Earth and Environmental Sciences Library

Series Editors

Abdelazim M. Negm, Faculty of Engineering, Zagazig University, Zagazig, Egypt

Tatiana Chaplina, Antalya, Türkiye

Earth and Environmental Sciences Library (EESL) is a multidisciplinary book series focusing on innovative approaches and solid reviews to strengthen the role of the Earth and Environmental Sciences communities, while also providing sound guidance for stakeholders, decision-makers, policymakers, international organizations, and NGOs.

Topics of interest include oceanography, the marine environment, atmospheric sciences, hydrology and soil sciences, geophysics and geology, agriculture, environmental pollution, remote sensing, climate change, water resources, and natural resources management. In pursuit of these topics, the Earth Sciences and Environmental Sciences communities are invited to share their knowledge and expertise in the form of edited books, monographs, and conference proceedings.

Vasiliy Tsygankov

Persistent Organic Pollutants in the Ecosystems of the North Pacific

 Springer

Vasiliy Tsygankov
School of Advanced Engineering Studies
"Institute of Biotechnology, Bioengineering
and Food Systems"
Far Eastern Federal University
Vladivostok, Russia

ISSN 2730-6674   ISSN 2730-6682 (electronic)
Earth and Environmental Sciences Library
ISBN 978-3-031-44895-9   ISBN 978-3-031-44896-6 (eBook)
https://doi.org/10.1007/978-3-031-44896-6

This Springer imprint is published by the registered company Springer Nature Switzerland AG
The registered company address is: Gewerbestrasse 11, 6330 Cham, Switzerland

Paper in this product is recyclable.

Preface

Nowadays, millions of new chemical compounds are synthesized each year to meet increasing demands in various industries. As manufacturers strive to introduce them as soon as possible, this often leads to undesirable consequences for the environment and human health, because a substantial portion of such substances are persistent organic pollutants (POP) that have not been tested for toxicity. The wide range of POPs includes also such organochlorine compounds (OCC) as dichlorodiphenyl-trichloroethane (DDT) and its metabolites (DDD and DDE), hexachlorocyclohexane (HCH) isomers, and congeners of polychlorinated biphenyls (PCB). DDT and HCH that reached their peak of popularity in the second half of the twentieth century were widely applied in agriculture worldwide due to their insecticidal properties, high effectiveness, and low cost. Subsequently, these substances were used in public health to control vectors of malaria and other dangerous diseases and also for the treatment of certain parasitic infections, e.g., scabies and pediculosis. PCBs, having good dielectric properties and high thermal, chemical, and electrical stability, were actively used as coolants or insulators for manufacturing transformers, capacitors, and other electrical equipment. However, all these compounds have been banned in many countries and are listed on the Stockholm Convention on Persistent Organic Pollutants (2002) as very toxic, environmentally resistant, and having high bioaccumulation and biomagnification potentials.

Despite the general ban on production and use, several countries in Asia, primarily India, and Africa have retained the right to use DDT to control vectors of dangerous diseases until an equivalent low cost and effective alternative is found. HCH is still added as an agent to some drugs against scabies mites and lice. Polychlorinated biphenyls are still contained in obsolete electrical equipment.

Although there is an increasing interest of the scientific community in the issues of accumulation, toxicity, and distribution of POPs, some territories where DDT, HCH, and PCBs were widely used in the past have been studied fragmentally and to an insufficient extent. This is especially true for the territory of Russia, in particular, the Far East having common borders with the countries that are potential producers and consumers of these xenobiotics. The remoteness of this region largely explains the slow progress of State's measures for the removal and disposal of hazardous substances to prevent their entry into the environment in Russia.

The Russian Far East is known as a land of fish, forests, and minerals. Its rich fish resources are formed under unique natural conditions in the seas of the Pacific (the Sea of Japan, Sea of Okhotsk, and Bering Sea) and Arctic Oceans. The biological productivity of these seas has a major effect on the socioeconomic well-being of the region. The aquatic resources harvested here constitute an important food supply for the populations of not only Russia but also the neighboring countries (Japan, South Korea, China, USA, and Canada).

In this book, we have systematized our results and those obtained by other researchers on the fate of POPs in the abiotic and biotic components of the aquatic environment and the possible health risks to residents of coastal areas exposed to these hazardous pollutants. Thus, we have identified convenient matrix organisms (certain species of fish, birds, and mammals) that indicate contamination in the region; shown possible ways of sea-to-land transport of xenobiotics; found targets of POP effects on living organisms; estimated time periods of exposure to the pollutants in the biosphere; compiled a list of *priority* toxicants for the region based on qualitative screening analysis; assessed the risks that the use of contaminated aquatic organisms poses to human health; and measured the levels of POPs in the human body.

I am deeply grateful to the authors who made substantial contributions to this collective book, for their valuable and critical comments, and also appreciate their moral support.

Vasiliy Tsygankov
Dr. Sci. Biol., Associate Professor
Dean of the Industrial Biotechnology
and Bioengineering Faculty, School
of Advanced Engineering Studies
"Institute of Biotechnology
Bioengineering and Food Systems"
Far Eastern Federal University
Vladivostok, Russia

Contents

Chapter 1
The *Dirty Dozen* of the Stockholm Convention and Other Persistent Organic Pollutants: A Review

Abstract This chapter provides general information about the Stockholm Convention and the stages of Convention's development and implementation. The major persistent organic pollutants (POP) are considered. The physical and chemical properties, distribution in the environment, metabolism and degradation, and toxicity of POPs such as organochlorine pesticides (OCP), polychlorinated biphenyls (PCB), and polycyclic aromatic hydrocarbons (PAH) are described. The authors of the present book study these POPs in the environment.

Keyword *Dirty Dozen* · Stockholm convention · POP · OCP · PCB · PAH · Physical and chemical properties · Distribution in the environment · Metabolism and degradation · Toxicity

1.1 The *Dirty Dozen* of the Stockholm Convention

Of all chemicals that occur in the environment as a result of human activity (industrial production, agriculture, infection control, etc.), the most hazardous are persistent organic pollutants (POP). Over the past few decades, these highly toxic compounds have done unprecedented harm to wildlife and human health by causing cancer diseases, damaging the nervous, reproductive, and immune systems of organisms, and provoking some congenital pathologies in various populations.

Trace amounts of these substances can be found in all people. Moreover, they can build up in the human body through bioaccumulation.

Although some of them have been banned in many developed and a number of developing countries since the 1970s, the fact that these substances can *jump* through the atmosphere and cross the boundaries of biotic and ecological environments necessitated creating a legally binding treaty for all nations. Such a treaty should be binding for all parties and, simultaneously, take into account specific socio-economic and political conditions in certain countries. Under the current circumstances, there is no alternative other than adopting such an international legal instrument to regulate the use of POPs.

In May 1995, the Governing Council of the United Nations Environment Programme (UNEP), in its Decision 18/32, determined the need for an international evaluation of the original list of POPs that required urgent regulation measures. It included aldrin, chlordane, DDT, dieldrin, dioxins, endrin, furans, hexachlorobenzene (HCB), heptachlor, mirex, polychlorinated biphenyls (PCB), and toxaphene. The UNEP Council ordered the Intergovernmental Forum on Chemical Safety (IFCS) to develop, no later than 1997, recommendations for international measures aimed at reducing the risks that the 12 listed POPs (subsequently referred to as the *Dirty Dozen*) posed to human health and the environment. In June 1996, IFCS submitted the required document to the UNEP Governing Council and the World Health Assembly for consideration.

In February 1997, the UNEP Governing Council proposed to convene an Intergovernmental Negotiating Committee (INC) in order to develop a legal mechanism for holding international events, initially focused on the 12 compounds. The Committee was also ordered to create an expert group that would develop criteria and procedures for identifying other POPs subject to future international measures. The first INC session was held in Montreal, Canada, June 1998. As a result, a Criteria Expert Group (CEG) was formed. The following INC sessions took place in Nairobi (Kenya, January 1999), Geneva (Switzerland, September 1999), Bonn (Germany, March 2000), and Johannesburg (South Africa, December 2000). At two sessions, INC was represented by CEG experts: in Bangkok, Thailand, October 1998, and in Vienna, Austria, June 1999. In June 2000, INC held a meeting of representatives from 18 countries in Vevey, Switzerland, to discuss financial issues.

The Convention was adopted and opened for signature at a conference held in Stockholm, Sweden, May 22–23, 2001. On May 23, 2001, a total of 92 states signed the Convention. It remained open for signature at the UN Headquarters in New York from May 24, 2001 to May 22, 2003. For this time, the number of signatories of the treaty increased to 151. The Convention was announced to enter into force 90 days after the submission of ratification documents by the governments of 50 states.

Initially, the participating countries ratified and signed 30 articles and six annexes. UNEP and INC were authorized to monitor the implementation of international policy in compliance with the Convention and take temporary measures until the treaty enters into force. On February 17, 2004, France became the 50th state to ratify the treaty. On February 18, 2004, in Geneva/Nairobi, the UNEP Council announced the Convention to become legally binding from May 17, 2004. By July 23, 2005, a total 103 countries ratified the treaty.

The governments immediately began taking actions within the framework of the treaty at the first meeting of the Conference of the Parties (COP1) in Punta del Este, Uruguay, May 2–6, 2005. As a result, a wide range of decisions were made, including the approval of the National Implementation Plans (NIP) and Technical Assistance Guidelines and the establishment of the POP Review Committee (POPRC). The priority task of the Conference was to evaluate developing countries' feasibility to comply with the regulations without technical assistance, in particular, the best available technologies and best environmental practices, for reducing emissions of dioxins and furans, two toxic groups of POPs.

The major goal of the Convention was to immediately ban most of the 12 POPs. Nevertheless, according to the decision of the World Health Organization (WHO), DDT can be used to control insect vectors of infections, since in many countries this pesticide remains a key insecticide applied for controlling malaria mosquitoes. This allows the governments of these states to keep protecting their populations from malaria, a devastating infectious disease of tropical regions, until an economically viable and environmentally friendly, chemical or non-chemical alternative/ substitute of DDT is found. Meanwhile, the Convention takes every measure to promote research aimed at finding and developing more effective agents to control malaria.

In addition to the ban on the use of such chemicals, the treaty also focuses on the liquidation of unnecessary and obsolete storage facilities for pesticides and toxic compounds that are still increasing in number in some of developing countries. By obliging the governments of the states to stop the production and emissions of these substances into the environment, the Stockholm Convention will greatly contribute to the protection of human health and the environment, strengthen the authority and enhance the effectiveness of international environmental law.

Another key goal of the Convention is to complete development of the guidelines to provide "best environmental practices and best available techniques" for the reduction and elimination of emissions of dioxins and furans (apparently the most toxic of all POPs) from a wide range of industrial and other sources such as open dumpsites of municipal and other waste in developing countries. The governments of the states that are parties of the Convention should also reduce the use of PCB-containing electrical equipment by introducing PCB substitutes and dispose PCBs in an environmentally friendly way no later than 2028.

Fortunately, alternatives to POPs do exist. The high cost and the lack of public competence and suitable infrastructure often prevented governments from making this choice. The Convention states that appropriate solutions should meet the specific properties and applications of each substance, as well as the climatic and socio-economic needs of each country.

To guarantee implementation of the decisions, the funding states have committed to allot hundreds of millions of dollars over the following few years. The head organization of the temporary financial mechanism of the Convention is the Global Environment Facility (GEF). Resources have already been attracted to support several POP research projects such as the Africa Stockpiles Programme in more than 100 countries. Supported by the alliance of developed and developing countries and having industrial and environmental groups in the management, the Stockholm Convention guarantees the world without POPs for future generations.

Below is a brief description of each of the POPs from the *Dirty Dozen* of such substances (Table 1.1).

Table 1.1 Characteristics of the persistent organic pollutants included in the *Dirty Dozen* of the Stockholm Convention [1, 2]

Compound (CAS registry number)	IUPAC name	2D structure	General information
Aldrin (309-00-2)	(1R,4S,4aS,5S,8R,8aR)-1,2,3,4,10,10-hexachloro-1,4,4a,5,8,8a-hexahydro-1,4:5,8-dimethanonaphthalene		Aldrin is usually applied to control termites, worms, snout beetles, and locusts, as well as to protect crops of cereals and potatoes from a variety of soil insects. It is easily metabolized to dieldrin by plants and animals. Aldrin binds to bottom sediments and soils and, therefore, rarely leaches into groundwater, but can evaporate from the soil surface and be redistributed by air currents, thus, polluting territories remote from the sources of entry. Acute aldrin poisoning can cause mortality of birds, fish, and humans. Chronic intoxication occurs through contaminated food. This compound is banned in most developed countries, but it continues to be used as a termiticide in many developing countries of Asia and Africa. Although aldrin is highly toxic, it has a selective effect. It can often cause liver damage [3]. Furthermore, aldrin is a carcinogen and can induce neurological and reproductive disorders. Aldrin is banned in a number of countries including Bulgaria, Ecuador, Hungary, Singapore, Switzerland, and its use is strictly regulated in Argentina, Canada, Japan, New Zealand, the USA, etc.

(continued)

Table 1.1 (continued)

Compound (CAS registry number)	IUPAC name	2D structure	General information
Dieldrin (60-57-1)	(1aR,2R,2aS,3S,6R,6aR,7S,7aS)-3,4,5,6,9,9-hexachloro-1a,2,2a,3,6,6a,7,7a-octahydro-2,7:3,6-dimethanonaphtho[2,3-b]oxirene	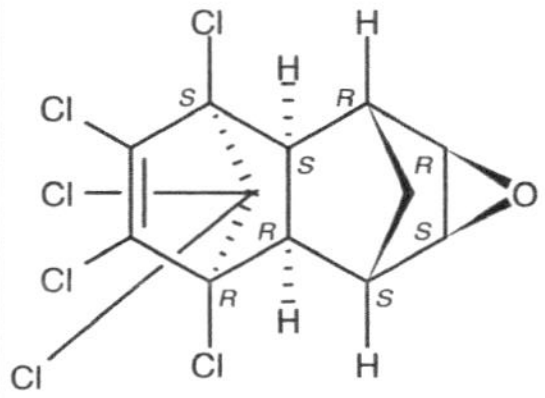	Dieldrin was mainly used in agriculture to kill soil insects and some insect vectors of infections. In the environment and living tissues, the pesticide aldrin quickly turns into dieldrin, which results in higher concentrations of the latter in the environment than those of aldrin. Residual amounts of dieldrin in tissues are, apparently, an additive effect of aldrin and a direct effect of dieldrin. Residual dieldrin is found in the air, water, soil, and also in birds, mammals, and humans that ingest it with food. Dieldrin is banned in many of EU countries and Singapore and is strictly regulated in Argentina, Canada, Austria, Colombia, India, New Zealand, the USA, etc.
Endrin (72-20-8)	(1aR,2R,2aR,3R,6S,6aS,7S,7aS)-3,4,5,6,9,9-hexachloro-1a,2,2a,3,6,6a,7,7a-octahydro-2,7:3,6-dimethanonaphtho[2,3-b]oxirene	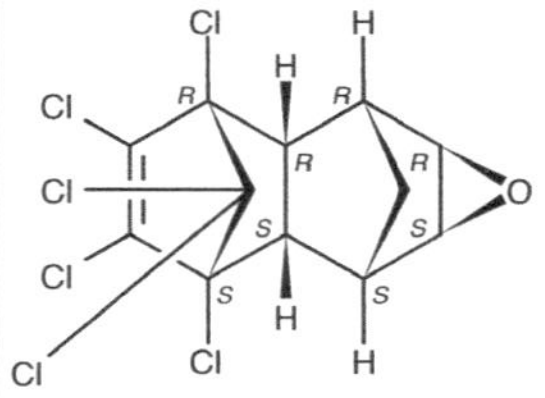	Endrin is a rodenticide applied to kill mice and voles. It is sprayed on leaves of crops such as cotton and cereals. Unlike other organochlorine compounds (OCC) with similar structures, endrin is rapidly metabolized by animals and does not build up to high concentrations in their adipose tissues. However, endrin is even more toxic than the closely related aldrin and dieldrin, being especially highly toxic to fish. It can enter the human body with food. After the treatment of cereals with endrin, rats, mice, and hamsters showed brain damage and also hepatic and embryonic disorders. Endrin is banned in many countries including Finland, Israel, the Philippines, and Singapore and is strictly regulated in Argentina, Canada, Chile, India, Japan, Pakistan, the USA, etc.

(continued)

Table 1.1 (continued)

Compound (CAS registry number)	IUPAC name	2D structure	General information
Chlordane (57-74-9)	1,2,4,5,6,7,8,8-octachloro-3a,4,7,7a-tetrahydro-4,7-methanoindane	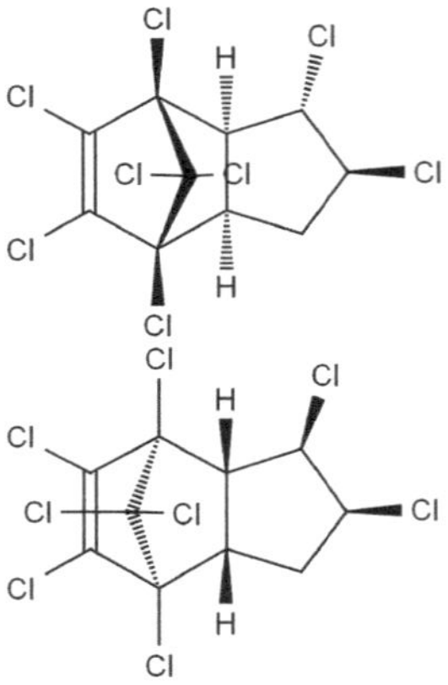	Chlordane is a broad-spectrum insecticide used to protect agricultural land and control termites. Chlordane is insoluble in water, soluble in organic solvents, and is semi-volatile. It binds to bottom sediments and builds up in adipose tissues of organisms. Its toxic effect on humans is mainly airborne. Acute toxic effects are species-specific. Chlordane is considered a carcinogen, can cause minor neurological disorders, and can mimic sex steroids or change their levels in individuals exposed to it. Chlordane has been banned in a number of countries including Brazil, the Netherlands, the Philippines, Singapore, Spain, and Sweden. Its application is strictly limited to non-agricultural use in Argentina, Canada, China, New Zealand, the USA, etc.

(continued)

Table 1.1 (continued)

Compound (CAS registry number)	IUPAC name	2D structure	General information
DDT (50-29-3)	1,1′-(2,2,2-Trichloroethane-1,1-diyl)bis(4-chlorobenzene)	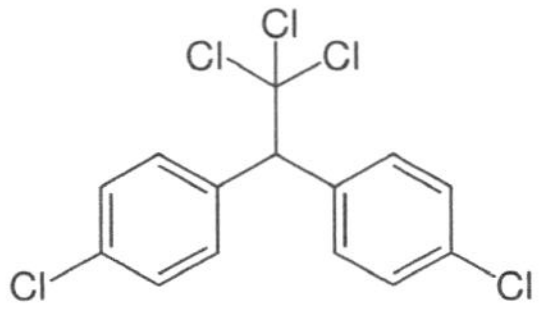	DDT is perhaps the most notorious POP. It was first applied during World War II to protect military troops and civilians from malaria, typhoid fever, lice, and other insect-borne diseases. After the war, DDT was widely applied in agriculture as an insecticide. It is highly persistent in soil. Being the very first, well-known, and one of the most widespread pesticides, DDT caused world-wide pollution of water and soil resources, which led to serious deterioration of human and animal health. Since DDT is a highly effective insecticide, WHO cannot recommend that state governments completely ban it and insists on limiting its use only to public health protection measures and indoor applications. DDT is easily metabolized into dichlorodiphenyldichloroethylene (DDE), a stable and equally toxic compound. Food is the main route of entry of DDT and DDE into living organisms. Both compounds are fat-soluble and build up in adipose tissues. DDT causes reproductive and embryonic disorders, as well as disorders of the immune system, in animals and humans. Like other OCCs, DDT affects the nervous system. Chronic DDT exposure leads to liver and kidney failure. By 1995, the use of DDT was banned in 34 countries and strictly regulated in another 34 countries [4]. After the ban, residual levels of DDT continue to be detected in all environmental and biota samples, including breast milk, which raises great concern about its potential impact on children

(continued)

Table 1.1 (continued)

Compound (CAS registry number)	IUPAC name	2D structure	General information
Heptachlor (76-44-8)	1,4,5,6,7,8,8-Heptachloro-3a,4,7,7a-tetrahydro-1H-4,7-methanoindene	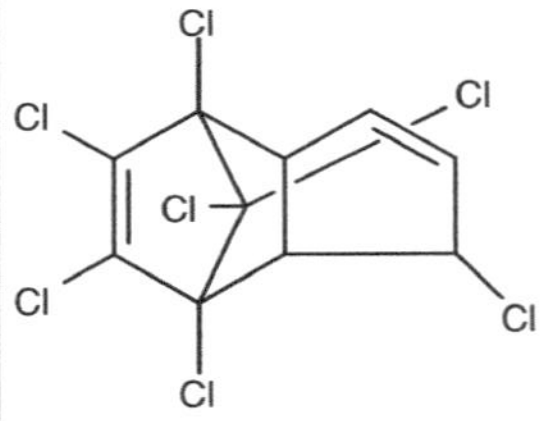	Heptachlor was used to kill cotton pests, locusts, and malaria mosquitoes. It is a gastric and contact insecticide, highly volatile and extremely insoluble in water, but soluble in organic solvents. It binds strongly to bottom sediments, is distributed in the atmosphere, and bioconcentrated in adipose tissues. Heptachlor epoxide, a metabolite of heptachlor in living tissues, is also toxic and bioaccumulative as parent compound. In rats, heptachlor was observed to affect progesterone and estrogen levels. It is highly toxic to humans, causes overexcitation of the central nervous system and liver damage. It probably has carcinogenic properties. Food is the major route of entry into the human body. The use of heptachlor has been banned in more than 12 countries and is strictly regulated in the EU, Canada, Japan, New Zealand, the USA, etc.

(continued)

Table 1.1 (continued)

Compound (CAS registry number)	IUPAC name	2D structure	General information
Hexachlorobenzene (118-74-1)	Hexachlorobenzene	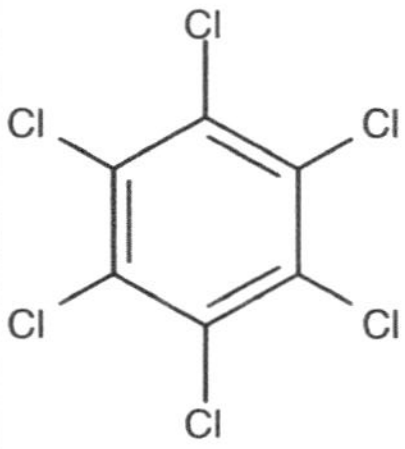	Hexachlorobenzene is a fungicide that was applied for seed treatment in the late 1940s. It is a by-product of the synthesis and production of industrial chemicals such as carbon tetrachloride, perchloroethylene, trichloroethylene, and pentachlorobenzene, and is also an impurity in a number of pesticide formulations. Hexachlorobenzene is detected in all types of foods. It has been found to be fatal for some animals at high doses; at lower concentrations, it causes skin lesions in humans and serious reproductive disorders in many animals. Being very stable, this compound is transported over long distances and, therefore, has a high distribution coefficient in the environment, which leads to its high bioconcentration. Use of hexachlorobenzene is banned in Austria, Belgium, Denmark, the EU countries, the Netherlands, Panama, the UK, and is also strictly regulated in Argentina, New Zealand, Sweden, etc.

(continued)

Table 1.1 (continued)

Compound (CAS registry number)	IUPAC name	2D structure	General information
Mirex (2385-85-5)	Dodecachlorooctahydro-1H-1,3,4-(epimethanetriyl)cyclobuta[cd]pentalene	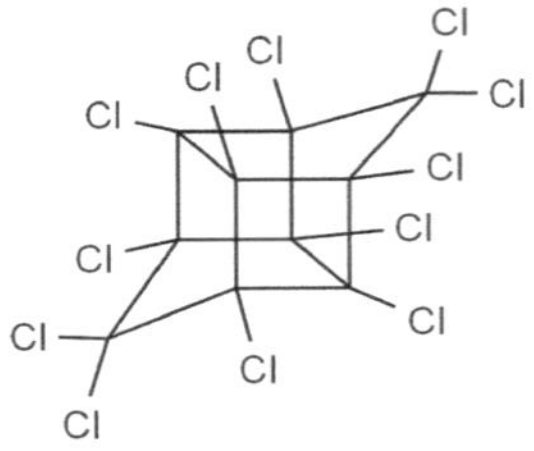	Mirex is a potent insecticide of low contact activity. It is mainly applied against ants, wasps, bugs, and termites, and in industry as a flame retardant in plastic, rubber, dye, and electrical products. When exposed to light, mirex decomposes to form photomirex, which is more toxic. Mirex is one of the most stable and persistent pesticides, very resistant to decomposition, insoluble in water, showing bioaccumulative potential, and binding strongly to bottom sediments. This compound destroys the endocrine system of animals, suppresses immunity, and has carcinogenic and teratogenic properties. Data on the adverse effects of mirex on humans are limited. This substance enters the body mainly with food, in particular, fish and meat of domestic and wild animals

(continued)

Table 1.1 (continued)

Compound (CAS registry number)	IUPAC name	2D structure	General information
Polychlorinated biphenyls (each PCB congener was assigned CAS no.)	1,1′-Biphenyl, chloro derivs	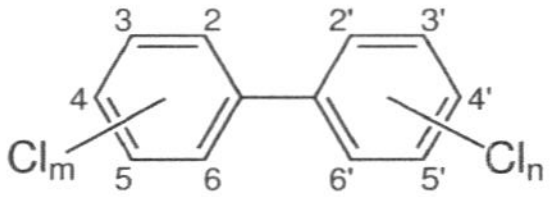	Polychlorinated biphenyls (PCB) are mixtures of chlorinated hydrocarbons. Since the 1930s, PCBs have been widely applied for industrial purposes, e.g., in the production of electrical transformers and capacitors, dyes, transfer papers, and plastics. Theoretically, there exist 209 PCB isomers, from three monochlorinated isomers to fully chlorinated decachlorobiphenyl. With increasing number of chlorine atoms, the solubility in water and vapor pressure decreases, while the solubility in lipids increases. Mixtures of various commercial brands in different countries actually contain different numbers of isomers. Usually PCBs are released into the environment in the form of an unpurified mixture containing also other reagents. They have low volatility and relative persistence, are chemically stable and resistant to heat, which poses a threat to the environment. PCBs without ortho-substituents (coplanar) exhibit dioxin-like toxicity. There are 13 known such compounds, with the rest being non-planar. PCBs are insoluble in water, tend to adsorb to organic particles in the environment and bioaccumulate in animal adipose tissues. Food is the major route of their entry into the human body. The US Environmental Protection Agency classifies PCBs as carcinogenic substances hazardous to humans. These are neurotoxic and cause intrauterine fetal damage. Visible changes in the nervous system caused by PCB have been recorded from animals, including mice, rats, apes, and quails [5–7]. Furthermore, PCBs show an evident hormone-suppressing effect

(continued)

Table 1.1 (continued)

Compound (CAS registry number)	IUPAC name	2D structure	General information
Polychlorinated dibenzodioxins	… chlorodibenzo-*p*-dioxin		Polychlorinated dibenzo-*para*-dioxins (PCDD, dioxins) and polychlorinated dibenzofurans (PCDF, furans) represent two groups of planar tricyclic compounds with very similar chemical structures and properties. Neither dioxins nor furans are produced in industry and have any commercial use. These are by-products of incomplete combustion of medical and household waste and also result from the production of several chlorinated compounds. There are 75 different dioxin and 135 furan isomers. At least 20 compounds are considered potentially toxic. In approximately 90% of cases, they enter the human body with food, in particular animal meat. Daily exposure leads to the buildup of PCDDs and PCDFs in lipids, breast milk, and blood. Infants receive higher doses with heavily contaminated mother's milk [8, 9]. The total toxicity of dioxin-containing mixtures is usually expressed in terms of toxic equivalent (TEQ) to a certain quantity of pure 2,3,7,8-tetrachlorodibenzo-*p*-dioxin (TCDD), the most toxic and best studied dioxin [10]. These substances have been proven to cause chloracne in humans. Dioxins can directly affect growth and differentiation of cancer cells through gene dysregulation; they are also well-known endocrine and immune suppressors
Polychlorinated dibenzofurans (each compound was assigned CAS no.)	… chlorodibenzofuran		
Toxaphene (8001-35-2)	1,4,5,6,7,7-hexachloro-2,2-bis(chloromethyl)-3-methylenebicyclo[2.2.1]heptane		Toxaphene is a somatic and contact insecticide used to control pests of cotton, fruits, etc., and also ticks parasitizing agricultural and domestic animals. It is not toxic to plants. Its low solubility in water, high persistence, and volatility contribute to the transport of toxaphene, as well as other pesticides, over long distances. Although the use of toxaphene stopped about 20 years ago, in some countries it is still allowed to be applied in small quantities. Toxaphene induces the activity of modified enzymes in the liver and causes dose-dependent adverse changes in the kidneys, thyroid gland, and nervous system. Some estrogenic effects of toxaphene and its association with the prevalence of cancerous tumors in animals have been reported [11]. This substance is banned in several countries including Egypt, India, South Korea, Singapore, the EU, and its use is restricted in Argentina, Pakistan, South Africa, Turkey, etc.

These compounds are listed in the Annexes of the Convention as follows:

1. Annex A (elimination of the production and use: aldrin, dieldrin, endrin, chlordane, mirex, toxaphene, and heptachlor);
2. Annex B (restriction of the production and use: DDT);
3. Annex C (reduction of the unintentional releases: hexachlorobenzene, PCB, and PCDD/PCDF).

After the fourth meeting of the Conference of the Parties, held on May 4–8, 2009, it was decided (SC-4/12) to include nine more organic compounds.

1. Annex A: α-HCH, β-HCH, lindane (γ-HCH), chlordecone, hexabromobiphenyl, hexa- and heptabromodiphenyl ethers, pentachlorobenzene, tetra- and pentabromodiphenyl ethers.
2. Annex B: perfluorooctane sulfonic acid and its salts, perfluorooctane sulfonyl fluoride.
3. Annex C: pentachlorobenzene.

Besides POPs, other priority pollutants are polycyclic aromatic hydrocarbons (PAH) including such a widely distributed compound as benzo(a)pyrene. PAHs are hazardous due to their transformation activity that can contribute to carcinogenic, teratogenic, or mutagenic changes in organisms. Depending on the conditions of exposure, they can cause mutagenesis, teratogenesis, growth inhibition, accelerated aging, toxicogenesis, and disruption of the immune system functions, which leads to both ontogenetic disturbances, changes in gene pool, and also undesirable deviations in cenoses [12].

For this reason, we decided that these compounds should also be considered in our book on persistent pollutants.

1.2 Organochlorine Pesticides (OCP)

1.2.1 Physical and Chemical Properties of OCPs

Organochlorine pesticides (OCP) are a large group of chemicals represented by halogen derivatives of alicyclic and aromatic hydrocarbons. These are anthropogenic substances, i.e. they are not formed in nature and enter the environment as a result of human industrial activities. In the 1940s–1960s, such OCPs as dichlorodiphenyltrichloroethane (DDT) and hexachlorocyclohexane (HCH) (Fig. 1.1) were most commonly and widely applied due to their pronounced insecticidal properties. OCPs are solid, crystalline substances that exhibit high thermal stability, low saturated vapor pressure, poor solubility in water, but good solubility in fats and lipids. The physical and chemical properties of compounds from the DDT group and HCH isomers are presented in Table 1.2.

In their chemical properties, OCPs are very inert under normal conditions and almost cannot be degraded by concentrated acids, alkalis, and water. An exception is

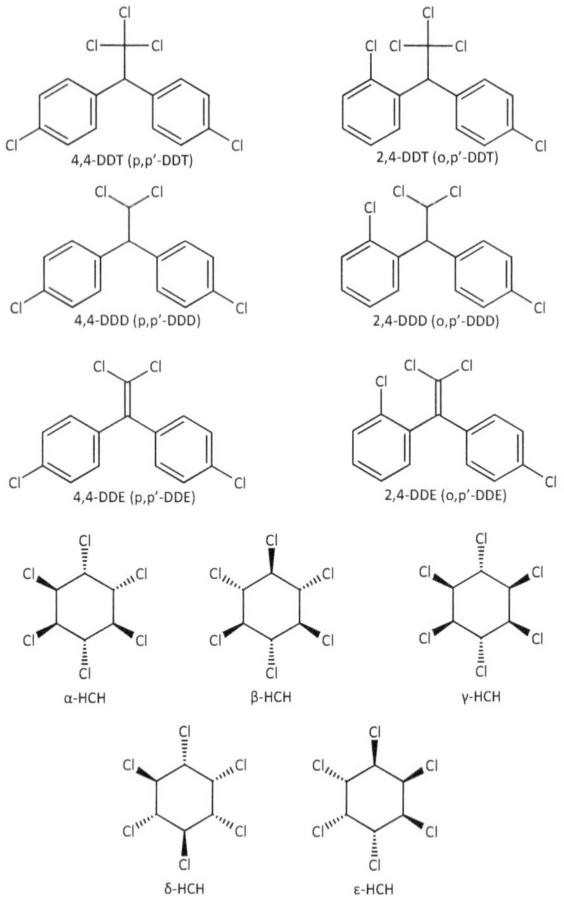

Fig. 1.1 Structural formulas of HCH isomers and DDT and its metabolites

4,4-DDT (p,p′-DDT)　　　2,4-DDT (o,p′-DDT)

4,4-DDD (p,p′-DDD)　　　2,4-DDD (o,p′-DDD)

4,4-DDE (p,p′-DDE)　　　2,4-DDE (o,p′-DDE)

α-HCH　　　β-HCH　　　γ-HCH

δ-HCH　　　ε-HCH

HCH isomers that are hydrolyzed to form trichlorobenzenes and trichlorophenols and release hydrogen chloride at a temperature of 100 °C and above. OCP degradation occurs in a reducing environment and is accelerated by catalysts at elevated temperatures. Thus, in an alcoholic solution of caustic alkali, DDT quantitatively passes into DDE (dehydrochlorination reaction). DDT decomposition is observed in case of heating a benzene solution of DDT with anhydrous aluminum chloride, as evidenced by the appearing orange color of the solution. In the presence of strong oxidizing agents (nitric acid or chromium oxides), DDT and its metabolites decompose with the formation of benzophenols [16].

When heated to the decomposition temperature, hexachlorobenzene (HCB) degrades with the release of toxic chloride vapors; at 65 °C, it violently reacts with dimethylformamide (DMF). HCB is produced industrially through chlorination of benzene with excess of chlorine in the presence of a catalyst, aluminum chloride [17].

Table 1.2 Physical and chemical properties of HCH isomers and DDT and its metabolites [13–15, PubChem]

Compound	M_r	Melting point, °C	Solubility, µg/L		Vapor pressure, Pa	H	K_{OW}	LD_{50}
			In water	In organic solvents				
α-HCH	290.83	159–160	10	1.8 g/100 g ethanol 6.2 g/100 g ether	6.0×10^{-3}	6.9×10^{-6}	3.8	500–600
β-HCH		314–315	5	1.1 g/100 g ethanol 1.8 g/100 g ether 1.9 g/100 g benzene	4.8×10^{-5}	4.5×10^{-7}	3.78	6000
γ-HCH		112.5	17	6.4 g/100 g ethanol 20.8 g/100 g ether 28.9 g/100 g benzene	5.6×10^{-3}	3.5×10^{-6}	3.72	125–225
δ-HCH		141–142	10	24.4 g/100 g ethanol 35.4 g/100 g ether 41.4 g/100 g benzene	4.7×10^{-3}	2.1×10^{-7}	4.14	–
ε-HCH		112.5	4	–	6.7×10^{-2}	5.1×10^{-6}	4.26	–
2,4-DDT	354.49	74.2	0.085	–	1.5×10^{-5}	5.9×10^{-7}	6.79	250–400
2,4-DDD	320.05	76–78	0.1	Ethanol, isooctane, CCl4	2.6×10^{-4}	8.2×10^{-6}	5.87	400
2,4-DDE	318.03	–	0.14	–	8.3×10^{-4}	1.8×10^{-5}	6.0	2500–3400

(continued)

Table 1.2 (continued)

Compound	M_r	Melting point, °C	Solubility, µg/L		Vapor pressure, Pa	H	K_{OW}	LD_{50}
			In water	In organic solvents				
4,4-DDT	354.49	109	0.025	Acetone, ethanol	2.1×10^{-5}	8.3×10^{-6}	6.91	250–400
4,4-DDD	320.05	109–110	0.09	Acetone, ethanol	1.8×10^{-4}	4.0×10^{-6}	6.02	400–3400
4,4-DDE	318.03	89	0.12	Acetone, ethanol	8.0×10^{-4}	2.1×10^{-5}	6.51	2500–3400

M_r is relative molecular weight;

K_{OW} is octanol/water partition coefficient;

H is Henry's law constant (for describing processes in the water–air system), atm·m^3/mol;

LD_{50} is the lethal dose at which 50% of experimental animals (rats, with intragastric injection) die, mg/kg body weight

The formula of a HCH technical mixture included various HCH isomers differing in the cycle conformation. In a technical mixture obtained through photochemical chlorination, α-HCH made up 55–70%; β-HCH, 5–14%; and γ-HCH, 9–13% [18]. Furthermore, mixtures with different compositions were produced, containing 25% and 90–99% γ-HCH (lindane) as an active agent [19].

The DDT technical mixture was based on dichlorodiphenyltrichloroethane, represented by two isomers, *p,p'*-DDT [1,1,1-trichloro-2,2-bis(*p*-chlorophenyl)ethane] and *o,p'*-DDT [1,1,1-trichloro-2-(*p*-chlorophenyl)-2-(*o*-chlorophenyl)ethane] that differ in the position of chlorine atoms in benzene rings. The formula of the DDT technical mixture, in addition to these two isomers, included *p,p'*-DDD [1,1-dichloro-2,2-bis(*p*-chlorophenyl)ethane], *o,p'*-DDD [1,1-dichloro-2-(*o*-chlorophenyl)-2-(*p*-chlorophenyl)ethane], *p,p'*-DDE [1,1-dichloro-2,2-bis(*p*-chlorophenyl)ethylene], and *o,p'*-DDE [1,1-dichloro-2-(*o*-chlorophenyl)-2-(*p*-chlorophenyl)ethylene]. A usual composition of technical DDT was as follows: *p,p'*-DDT, 77.1%; *o,p'*-DDT, 14.9%; *p,p'*-DDD, 0.3%; *o,p'*-DDD, 0.1%; *p,p'*-DDE, 4.0%; *o,p'*-DDE, 0.1%; and trace concentrations of other compounds [20].

1.2.2   Patterns of OCP Distribution in the Environment

After entering the biosphere, persistent organic compounds become involved in various physical and chemical processes. The resistance of POPs to photochemical, chemical, and biological degradation in the atmosphere, the aqueous phase, and soil results in their long-term presence and circulation in the environment. Although the POP group under consideration is characterized by low vapor pressure values, they, nevertheless, exhibit a pronounced capability of passing into the vapor/gas phase, i.e. evaporate into the atmospheric air, e.g., from the surface of soil, water, etc. and circulate between the components of the environment [16, 21, 22]. They are transferred within various components of the biosphere (soil, aquatic systems, and atmosphere) and from one subsystem to another. Currently, POPs are distributed ubiquitously, which is evidenced by findings of these compounds in both abiotic and biological samples from various parts of the globe. The facts of their detection in regions geographically remote from possible sources of emission such as the Arctic and Antarctic indicate the global distribution of POPs [23–26]. The lipophilicity of POPs leads to their bioaccumulation, i.e. buildup in plants or animals (mainly in adipose tissues) and the increase in concentrations of these compounds during the transition from the lowest to the highest trophic levels of a food chain (biomagnification) [27–29].

The behavior of OCPs, as well as PCBs, in the environment and the pattern of distribution in various components of natural ecosystems are determined by the physical and chemical properties of these compounds (Table 1.2). The patterns of POP distribution in the environment are described by various parameters, of which the most important are the solubility in water, octanol/water partition coefficient, Henry's law constant, and saturated vapor pressure [30]. Values of these parameters for the group of the POPs considered here are presented in Table 1.2.

The octanol/water partition coefficient is an indicator of lipophilicity of a compound, which characterizes the distribution of this substance between water and organic matter and is calculated as the ratio of the substance's equilibrium concentration in octanol (C_O) to the concentration of the substance in water (C_W):

$$K_{OW} = C_O/C_W.$$

Substances with higher K_{OW} values exhibit more hydrophobic properties and are more easily bound by organic matter of the matrix (soils, bottom sediments, living organisms, etc.).

The Henry' law constant is of major importance for describing partition processes in the water/air system. According to Henry's law, the ratio of a substance's partial pressure in air (P_A) to the concentration in water (C_W) is equal to the ratio of the substance's saturated vapor pressure (P°) to the solubility of this substance (S):

$$P_A/C_W = Po/S = H,$$

where H is the Henry' law constant.

Substances with higher Henry's law constant values show an easier transition into the gas phase.

Saturated vapor pressure is the maximum pressure of vapor of a compound when it passes into the gas phase from a solid (sublimation) or liquid state (evaporation) at certain temperature. Leading researchers from Norway and Canada [21], based on vapor pressure values, categorized compounds into groups depending on behavior (partition between the gas phase and aerosols) in the atmosphere. According to them, compounds with vapor pressures in a range from 10^{-2} to 10^{-4} Pa at 25 °C, i.e.

HCHs and some PCBs, can be in the atmosphere in both the gas phase and aerosols. Compounds with vapor pressures lower than 10^{-4} Pa at 25 °C (DDT group and higher chlorinated PCBs) can be in the atmosphere only in the aerosol-sorbed phase.

The extreme persistence of OCPs has resulted in their ubiquitous distribution in the biosphere through dissolution, sorption, bioaccumulation, and evaporation processes. The high K_{OW} values and the low solubility in water of DDT group compounds and HCH isomers (Table 1.2) contribute to binding of these substances in the environment by soil particles, sediments, and particles suspended in the water and air [31, 32]. Sorption on solid particles is important for partition in the environment. Heavy, rich in humus soils and bottom sediments with high organic carbon contents have maximum adsorption capacities [16]. In water bodies, OCPs are absorbed by allochthonous and autochthonous particles and precipitate onto the bottom. OCPs enter the atmosphere from soils and water bodies as a result of evaporation and wind erosion. Furthermore, these can be involved in repeated evaporation–condensation cycles and transported over long distances from their sources, thus, causing both regional and global pollution of the biosphere. Like other POPs, OCPs undergo global redistribution and are transported to polar regions, being found in various environmental components in the Arctic [25, 33–35]. Processes of POP redistribution in the environment depend on temperature. The vapor pressure of any substance increases with increasing temperature. A rise in the ambient temperature contributes to the transition of a substance into the vapor/gas phase; the condensation process occurs as the temperature decreases (Fig. 1.2) [21].

Thus, POP concentrations in the atmospheric air show seasonal relationships, being higher in the warm than in the cold season. The water temperature on the planet varies from about + 30 °C at the equator to – 1.7 °C in polar regions, while the air temperature varies in a wider range, from – 90 to + 50 °C, depending on season and latitude [25]. Such a temperature gradient causes POPs to migrate from

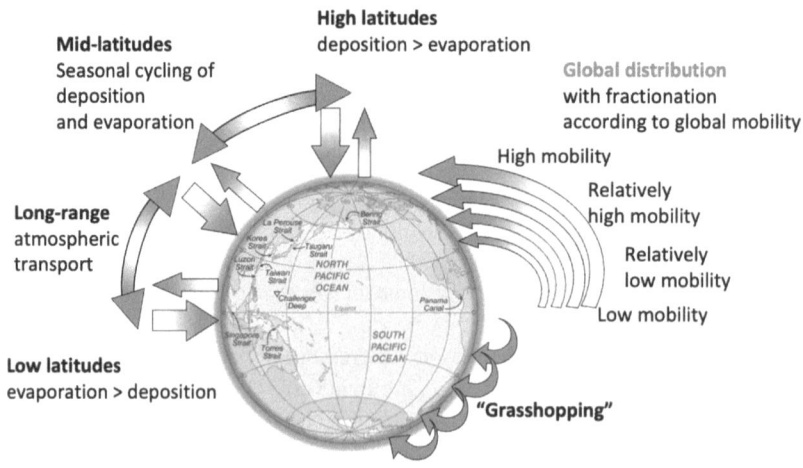

Fig. 1.2 Global transport of persistent organic pollutants [21]

warm regions of the planet to cold ones, with the partition (fractionation) of these compounds depending on their physical and chemical properties [21, 30]. To date, a sufficient number of publications have confirmed the phenomenon of POP redistribution on a global scale [24, 36, 37]. Various compounds, depending on their physical and chemical properties, have different potentials for transition into the vapor/gas phase and migration from places of initial entry into the natural environment. A simultaneous assessment and analysis of POP distribution in bottom sediments and water has shown that DDT and PCBs have a low potential for partition and migration from original sources. The distribution of POPs between water and bottom sediments is determined by the physical and chemical properties of these compounds and can be described by the octanol/water partition coefficient (K_{OW}). Persistent compounds with low K_{OW} values are present in water bodies mainly in the aqueous phase (to a lesser extent, in bottom sediments), from which they can pass into the vapor/gas phase and be involved in the atmospheric transport.

1.2.3 Metabolism and Degradation of OCPs

The major mechanisms of OCP degradation in the environment can conditionally be divided into abiotic (photochemical reactions) and biotic processes of metabolic decomposition with the involvement of living organisms.

Photochemical degradation of OCPs, whose molecules contain aromatic groups and unsaturated chemical bonds, occurs through the absorption of solar energy in the ultraviolet (UV) and visible regions of the spectrum (Fig. 1.3) [38, 39].

The rate of photochemical degradation and the composition of the final products of this reaction depend on the environment where this process occurs. As studies showed, after being exposed to UV light ($\lambda = 254$ nm) for 48 h, 80% of DDT decompose, and DDE (major portion), DDD, and ketones are found among the resulting products. Further experiments showed that DDD is very resistant to UV radiation, and DDE gradually turns into a number of compounds including polychlorinated biphenyls (PCB). Most PCB components being in the gas phase in the atmospheric air are not subject to significant changes with light exposure. PCBs are products of industrial activities. The choice of PCBs in technologies is based on their chemical inertia, incombustibility, and resistance to temperatures of about 500 °C. Being effective pesticides, these compounds are used as fungicides to protect trees from microbiological attacks, etc. The potential of marine organisms to accumulate PCBs is assessed on the basis of congeners that differ in the number and position of chlorine atoms in the molecule [40].

Lindane (γ-HCH) is isomerized into α-HCH under UV light [38, 39, 41].

The chemical degradation of most OCPs usually occurs in a reducing environment (for some substances, degradation is possible also under oxidizing conditions) and is intensified by catalysts and at elevated temperatures. An increase in pH, oxygen content, and the presence of heavy metal salts accelerate the OCP degradation as well [42–44].

Fig. 1.3 Diagram of photochemical transformations of DDT [38]

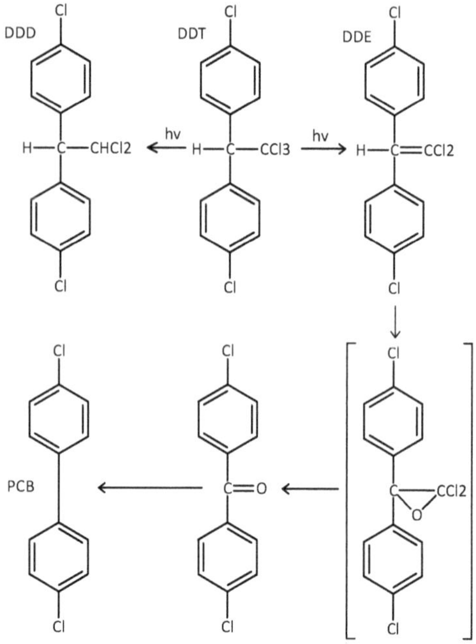

In an oxidizing environment, hydrolytic transformations occur with the replacement of halogen atoms by a hydroxyl group. The rate of hydrolysis depends on the pH value of the environment and the temperature. Thus, in acidic soils and at low temperatures, OCPs persist much longer [45].

Degree of OCP accumulation in animals is estimated as the ratio of two processes, absorption and excretion. After entering the body, chemical compounds undergo changes in two directions: (1) oxidation, reduction, and hydrolysis; (2) conjugation, which is the formation of complexes with some low-molecular-weight compounds in cells.

The major organ of metabolism of xenobiotics in some marine organisms (fish and mammals) is the liver, mainly due to the diversity and high activity of biotransformation enzymes acting in it.

The products of the first stage of metabolism enter the bloodstream and can have an effect on the organs and systems. The liver also releases products of the second stage of metabolism into the blood. From the bloodstream, the transformation products can be captured by the kidneys, lungs, other organs, and also recaptured by the liver for excretion with bile. This way, the metabolites enter the intestine where they are partially reabsorbed and reenter the liver (enterohepatic recirculation).

The general trend of such metabolism is the transformation of exogenous substance into a more polar compound with the subsequent binding of the resulting product to a highly polar fragment, which facilitates its excretion. In plants that lack excretion systems, exogenous substances (or their metabolites) are usually conjugated with carbohydrates and deposited in plant's parts unrelated to the general metabolism [40].

Insects have a smaller set of hydrolases than mammals. Thus, there is no rapid neutralization in insects and, therefore, they can accumulate OCPs and organophosphorus pesticides to lethal concentrations [40].

Oxidative processes are the most common trends of transformation of chemical compounds in organisms. These are often accompanied by detoxification of xenobiotics. The major types of oxidative processes are categorized as oxygenase-based, oxidase-based (dehydrogenase-based), and peroxidase-based. Oxidative processes in plant cells are mainly catalyzed by peroxidases. Reductive processes in organisms have not been studied as comprehensively as oxidative ones. Under anaerobic conditions in soil, cyclic and aromatic hydrocarbons are known to degrade with the formation of cyclohexanone as an intermediate product. An assumption has been made that nitro compounds are restored by nitro- and azoreductases in plants. In animals, reduction of sulfides also occurs [40, 46].

Hydrolysis is accelerated by the action of enzymes, hydrolases. Hydrolysis activity depends on the characteristics of substituents in the molecule of the compound cleaved: the bulky substituents near the group cleaved off sterically hamper hydrolysis and, therefore, OCPs having such substituents in their structure are resistant to hydrolysis and build up in adipose tissues.

The enzymatic oxidation of OCPs by microorganisms leads to formation of various metabolites, which may turn out to be both harmless substances and more dangerous and toxic than the original compounds.

In animals, the OCP metabolism is controlled by enzymes mainly in the liver and, to a lesser extent, in the intestines and kidneys. Their activity varies significantly between members of different taxonomic groups, being much higher in mammals and birds than in fish and amphibians. In living organisms, the major trends of DDT degradation can be represented by the following diagram (Fig. 1.4).

Fish lack some of enzymes for xenobiotic metabolism, and the system of production of water-soluble metabolites and conjugates that could be excreted from the body is not developed in them. Therefore, fish remove lipophilic compounds into the surrounding water by passive diffusion. However, this does not protect them from high levels of contamination by compounds such as DDT and aldrin that enter from the water and build up in large quantities. Unlike other abiotic components of ecosystems, the half-life of DDT in fish is approximately 30 days [16]. DDE and DDD are less toxic to fish than the original compound. Lipophilic OCPs that build up in the adult fish liver and are transferred to the gonads have a negative impact on the reproductive function of fish [49].

The DDT accumulation depends on the lipid content of the organs and tissues of aquatic organisms. For example, the relationship of DDT accumulation with the organs can be arranged in the following order: fat → liver → muscles. The increased lipid content in fish from pesticide-contaminated waters can be regarded as obesity and adaptation to habitat conditions [50].

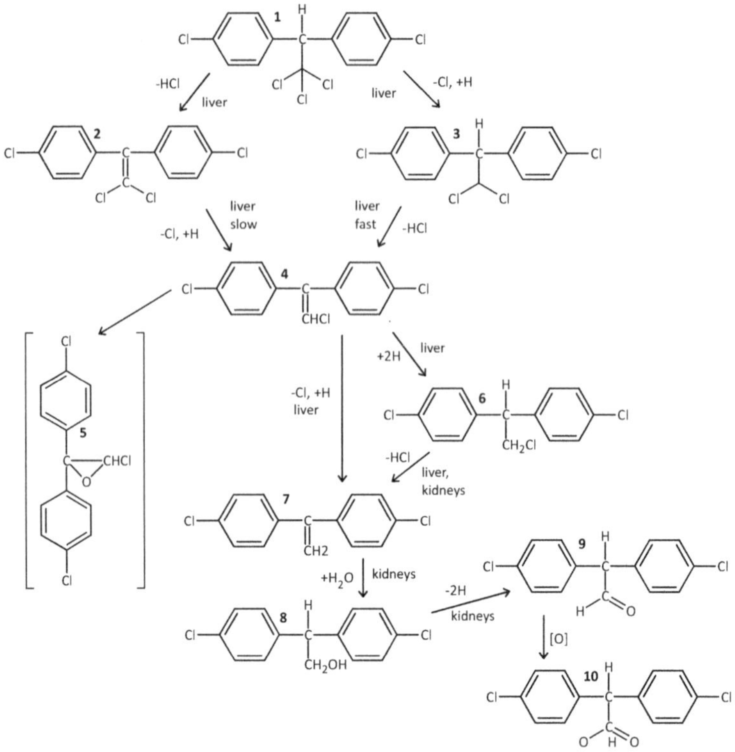

Fig. 1.4 Diagram of metabolic transformations of DDT in animals [15, 47, 48]: 1—DDT; 2—DDE; 3—DDD; 4—dichlorodiphenylchloroethylene; 5—dichlorodiphenylchloroethylene epoxide; 6—dichlorodiphenylchloroethane; 7—dichlorodiphenylethylene; 8—dichlorodiphenylethanol; 9—dichlorodiphenylethanal; 10—dichlorodiphenylethane (dichlorodiphenylacetic) acid

HCH is less persistent than DDT and is more rapidly degraded microbiologically. Studies conducted on model water bodies to determine HCH content in various ecosystem components have identified the major factors affecting the metabolism of this pesticide. According to the experimental data [43], at one day after the introduction of γ-HCH into a model water body, concentrations of this compound in the water decreased from 0.50 to 0.27 mg/L; in silt, increased from 0.0 to 0.34 mg/kg; in higher aquatic plants, from 0.0 to 2.3 mg/kg. The γ-HCH accumulation coefficient (K_A) for sediments was 1.3–4.0; for aquatic plants, from 8.0 to 67.0.

The accumulation coefficient K_A is calculated by the following formula:

$$K_A = \frac{C_O}{C_W},$$

where C_O is the concentration of the pesticide in aquatic organisms, mg/kg; C_W is the concentration of the pesticide in water, mg/L.

As was shown using the example of ecosystems in the northwestern Pacific Ocean, α-HCH dominates the HCH isomers in surface waters and zooplankton, reaching 55–56%; γ-HCH amounts to 35–40%; the proportion of β-HCH is the smallest, less than 10%. In cephalopods, the proportion of β-HCH is higher, about 20%. In tissues of striped dolphins, the proportions of the isomers are different: β-HCH dominates, up to 80%; α-HCH has the second largest proportion, 10–15%; γ-HCH is at a minimum level [51].

After HCH was introduced to soil, the concentration of α-HCH in samples decreased by 30% in anaerobic and by 55% in aerobic conditions within 20 weeks [52].

When lindane (γ-HCH) was applied on the soil surface, the largest portion of the substance vanished from the soil within the first days, and only 10% remained after three months [53].

The processes of γ-HCH transformation by a culture of the bacterium *Pseudomonas aeruginosa* with the formation of α-, β-, γ-HCH, as well as chlorocyclohexane and tetrachlorobenzene, occurred within the first three days. At high γ-HCH concentrations (400 mg/kg), the growth of Gram-positive bacteria was observed to increase [52].

The major trends of lindane metabolism in the environmental components are shown in Fig. 1.5.

The γ-HCH degradation leads to the formation of pentachlorocyclohexane, which, depending on environmental conditions, is transformed mainly into the following

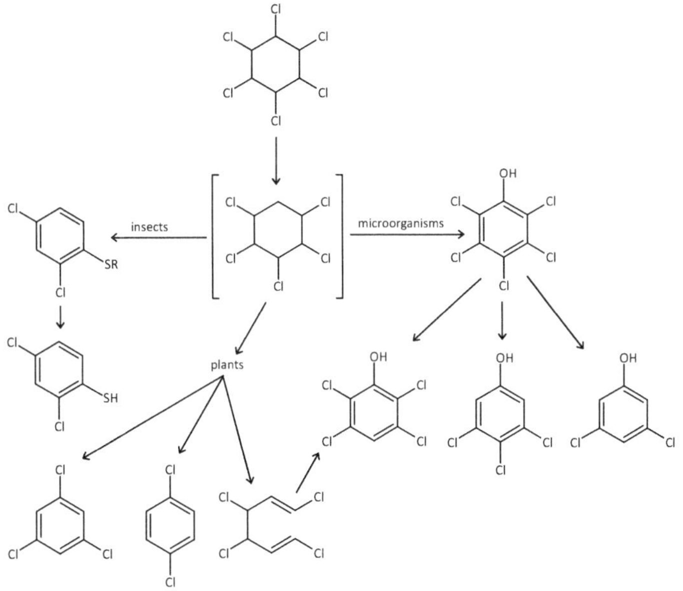

Fig. 1.5 Diagram of lindane metabolism in the components of the environment [16, 54]

compounds: di- or trichlorobenzene (in plants); di-, tri-, and tetrachlorophenol (under the effect of microorganisms); and thiophenols (in insects). Pentachlorophenol, an intermediate product, refers to highly toxic substances (LD$_{50}$ = 29–36 mg/kg) [55]. This compound probably degrades rapidly, unlike the products of its further transformation. Di-, tri-, and tetrachlorophenols, di- and tetrachlorobenzenes, categorized as medium- and highly toxic compounds, were found in various animals and plants [54]. Hexachlorobenzene, which is formed as a result of lindane decomposition, exhibits low toxicity, but is highly persistent and carcinogenic to mammals [56]. Pentachlorobenzene, formed through further degradation, has the same properties.

Concentrations and proportions of HCH isomers in the components of the marine environment, open and enclosed water bodies, and living organisms depend on a multitude of factors: the physical and chemical properties of water and sediments, illumination, the specific features of biotransformation processes, and also the duration of exposure to the pesticides in the environment. The transition of OCPs dissolved in river water to suspension occurs largely in the ecosystems of river deltas and estuaries, at the river–sea interface, where the salinity sharply increases. The solubility of OCPs in water drops with increasing salinity, and they pass into the suspended state [57]. The increased OCP content of the organs of riverine and estuarine species, as compared to marine ones, is associated with this solubility variation [50].

The effects of these and many other factors cause the most stable isomer, β-HCH, to be accumulated over time and with the transfer up the food chain [51]. The ratio of α- and γ-HCH concentrations is used to estimate the time period since the entry of pesticides into the ecosystem. A high value of the ratio, more than unity, indicates the long-term presence of OCPs in the environment; a value below unity, i.e. the dominance of γ-HCH, is evidence of recent (fresh) entry [16].

DDT exists in the form of the main product and its metabolites, DDD and DDE. The time period of exposure of the environmental components to DDT is estimated as the ratio of the DDT concentration to that of DDE, a product of DDT degradation. High values of the DDT/DDE ratio indicate a recent entry of DDT into the environment; low values are evidence of a long-term presence in the system and the gradual transformation into DDE.

The relative stability of DDE and β-HCH in the environment is determined by both chemical and physical factors. The persistence of DDE is due to the low reactivity of the conjugate π–π and π–*n* bond system characteristic of vinyl halogen compounds. The persistence of β-HCH is explained by the low solubility of the compound in aqueous systems and, accordingly, low availability for further chemical and biochemical transformation.

The major biological mechanisms of OCP destruction in the environment are processes of metabolic degradation by microorganisms. The soil microorganisms *Aerobacter aerogenes*, *Pseudomonas fluorescens*, *E. coli*, and *Klebsiella pneumoniae* can decompose DDT under aerobic and anaerobic conditions, forming 4-chlorobenzoic acid and DDE, respectively, as the main metabolites. DDE, the main metabolite of DDT, is extremely resistant to biodegradation. In forest soils, the half-life of DDT is 20–30 years [20].

Thus, concentrations and proportions of OCP isomers and derivatives in the ecosystem continuously vary and depend on the physical and chemical properties of water and soils, as well as on the duration of exposure of the environment these pesticides.

1.2.4 Toxicity of OCPs

The most commonly used characteristic of the hazard posed by a chemical is quantitative data on its acute toxicity for some laboratory animals. However, for adequate assessment of the impact on populations, communities, and ecosystems, more extensive experiments are required, involving not only animals, but also microorganisms and plants. Thus, in the EU countries, a set of standard tests are usually performed for these proposes that provide an essential minimum of information to make a conclusion about the toxicity of the substance under consideration. For new pesticides, the following tests of lethal (L) or effective (E) doses (LD, ED) and concentrations should be carried out [40]:

(1) LD_{50} level for mammals (orally on rats);
(2) LC_{50} for mammals (through inhalation on rats);
(3) LC_{50} for earthworms;
(4) ED_{50} for soil bacteria;
(5) LD_{50} orally for birds;
(6) LC_{50} orally for bees;
(7) LC_{50} within 96 h for fish;
(8) LC_{50} within 48 h for water fleas *Daphnia*;
(9) EC_{50} in experiments with microalgae growth inhibition.

Estimates are obtained for each of the tests, which are summed up to calculate the efficiency factor (EF). The latter, as well as the exposure factor (E), can be found for each environment separately. Then the risk factors are calculated for these environments:

$$X = EF_s E_s (\text{soil}),$$
$$Y = EF_w E_w (\text{water}),$$
$$Z = EF_a E_a (\text{air}),$$

which are normalized into solubility factors S_x, S_y, and S_z (maximum values of X, Y, and Z are assumed to be equal to 1; minimum values, to 0).

Preliminary conclusions about the toxicity or risk are drawn based on the values obtained: the risk is considered high at $S_i \geq 0.55$, potentially significant at $0.55 < S_i \geq 0.30$, and insignificant or absent at $S_i < 0.30$ [40].

The hazard that organochlorine compounds pose to terrestrial and aquatic organisms became clearly evident as early as in the middle twentieth century. The danger

Table 1.3 Importance of safety standards: safe levels of organochlorine compounds in the environment [16, 42, 56, 61]

Standard/Environment	HCH	DDT	PCB
MPC/Air (mg/m^3)	0.001	0.001	0.001
MPC/Water (mg/dm^3)	0.002	0.1	0.001
MPC/Soil (mg/kg)	0.1	0.1	0.00006

of insecticides to fish and aquatic invertebrates was revealed by studies in the 1940s and 1950s. As was found in the 1960s, the toxic effect of these substances on the most sensitive aquatic organisms becomes acute in a concentration range from 10^{-3} to 10^{-12} g/L [58, 59]. Such a high sensitivity to low concentrations is explained, on the one hand, by the extreme toxicity of these compounds and, on the other, by the specific pattern of their effects on the life functions. They easily affect any arthropods, in particular crustaceans, that represent the major part of marine and freshwater zooplankton.

Mass mortality of common aquatic species reduces ecosystem's self-purification capacities, since the vital activity of aquatic organisms such as bacteria, algae, crustaceans, and mollusks provides the transformation of organic matter in a water body.

The large-scale application of pesticides necessitates a thorough study of the harmful effects that they exert on the human body, taking into account both immediate and long-term consequences [8, 60]. In this regard, strict regulation of the levels of these compounds in food products is necessary.

Environmental pollution control is based on sanitary and hygienic standards that regulate maximum permissible concentrations (MPC) of harmful chemicals in the air, water, soil, and food. Such regulation consists in determining the minimum concentrations of pollutants that guarantee safety for the human health and environment. Safe levels of OCPs in the air, water, soil, and animal feed are indicated in many studies and regulatory documents (Table 1.3).

Table 1.4 presents the hygienic standards (MPC of toxic substances) of food safety for humans established in the Russian Federation. Manufacture, import, and distribution of food products are required to comply with these standards [62].

For some of food items (fruits and vegetables, milk), the current regulatory standards for DDT and its metabolites (DDE and DDD) adopted by various international organizations are the same as those in Russia, while for others (eggs, meat, cereals), the standards are approximately fivefold higher (Table 1.5).

The MPCs of DDT and its metabolites established in the Russian sanitary regulations and standards (SanPiN 2.3.2.1078-01) are lower for some food items than the US EPA standards (for eggs and meat). The HCH level is not determined in the international standards, since its level is considered low.

Table 1.4 MPC of organochlorine pesticides and polychlorinated biphenyls in food products, mg/kg wet weight [62]

Standard	HCH isomers	DDT and its metabolites	PCB congeners	Notes
MPC, meat	0.1	0.1	–	
MPC, eggs	0.1	0.1		
MPC, milk	0.05	0.05		
MPC, fish	0.2	0.2	2.0	Marine
	0.2	2.0		Sturgeon, salmon, herring
	0.03	0.3		Freshwater
MPC, meat	0.2	0.2		Marine animals
MPC, fish roe	0.2	2.0		
MPC, fish liver	1.0	3.0	5.0	
MPC, cereals	0.5	0.02	–	
MPC, vegetables	0.1	0.1		
MPC, fruits	0.05	0.1		

Table 1.5 Current standards for DDT and its metabolites (DDE and DDD) adopted by various international organizations [63]

Source of data or organization	Standard for DDT and metabolites
FAO/WHO, permissible daily intake	20 μg/kg
WHO, guidelines on drinking water	1 μg/L
WHO guidelines on DDT, milk (in fat)	1 μg/g fat
US EPA*, minimum risk level (MRL)	0.5 ng/kg/day
US EPA, fruits and vegetables	0.1–0.5 mg/kg
US EPA, eggs, cereals	0.5 mg/kg
US EPA, milk	0.05 mg/kg
US EPA, meat	5.0 mg/kg

*US EPA is the Environmental Protection Agency of the United States

1.3 Polychlorinated Biphenyls (PCB)

1.3.1 Physical and Chemical Properties of PCBs

PCBs were used as electrically conductive liquids in electrical equipment, lubricants, and coolants, and also as components in the production of plasticizers, pesticides, dyes, and varnishes [16, 64]. PCBs are usually heavy, high-boiling, oil-like liquids with dielectric properties. PCBs are extremely inert and poorly soluble in water [65]. The physical and chemical properties of PCBs are presented in Table 1.6.

Table 1.6 Physical and chemical properties of polychlorinated biphenyls [2, 65]

Group of congeners	Molecular weight	Vapor pressure, Pa	Solubility in water, mg/L	log K_{OW}
Monochlorobiphenyl	188.7	0.9–2.5	1.21–5.5	4.3–4.6
Dichlorobiphenyl	223.1	0.008–0.60	0.06–2.0	4.9–5.3
Trichlorobiphenyl	257.5	0.003–0.22	0.015–0.4	5.5–5.9
Tetrachlorobiphenyl	292.0	0.002	0.0043–0.010	5.6–6.5
Pentachlorobiphenyl	326.4	0.0023–0.051	0.004–0.02	6.2–6.5
Hexachlorobiphenyl	360.9	0.0007–0.012	0.0004–0.0007	6.7–7.3
Heptachlorobiphenyl	395.3	0.00025	0.000045	6.7–7
Octachlorobiphenyl	429.8	0.0006	0.0002–0.0003	7.1
Nonachlorobiphenyl	464.2	–	0.00018–0.0012	7.2–8.16
Decachlorobiphenyl	498.7	0.00003	0.000001	8.26

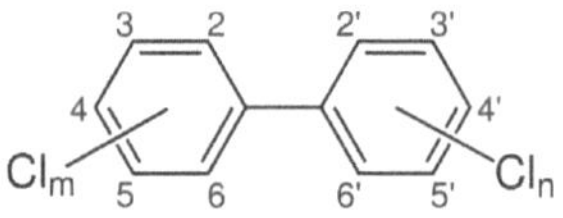

Fig. 1.6 General structural formula of polychlorinated biphenyls (PCB)

PCBs differ from each other in the *n* number (from 1 to 10) and the position of chlorine atoms in the molecule. Theoretically, there can be a total of 209 PCB congeners (Fig. 1.6) differing in the number and position of chlorine atoms in the molecule, but only about 130 of them have been recorded from the environmental components to date. The numbering of PCB congeners now adopted by IUPAC was proposed in 1980 [66].

The PCB congeners exhibit different physical and chemical properties, which eventually cause their different environmental behaviors and toxicities. Composition of technical mixtures depends on the conditions of synthesis and usually includes an extremely wide range of PCBs. These have a low solubility in water, which decreases with increasing number of chlorine atoms in the molecule [66].

Two benzene rings in the biphenyl molecule can be oriented on the same plane or at an angle to each other (up to 90°). The number and positions of substituents determines the degree of rotation of the benzene rings relative to the bond axis. Thus, the absence of chlorine atoms at *ortho* positions or the presence of only a single substituent at *ortho* position (mono-*ortho* substituted biphenyls) contributes to the retention of a flat structure, and such PCBs are referred to as planar or coplanar congeners [65, 67]. Previously, it was suggested that the PCB toxicity increases with increasing number of chlorine atoms in the molecule, but studies have eventually shown that persistence, bioaccumulation capacity, and toxicity of certain isomers depend on the positions of chlorine atoms in the molecule. The PCBs with no chlorine atoms at *ortho* positions and mono-*ortho* substituted PCBs (coplanar PCBs) have been found to be the most toxic ones [66].

In the environmental components, PCBs are present in the form of mixtures of various isomers with different toxicities. Therefore, to assess the overall PCB toxicity in this system, a set of toxic indicators is used, which have been selected on the basis of toxicological studies for various *ortho* unsubstituted and also mono-*ortho* and di-*ortho* substituted PCBs [68].

As a result of combustion, PCBs can form even more toxic substances such as hydrogen chloride, dioxins, and dibenzofurans. It was reported that pyrolysis of technical PCB mixtures causes the formation of some dibenzofurans. The latter are also by-products of technical synthesis of PCBs [65].

1.3.2   Distribution, Metabolism, and Degradation of PCBs

PCBs are found ubiquitously in the environment. Atmospheric transport plays a crucial role in the global spread of these compounds. The patterns of distribution of PCB congeners in the atmosphere vary depending on the number of chlorine atoms in the molecule [21]. Thus, monochlorobiphenyls are found mainly in the atmosphere; *ortho*-substituted PCBs with 1–4 chlorine atoms are transported to the poles through repeated evaporation–condensation cycles between air and water/soil. PCBs with 4–8 chlorine atoms are found in mid-latitudes, while PCBs with 8–9 chlorine atoms remain close to sources of their initial entry into the environment. In the atmosphere, PCBs stay in the form of vapors and adsorbed to aerosol particles. In the vapor form, PCBs are more mobile and can be transported by air masses over longer distances than the aerosol-sorbed form. PCBs with vapor pressures of more than 10^{-4} mmHg (mono- and dichlorobiphenyls) are present in the atmosphere almost only in the vapor form; PCBs with values of less than 10^{-8} mmHg (tri-, tetra-, penta, hexa-, and heptachlorobiphenyls) are in the aerosol-sorbed form; and PCBs with values from 10^{-4} to 10^{-8} can be both in the vapor form and as part of aerosol particles [69]. Thus, lower chlorinated PCBs more easily become subject to atmospheric transport.

PCBs enter water bodies from the atmosphere with wet and dry precipitation, and also with leachates from soils. In water bodies, PCBs are redistributed between water and bottom sediments. Higher chlorinated PCBs without substituents at *ortho* positions and characterized by low solubility in water and a high octanol/water partition coefficient (K_{OW}) tend to be adsorbed in bottom sediments, whereas lower chlorinated PCBs with higher solubility in water and low K_{OW} are found predominantly in the water column. In the upper layers of bottom sediments, PCBs are more easily leached and involved in redistribution processes [5, 70, 71].

Persistence of PCBs in the environment depends on the number of chlorine atoms and their positions in the molecule. In the atmosphere, the prevailing process of PCB transformation is reaction with hydroxyl radical, whereas the role of photolysis in the PCB degradation in the atmosphere has turned out to be insignificant. The estimated half-life values for various PCBs subject to reaction with hydroxyl radical increase with increasing number of chlorine atoms in the molecule from two days for biphenyl to 75 days for hexachlorobiphenyl. The total rate of degradation of PCBs as a result of this reaction is 8300 t per year [65, 72].

In water, photolysis is the prevailing mechanism of abiotic degradation of PCBs, while the contribution of hydrolysis and oxidation processes is insignificant [65]. During photolysis, the chlorine–carbon bond is breaking and the chlorine atoms are gradually replaced by hydrogen [73]. The PCB biodegradation in surface water is usually an aerobic process that depends on the structure of isomers and environmental conditions. It is known that lower chlorinated PCBs (mono- and disubstituted) more easily undergo biodegradation than higher chlorinated ones [64].

PCB biodegradation is the major mechanism of PCB decomposition in bottom sediments and soil that can occur under both anaerobic and aerobic conditions. The aerobic PCB degradation usually has two stages: the transformation of PCBs into the respective benzoic acids and then the mineralization of chlorobenzoates into carbon dioxide and inorganic chlorides. Under anaerobic conditions, the reductive dechlorination of PCBs occurs, which causes chlorine atoms to be cleaved off without destroying the benzene rings, and less toxic mono- and dichlorobiphenyls are formed. The rate of PCB biodegradation is also determined by the position and number of chlorine atoms in the molecule. An increase in the number of chlorine substitutes in the molecule in general and at *ortho* positions in particular significantly inhibits the PCB biodegradation (Fig. 1.7) [64, 74, 75].

The global redistribution of PCBs was recorded using several semi-permeable membrane devices for sampling atmospheric air that were installed at certain distances from each other between southern England and northern Norway [24].

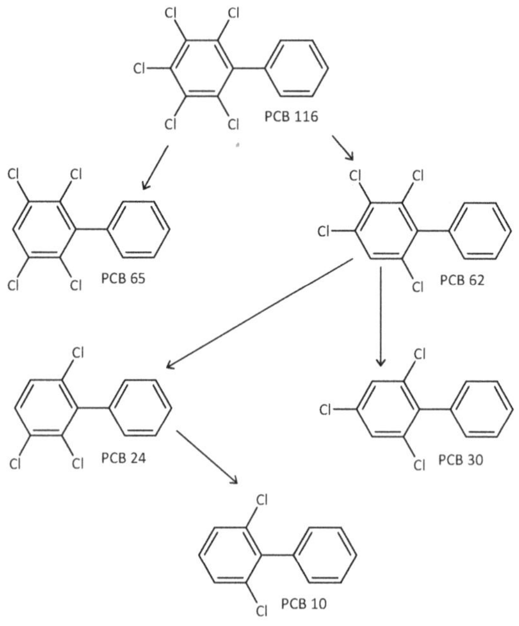

Fig. 1.7 Diagram of metabolic transformations of PCB by microorganisms [76]

In the air taken with these samplers, the qualitative composition of PCB isomers varied from south to north, with an increase in the proportion of lower chlorinated PCBs at more northerly points.

PCBs are usually considered persistent organic pollutants. However, many PCB congeners, including chiral PCBs, are biotransformed through complex, species-dependent metabolic pathway to hydroxylated, methylsulfonylated, sulfated, glucuronidated, and other metabolites [77]. In chiral PCBs, both phenyl rings have an asymmetric substitution pattern relative to the axis formed by the central C–C bond of the biphenyl system. These compounds make up to 6% of technical PCB mixtures in weight, and are subject to substantial atropoisomeric enrichment in wild animals, laboratory animals, and humans. The oxidation of chiral PCBs, especially those with a 2,3,6-trichloro-substituted structure in one benzene ring, into hydroxy-PCBs has been extensively studied using recombinant enzymes, liver microsomes, isolated hepatocytes, liver, hippocampus, and skin sections from mammalians [78]. To the best of our knowledge, the PCB oxidation has not been investigated in amphibians, fish, or birds to date. Indirect evidence, e.g., from toxicokinetic studies, suggests that PCBs are also subject to biotransformation in a multitude of animal species.

For example, PCB 136 (Fig. 1.8) can be oxidized by direct insertion of oxygen into a *meta* C–H bond or form epoxide intermediates. PCB 136 epoxides can either rearrange to form HO-PCB, undergo 1,2-shift to form 3-HO-PCB 150, from a dihydrodiol epoxide and react with cellular nucleophiles, such as glutathione (GSH). HO-PCBs can be further oxidized to dihydroxylated metabolites, such as 4,5-HO-PCB 136. Alternatively, mono- and dihydroxylated PCB 136 metabolites may undergo phase II metabolism to glucuronide or sulfate conjugates. Similar to other dihydroxylated PCB metabolites, 4,5-HO-PCB 136 can likely be oxidized via a semiquinone radical to PCBs quinones, which subsequently will react with cellular nucleophiles via a chlorine displacement reaction. Glutathione conjugates formed from PCB 136 epoxides will be metabolized in several steps to the corresponding methylsulfonyl metabolites. Metabolites shown in black have been detected in in vitro and/or in vivo studies, whereas experimental evidence suggests that metabolites shown in blue are likely formed in mammals [77].

1.3.3   Toxicity of PCBs

PCBs have a low toxicity with single exposure and a high cumulative potential in case of long-term exposure. The major routes of entry into the human body are through the skin, lungs, as well as via food chains in polluted areas.

Coplanar isomers, which have no more than one chlorine atom at *ortho* position (2,2′,6,6′), are considered the most hazardous ones. In terms of toxicity, these are identical to polychlorinated dibenzo-*p*-dioxins and polychlorinated dibenzofurans. PCB toxicity is measured on a toxicity scale similar to the scale for dioxins/furans. Toxicity equivalents have been developed for 12 PCB congeners.

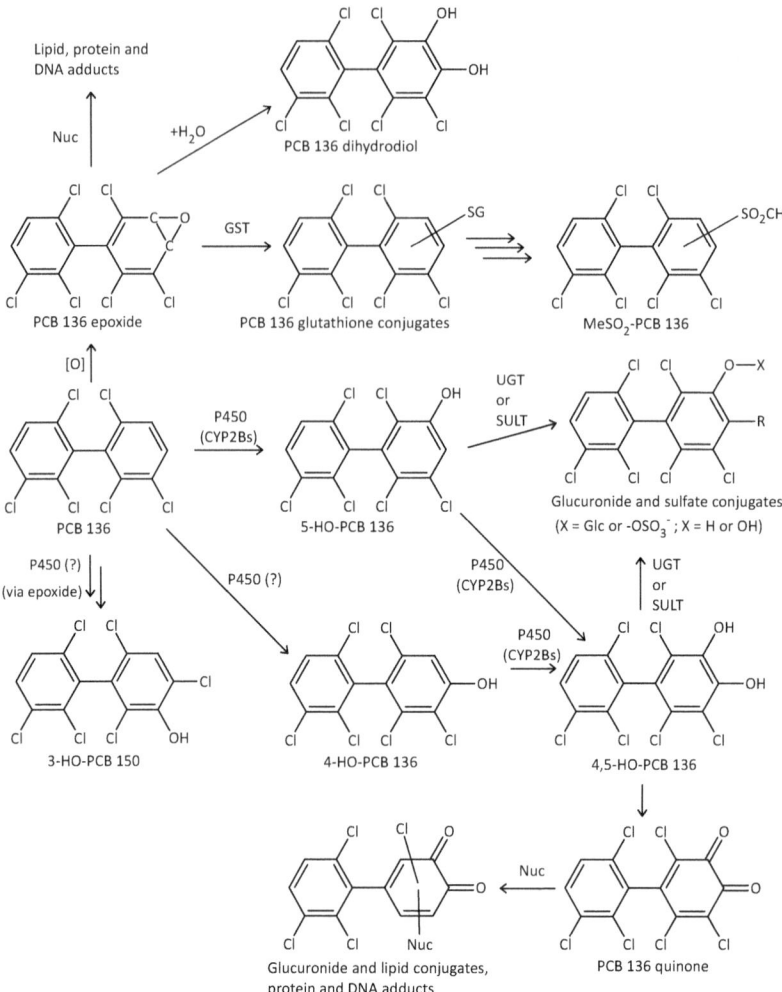

Fig. 1.8 Simplified diagram of metabolism of PCB 136 atropisomers [77]: Glc—glucuronide; GS—glutathione residue; GST—glutathione transferase; Nuc—cellular N- or S-nucleophile; P450—cytochrome P450 enzyme; SULT—sulfotransferase; UGT—UDP glucuronosyltransferase

A characteristic behavioral feature of PCBs in the environment is their very slow degradation. Having pronounced lipophilic properties, PCBs exhibit a high potential of bioaccumulation in fat-containing components (with the accumulation factor in some of biological components of the environment reaching dozens of millions). High PCB contents have been recorded from breast milk and human adipose tissues. PCBs are capable of penetrating the placenta and building up in fetal tissues [68, 79].

First cases of PCB poisoning of humans exposed to these compounds at various industries were recorded as early as in the 1930s. In 1963, the mass poisoning events at Japanese enterprises that produced capacitors became widely known. There is also an example of fatal PCB poisoning due to the damaged sealing of the heat exchange equipment during the rice oil refining process. This case, which went down in history as the *Yusho incident*, occurred in Japan in 1968. The poisoning symptoms were the enlargement and hypersecretion of the mammary glands, nausea, vomiting, hepatomegaly and hepatic porphyria, pathological changes in the peripheral nervous system and blood composition, and impaired adrenal function [80].

Of the symptoms of occupational PCB poisoning in workers exposed to these compounds during production activities, the most frequently reported ones are chloracne (skin damage) and various neurological phenomena in the form of headaches, fatigue, and formication (or crawling) sensation in the extremities.

Acute toxicoses are manifested as skin lesions, disorders of liver, kidneys, lungs, and central nervous system. Upon entering the body, PCBs are well absorbed in the gastrointestinal tract, in the lungs, penetrate the skin, and are accumulated mainly in adipose tissue. In most adipose tissue samples, the PCB content was 1 mg/kg or less (level in blood, 0.3 µg/100 mL); large amounts (up to 700 mg/kg) were found in adipose tissue samples from humans exposed to PCBs during their professional activity (level in blood, 200 µg/100 mL) [80].

PCBs are classified as immunotoxicants. Being potent immune suppressors (factors causing *chemical* AIDS), they pose serious hazard to human health. PCBs have pronounced embryotoxic and potential carcinogenic effects (with LD_{50} varying from 0.79 to 11 g/kg). After entering the body of a fetus and child, PCBs contribute to the development of, respectively, congenital malformations and childhood pathologies (developmental delay, decreased immunity, and impaired hematopoiesis). Furthermore, PCBs cause a reduction in the number of implantation sites and in the number of newborns, and also prolonged pregnancy. The long-term administration of PCBs to monkeys before and during gestation and during lactation resulted in miscarriages, premature birth, and early postnatal death. However, the most adverse effect of PCBs on humans is the mutagenic effect, which has negative consequences for the health of subsequent generations [65, 66].

It has been found that this group of compounds can interfere with the hormonal mechanisms and cause endocrine breakdowns; moreover, PCBs can mimic or block the action of thyroid hormones.

The elimination half-life of PCBs in humans is 5 years.

1.4 Polycyclic Aromatic Hydrocarbons (PAH)

1.4.1 Physical and Chemical Properties

Polycyclic aromatic hydrocarbons (PAH) are considered a group of priority pollutants. PAHs pose a hazard because, showing transformation activity, they can contribute to carcinogenic, teratogenic, or mutagenic changes in organisms [81].

PAHs are high-molecular-weight organic compounds with benzene ring as the main structural element (Fig. 1.9). Besides unsubstituted PAHs, there are a multitude of polycyclic structures containing the functional group either in the benzene ring or in a side chain. These include halogen, amino, sulfo, and nitro derivatives, as well as alcohols, aldehydes, ethers, ketones, acids, quinines, and other aromatic compounds (Table 1.7). The structural formulas of some polycondensed heterocyclic compounds, which are structural analogs of PAHs, are shown in Fig. 1.10.

According to the principle of annelation (addition) of benzene rings used [82], all unsubstituted PAH structures can be considered as derivatives of molecules of naphthalene and biphenyl, the simplest polycondensed compounds that originate two large groups of PAHs: cata-annelated and peri-condensed hydrocarbons, respectively.

PAHs are crystalline compounds (except a number of naphthalene derivatives) with high melting and boiling points (Table 1.8).

The solubility of PAHs in water is low and varies significantly between different hydrocarbons. Their solubility in organic solvents decreases with increasing molecular weight. Both parameters depend on relative positions of condensed benzene rings in the molecule. The solubility of pyrene in water is about 1000-fold higher than that of 3,4-benzapyrene (0.11 µg/L), which is minimum among the PAHs studied. The salt composition does not have any marked effect on PAH solubility.

The PAH solubility in water increases in the presence of benzene, oil, petroleum products, detergents, etc. The higher the level of these substances in effluents and water bodies, the more toxic and carcinogenic these PAHs can be in the water in a dissolved state.

Aromatic hydrocarbons are characterized by electrophilic substitution reactions. They almost do not enter addition reaction. Nitration, sulfonation, halogenation, alkylation, etc. reactions that occur through the mechanism of electrophilic substitution cause the formation of, respectively, nitro, sulfo, and halogen derivatives, alkyl-substituted PAHs, and aromatic ketones.

Aromatic hydrocarbons are oxidized to form quinones and carboxylic acids. Less stable hydrocarbons more easily enter substitution, oxidation, and addition reactions. Therefore, acenes or hydrocarbons containing acene structures are more reactive than phenes with the same number of rings or than hydrocarbons containing phenolic structures. Some PAHs, including carcinogenic ones, degrade under effects of strong concentrated acids, high-frequency currents, ultrasound, and UV light.

1.4.2   Sources and Transformation of PAHs in the Environment

The major sources of PAH formation are associated with various technological processes; more than half of emission comes from energy production and industrial discharges from coal-fired plants. In large cities, exhaust gases make a substantial contribution to the total PAH level.

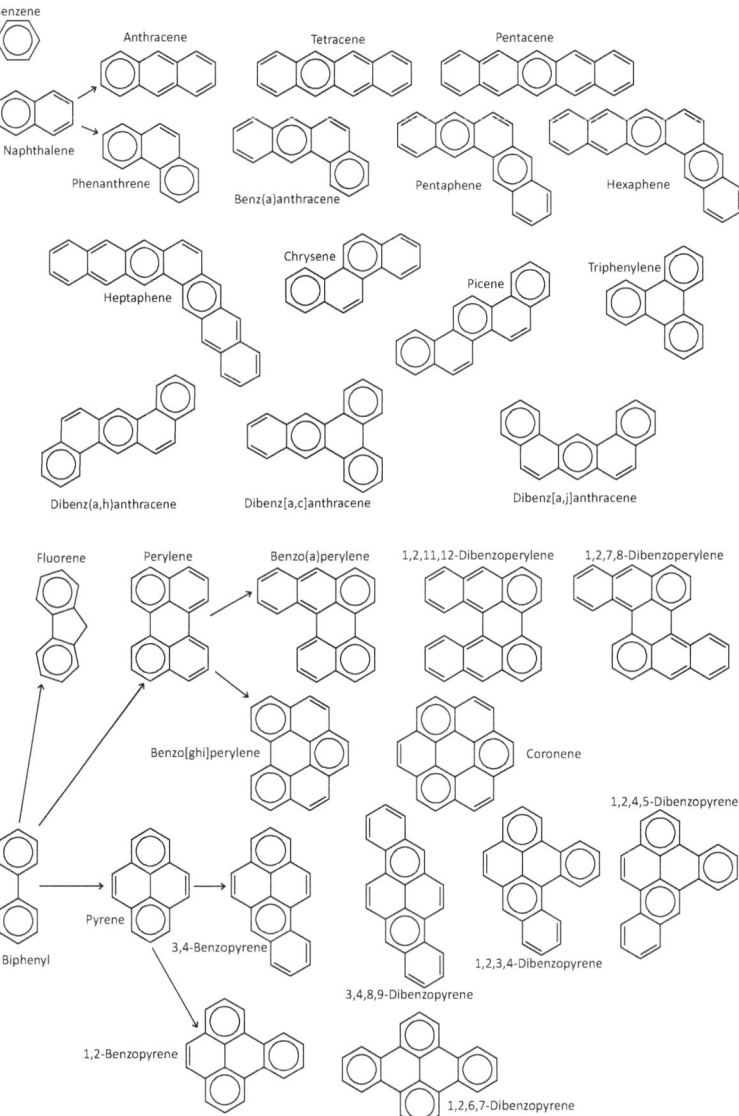

Fig. 1.9 Unsubstituted PAHs

PAHs are formed through free radical reactions from methane, other simple substituted and unsubstituted hydrocarbons, lipid peptides, carbohydrates, lignin, terpenes, nicotine, and leaf pigments. Several theories of hydrocarbon pyrosynthesis have been advanced, where ethylene, acetylene, and 1,3-butadiene act as intermediates. One of them is the well-known Badger's mechanism which consists in dimerization of 1,3-butadiene, accompanied by cyclization and dehydrogenation. The first stage of PAH

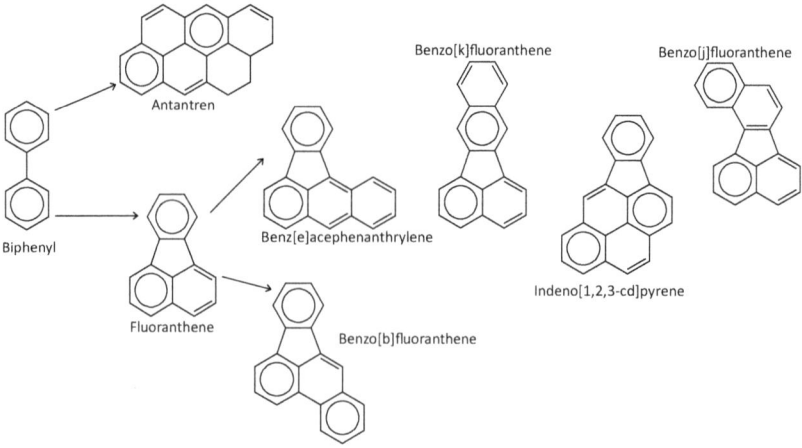

Fig. 1.9 (continued)

Table 1.7 Substituted derivatives of PAHs [2, 12]

Derivatives	Substituent	Substitution	
		In benzene core	In side chain
Alkyl derivatives	Alkyl radical	$Ar-C_nH_{2n+1}$	
Halogen derivatives	Halogen atom	ArHal	Ar–RHal
Nitro derivatives	Nitro group	$ArNO_2$	
Sulfo derivatives	Sulfo group	$ArSO_3H$	
Amino derivatives	Amino group		
Primary		$ArNH_2$	
Secondary		$(Ar)_2NH$	
Tertiary		$(Ar)_3N$	
Phenols	Hydroxyl group	ArOH	
Aromatic alcohols			Ar–ROH
Oxy derivatives	Oxygen atom	Ar=O	
		O=Ar=O	
Aromatic aldehydes	Aldehyde group	ArCHO	
Aromatic ketones	Keto group	$(Ar)_2CO$	
Aromatic acids	Carboxyl group	Ar–COOH	Ar–RCOOH
Phenolic acids	Carboxyl and hydroxyl groups	HOOC–Ar–OH	
Oxy acids			$Ar\text{-}R\text{-}COOH$ $\underset{OH}{\mid}$

Ar—aromatic radical; R—aliphatic radical

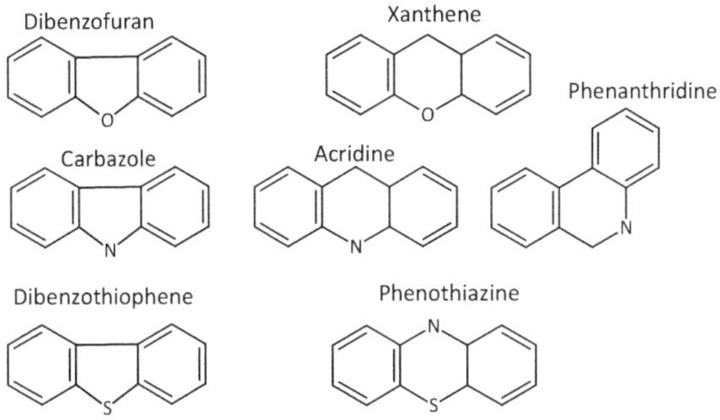

Fig. 1.10 Heterocyclic structural analogs of PAHs

Table 1.8 Physical and chemical properties of some PAHs [2, 12, 81]

Hydrocarbon	M_r	Temperature, °C		Solubility, µg/L		Vapor pressure, mmHg	K_{OW}	K_{OC}
		Melting point	Boiling point	In fresh water	In salt water			
Naphthalene	128	80	218	31,700	–	0.087	3.29	2.97
Acenaphthylene	152	92	–	16,100	–	0.029	4.07	1.40
Acenaphtene	154	96	–	3930	–	4.47×10^{-3}	3.98	3.66
Fluorene	166	116	293	1980	–	3.2×10^{-4}	4.18	3.86
Phenanthrene	178	100	340	1290	–	6.8×10^{-4}	4.45	4.15
Anthracene	178	218	340	73	–	1.75×10^{-6}	4.45	4.15
Fluoranthene	202	110	–	260	–	5×10^{-6}	4.90	4.58
Pyrene	202	156	399	95.8	78.9	2.5×10^{-6}	4.88	4.58
Tetraphene	228	158	396	0.91	0.63	2.5×10^{-6}	5.61	5.30
Chryzene	228	255	–	–	–	6.4×10^{-9}	5.9	–
1,2-benzopyrene	252	178	456	0.99	1.83	5.6×10^{-9}	6.06	6.74
3,4- benzopyrene	252	177	456	0.11	0.13	4.89×10^{-9}	6.11	5.98
1,12-benzoperylene	276	273	511	0.18	0.21	1.03×10^{-10}	6.50	6.20
1,2,5,6-dibenzanthracene	276	262	465	31.38	21.13	1×10^{-10}	6.84	6.52
1,2,3,4-dibenzanthracene	278	205	465	22.72	27.84	3.73×10^{-9}	6.41	–
1,2,7,8-dibenzanthracene	278	196	465	8.72	10.54	3.73×10^{-9}	6.54	–

M_r is relative molecular weight;
K_{OW} is octanol/water partition coefficient;
K_{OC} is organic carbon to water partition coefficient

formation is the thermal degradation of fuel resulting in relatively simple fragments of free radical molecules; the second stage is the recombination of radicals with the formation of PAHs [83].

The pyrolytic formation of PAHs is suggested to occur mainly at relatively high temperatures (650–900 °C) and low oxygen in flame. However, benzopyrene was also observed to form in case of low-temperature pyrolysis of wood within a narrow temperature range, with a maximum yield at 300–350 °C. Benzopyrene is formed in the low-temperature wood pyrolysis presumably by an ionic mechanism [84].

The composition and structure of PAHs that are formed during various pyrolytic processes depend both on the nature of source material and on the temperature. Polycyclic aromatic compounds formed through combustion mostly lack side substituents. As the temperature decreases, the amount of substituted PAHs formed grows. Alkyl-substituted PAHs, in particular, methylpyrenes, are also formed as a result of pyrolytic processes, although the proportion of structures with more complex side chains is very small. The soot formed during forest fires differs from the soot produced by anthropogenic sources by a high content of alkylated PAHs due to the lower ignition temperature of wood. Burning tobacco also gives a particularly high content of alkylated PAHs; indenes and isoprene are formed through pyrolytic reactions; the temperature of burning tobacco in cigarettes reaches 860 °C, but pyrolysis occurs mainly at a lower temperature, while hot gases are passing through organic material [12].

More than 200 PAHs (unsubstituted aromatic and alkyl-substituted hydrocarbons, nitrogen-, oxygen-, and sulfur-containing compounds) are found in the environment, primarily in the products of combustion of organic raw materials. The global emission of benzopyrene constitutes about 5000 t per year, of which coal combustion accounts for 61%; coke production, 20%; wood combustion, 4%; open fires of forests and crops, 8%; emissions from transport, 1%; oil and gas combustion, respectively, 0.09% and 0.06%. The percentages may differ between countries and cities.

Also, PAH concentrations may vary depending on the major activities of certain city or country. For example, the maximum benzapyrene concentrations exceeded 10 MPCs in Russia's 27 cities with populations of 9.6 million people each in 2016 (Table 1.9).

Petrogenic PAHs (from coal and oil) can enter the environment with emanations from producing horizons. In crude oil, PAHs with mainly two and three benzene rings make up from 0.2 to 7% and, according to some evidence, can even reach more than 30% [86]. In most cases, hydrocarbons, from benzene to phenanthrene, constitute about 90% of aromatic compounds in oil. Hydrocarbons of coal and oil usually contain more alkylated homologues than unsubstituted PAHs, with the number of alkyl radicals ranging from 2 to 5.

Organic fuel includes various proportions of PAHs: gasoline contains mainly monocyclic hydrocarbons and naphthalenes; diesel fuel includes polyarenes, from benzene to fluoranthene. Emissions from carburetor and diesel car engines can contain, according to various estimates, from 16 to 2300 mg/kg of polyarenes within the solid fraction of emissions. The major components of diesel engine emissions are fluoranthene and pyrene (from 46 to 88%); for carburetor engines, their content ranges within 15–17%. Benzofluorantenes, benzopyrene, and benzoperylene are also present in large quantities in emissions from gasoline engines.

Table 1.9 List of cities of the Russian Federation where high levels of benzopyrene pollution of atmospheric air were recorded (highest single concentrations of certain pollutants more than 10 MPC$_{h.s.}$) in 2016 [85]

City	Number of cases	Highest concentration MPC[1]	City	Number of cases	Highest concentration MPC[1]
Arkhangelsk	1	20.5	Novokuznetsk	7	28.8
Achinsk	2	12.7	Novosibirsk	3	14
Birobidzhan	3	24.6	Petrovsk-Zabaikalsky	3	35.2
Blagoveshchensk	2	13.4	Sayansk	1	10.3
Bratsk	9	80.3	Svirsk	1	14.9
Zima	6	53.3	Selenginsk	1	14.8
Irkutsk	1	18.1	Ulan-Ude	5	21.7
Kemerovo	1	13.9	Usolye-Sibirskoe	4	14.1
Krasnoyarsk	14	40.6	Chegdomyn	3	20.4
Kyzyl	4	17.5	Cheremkhovo	3	17.2
Lesosibirsk	6	33	Chernogorsk	3	16.3
Magnitogorsk	8	38.5	Chita	11	46.2
Minusinsk	4	36.9	Shelekhov	3	20
Nazarovo	2	14.4			

[1]The values are the highest single concentrations of pollutants divided by the highest single maximum permissible concentration (MPC$_{h.s.}$)

The major processes of transition of polyarenes from the atmosphere to the surface of land and water are dry precipitation, leaching by precipitation, and precipitation from the gaseous phase. The former two processes are of primary importance. The time of PAHs' presence in the atmosphere depends on the size of particles adsorbing them and also on weather conditions. Sub-micrometer particles can stay in the atmosphere for weeks, and, when entering water bodies, such particles remain suspended also for a long time. Almost all compounds entering the atmosphere with vehicle emissions precipitate within 50 m from the road and are leached from the land surface into water bodies, where they accumulate in bottom sediments [12, 87].

The decomposition of polyarenes in the environment occurs mainly through oxidation and biodegradation. The rate of degradation of these compounds is relatively low, especially for high-molecular-weight compounds, and can further decrease under low oxygen conditions.

PAHs can be subject to various chemical transformations and biological degradation in almost all environmental components.

Chemical transformations include photo-oxidation, interaction with oxidants, thermal reactions, etc. Chemical degradation, in particular photo-oxidation, is the major way of PAH transformation in the atmosphere; photo-oxidation can also have a substantial effect on the PAH degradation in an aquatic environment.

Biological degradation is associated with the involvement of PAHs in the metabolism of microorganisms, plants, and animals. For microorganisms, PAHs are a source of energy. For plant and animal organisms, PAHs, being involved in metabolic processes, in some cases become initial compounds for the synthesis of biopolymers [88]. Biological degradation of PAHs is the major mechanism of self-purification of soils, aquatic environments, and also plant and animal organisms.

Besides chemical and biological transformation, there are also physical processes such as leaching, weathering, etc. that provide the removal of PAHs from one natural environment and transition to another. Such processes are especially characteristic of atmospheric and soil conditions.

Biological degradation is a major route of PAH removal from the aquatic environment. Microbial populations play a primary role in the processes of biological self-purification of water bodies. For example, the PAH degradation by oil-oxidizing bacteria is usually considered the basic natural process of self-purification of the world's oceans from petroleum products [12].

The oxidation of aromatic hydrocarbons by microorganisms is an aerobic process, i.e. it requires molecular oxygen. The enzymatic cleavage of the aromatic core is suggested to follow the dihydroxylation. The vast majority of microorganisms cleave the aromatic ring through the formation of ortho-dioxy derivatives [88].

Many reactions of PAH photo-oxidation cause the formation of intrinsic peroxides in the solution, whose photolysis or pyrolysis leads, e.g., to the rupture of alkyl chains and rings. If intrinsic peroxides cannot be formed (due to their stereometry), ions emerge (Fig. 1.11).

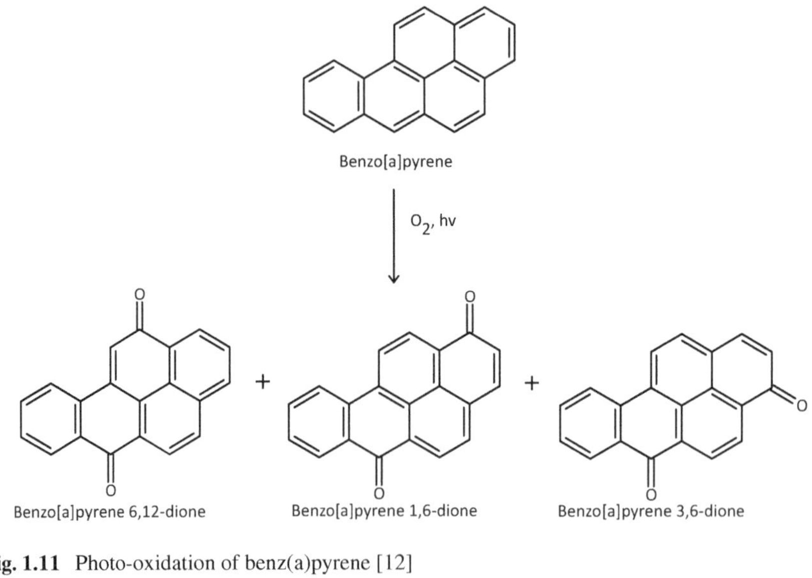

Fig. 1.11 Photo-oxidation of benz(a)pyrene [12]

Most bicyclic and tricyclic PAHs dissolved in water are not carcinogenous, but, when exposed to the UV radiation, they turn into compounds that become acutely toxic to various aquatic organisms.

1.4.3 Toxicity

Study of PAHs as chemically persistent and highly toxic compounds has a particular implication for environmental conservation and management. Toxicity is used as a collective term in this case. The same PAHs in some conditions can exhibit carcinogenic activity and cause tumor changes in the body (oncogenesis); in other conditions, they cause malformations (teratogenesis), poisoning (toxigenesis), mutations (mutagenesis), or stimulate or suppress the immune system [12].

Therefore, in the study [89], PAHs were suggested to be considered as chemical transformers of the biosphere. Some properties of PAHs affect the fate of living organisms, while others do the fate of future generations.

Carcinogenicity is one of the major indicators of the hazard that chemical compounds pose.

Carcinogenic PAHs with a pronounced blastomogenic effect caused tumors in 80–100% of experimental animals within a quite short time period. Weak carcinogens caused tumors in 20–30% of animals within significantly longer periods. Their carcinogenic activity towards humans has not been proven, but these compounds should be considered potentially hazardous [90].

Of the usual set of 13–16 unsubstituted PAHs present in the air and other natural environments, 3,4-benzopyrene and 1,2-5,6-dibenzanthracene exhibit the highest activity. These compounds caused tumors not only of subcutaneous tissues at the administration site, but also in the lungs of newborn mice, or after being injected into the trachea of adult mice or hamsters [12].

In the study of eight PAHs, the relative degrees of their carcinogenic activity were established as follows [91]: 3,4-benzopyrene > anthracene > 2,3-benzfluoranthene > (ortho-phenylene)pyrene > 11,12-benzfluoranthene and 10,11-benzfluoranthene. In the cases of 1,2-benzopyrene and 1,12-benzoperylene, no carcinogenic properties were found.

4-6-Cyclic PAHs substituted in a certain way show significant carcinogenic activity. Of particular interest are some methyl and dimethyl derivatives of PAHs, including 9,10- and 1,9-dimethyltetraphene and 20-methylcholanthrene, that are compounds with higher activity than 3,4-benzopyrene. These were identified in the atmospheric air only in few studies. Alkyl-substituted pyrene that are widely represented in anthropogenic emissions and in technogenically altered natural components of the environment, e.g., 1- and 3-methylpyrenes, show low carcinogenic and mutagenic activities [82, 92, 93].

Methyl groups introduced into the aromatic core at certain positions can increase the carcinogenic activity. Chrysene is not carcinogenic, and its methyl homologues are weakly (sometimes highly) active. Phenanthrene is not carcinogenic itself, but its

1,2,3,4-methyl derivative already exhibits low activity. When two methyl groups in this compound are substituted by a benzene ring, this leads to the formation of 1,2-dimethylchrysene, which shows higher activity, or 9,10-dimethyltetrafene, which is even more active. An increase in carcinogenic activity is also observed upon the transition of 3,4-benzphenanthrene to its 2-methyl derivative, and also in the case of antanthrene's transition to its 6-methyl and 2,6-dimethyl derivatives [82].

Heterocyclic analogs of PAHs, a number of compounds from the benzacridine group (which are not carcinogenic themselves, in contrast to their methyl homologues being very active), and other nitrogen-containing compounds (dibenzcarbazoles) also exhibit carcinogenic activity. For having this activity, both the position of the heteroatom in the ring and the positions of the substituents (if any) are important. Thus, alkyl derivatives of 1,2-benzacridine are more potent carcinogens than derivatives of 3,4-benzacridine. Some sulfur-containing PAH analogs are also active.

Heterogeneous polycyclic arenes (mainly nitrogen-containing) also pose a certain health hazard, although their pathogenic properties have not yet been sufficiently studied. In particular, nitroarenes are pollutants with a potential carcinogenic effect on humans and often contribute to increased mutagenic activity [12].

The carcinogenic activity of PAHs and their analogs is closely related to the property of these compounds to cause mutations at the genetic level. According to some evidence, the carcinogenic and mutagenic properties of chemical compounds, including PAHs, are recorded simultaneously in 85–93% of cases [94]; according to other data, in 60% of cases [95].

The major threat of the mutagenic effect is that carcinogenic substances are capable of interacting with macromolecular compounds in the cell (mainly with DNA) and, thus, increasing mutations in somatic and sex cells, which causes genetic pathologies in offspring and increases the prevalence of cancer in the current generation. Moreover, the accumulation of mutagens in the environment poses the hazard of increase in the mutation rate in other organisms, thus, probably contributing to acceleration of their evolution, which may be especially undesirable and dangerous in the case of pathogenic organisms.

The maximum permissible concentrations (MPC) of PAHs are closely associated with the harm that these compounds pose to animals and humans. In the Russian Federation, the standard safe level of benzopyrene, categorized as Class 1 hazardous substance, has been determined for almost all environmental components (including food). In addition, MPCs or approximately safe levels of impact (ASLI) have been established for six more PAH representatives (anthracene, acenaphthene, naphthalene, pyrene, phenanthrene, and dibenzoanthracene) (Table 1.10), however, so far only for the atmospheric air of populated places and the air of working zone [96].

Also, according to the "Carcinogenic Factors and Basic Requirements for the Prevention of Carcinogenic Hazard" sanitary regulations [103], benzoanthracene, benzopyrene, dibenzoanthracene, dibenzopyrene, and cyclopentapyrene are included in the list of factors carcinogenic to humans.

The level of MPC for benzopyrene adopted in different countries varies significantly. For example, in Russia, it is 20 µg/kg of soil (a general sanitary standard). In Germany, the Federal Soil Protection and Contaminated Sites Ordinance of July

Table 1.10 Sanitary and hygienic standards for PAHs in the environmental components in the Russian Federation

Polycyclic aromatic hydrocarbon	Soil[1,2], μg/g MPC	Air in populated places[3,4], mg/m^3	Air in working zone[5], mg/m^3 MPC	Water of water bodies used for economic, drinking, cultural, and household purposes[6], mg/L
Benzopyrene	0.02	*MPC$_{d.a.}$[3] = 0.000001	0.00015	0.00001
Anthracene	–	**ASLI[4] = 0.01	–	–
Acenaphthene	–	ASLI[4] = 0.07	10	–
Naphthalene	–	MPC$_{h.s.}$[3] = 0.000001	20	0.01
Pyrene	–	ASLI[4] = 0.001	0.03	–
Phenanthrene	–	ASLI[4] = 0.01	0.8	–
Dibenzoanthracene	–	MPC$_{d.a.}$[3] = 0.005	–	–

[1][97]; [2][98]; [3][99]; [4][100]; [5][101]; [6][102]; *MPC, maximum permissible concentration; **ASLI, approximate safe level of impact

12, 1999 established the differentiated standards for the benzo(a)pyrene content of soils depending on the category of use: for soils of children's playgrounds, 2000 μg/kg dry weight; for soils of urban park and recreational facilities, 10,000 μg/kg; for soils of industrial and commercial territories, 12,000 μg/kg [104, 105].

Acknowledgements The work was supported by the Ministry of Science and Higher Education of the Russian Federation, project no. FZNS-2023-0011.

References

1. CAS Common Chemistry. https://commonchemistry.cas.org/. Accessed 5 Jan 2023
2. PubChem PubChem. https://pubchem.ncbi.nlm.nih.gov/. Accessed 4 Jan 2023
3. Orris P (2000) Persistent organic pollutants and human health
4. Ritter L, Solomon K, Forget J et al (1995) Persistent organic pollutants an assessment report on: Ddt-Aldrin-Dieldrin-Endrin-Chlordane Heptachlor-Hexachlorobenzene mirex-toxaphene polychlorinated biphenyls dioxins and furans prepared. The International Program on Chemical Safety (IPCS) within the Framework of the Inter-Organization Program for the Sound Management of Chemicals (IOMC)
5. Burgess RM, McKinney RA, Brown WA (1996) Enrichment of marine sediment colloids with polychlorinated biphenyls: trends resulting from PCB solubility and chlorination. Environ Sci Technol 30:2556–2566. https://doi.org/10.1021/es9509137
6. Porpora M, Lucchini R, Abballe A et al (2013) Placental transfer of persistent organic pollutants: a preliminary study on mother-newborn Pairs. IJERPH 10:699–711. https://doi.org/10.3390/ijerph10020699

7. Hartwell SI, Apeti AD, Pait AS et al (2018) Benthic habitat contaminant status and sediment toxicity in Bristol Bay, Alaska. Region Stud Marine Sci 24:343–354. https://doi.org/10.1016/j.rsma.2018.09.009

8. Tsygankov VY, Boyarova MD, Kiku PF, Yarygina MV (2015) Hexachlorocyclohexane (HCH) in human blood in the south of the Russian Far East. Environ Sci Pollut Res 22:14379–14382. https://doi.org/10.1007/s11356-015-4951-3

9. Health Canada (2021) Federal contaminated site risk assessment in Canada: toxicological reference values (TRVs). Health Canada, Ottawa, Ontario

10. Chen M-W, Santos H, Que D et al (2018) Association between organochlorine pesticide levels in breast milk and their effects on female reproduction in a Taiwanese population. Int J Environ Res Public Health 15:931. https://doi.org/10.3390/ijerph15050931

11. ATSDR (2014) Toxicological profile for toxaphene. Department of Health and Human Services, Public Health Service, Atlanta, GA, U.S

12. Rovinsky FY, Teplitskaya TA, Alekseeva TA (1988) Background monitoring of polycyclic aromatic hydrocarbons. Gidrometeoizdat, Leningrad

13. UNEP (1989) DDT and its derivatives: environmental aspects. World Health Organization, Geneva

14. ATSDR (2002) Toxicological profile for Alpha-, Beta-, Gamma-, and Delta-Hexachlorocyclohexane. Department of Health and Human Services, Public Health Service, Atlanta, GA, U.S

15. ATSDR (2002) Toxicological profile for DDT, DDE, and DDD. Department of Health and Human Services, Public Health Service, Atlanta, GA, U.S

16. Rovinsky FY, Voronova LD, Afanasev MI et al (1990) Background monitoring of ground ecosystems contamination by organochlorine compounds. Gidrometeoizdat, Leningrad

17. ATSDR (2002) Toxicological profile for hexachlorobenzene. Department of Health and Human Services, Public Health Service, Atlanta, GA, U.S

18. Lobov VP, Efimov GA (1963) Pesticides. Gostekhizdat, Kyiv

19. Zhulidov A (2002) Levels of DDT and hexachlorocyclohexane in burbot (Lota lota L.) from Russian Arctic rivers. Sci Tot Environ 292:231–246. https://doi.org/10.1016/S0048-9697(01)01130-5

20. Tsydenova OV (2005) Organochlorine compounds in the ecosystems of Lake Baikal and its basin. Abstract of the dissertation for the degree of candidate of chemical sciences, Baikal Institute of Nature Management

21. Wania F, MacKay D (1996) Peer reviewed: tracking the distribution of persistent organic pollutants. Environ Sci Technol 30:390A-396A. https://doi.org/10.1021/es962399q

22. Wania F, Axelman J, Broman D (1998) A review of processes involved in the exchange of persistent organic pollutants across the air–sea interface. Environ Pollut 102:3–23. https://doi.org/10.1016/S0269-7491(98)00072-4

23. Van den Brink NW (1997) Directed transport of volatile organochlorine pollutants to polar regions: the effect on the contamination pattern of Antarctic seabirds. Sci Total Environ 198:43–50. https://doi.org/10.1016/S0048-9697(97)05440-5

24. Bidleman TF (1999) Atmospheric transport and air-surface exchange of pesticides. Water Air Soil Pollut 115:115–166. https://doi.org/10.1023/A:1005249305515

25. Macdonald RW, Barrie LA, Bidleman TF et al (2000) Contaminants in the Canadian Arctic: 5 years of progress in understanding sources, occurrence and pathways. Sci Total Environ 254:93–234. https://doi.org/10.1016/S0048-9697(00)00434-4

26. Negoita TG, Covaci A, Gheorghe A, Schepens P (2003) Distribution of polychlorinated biphenyls (PCBs) and organochlorine pesticides in soils from the East Antarctic coast. J Environ Monit 5:281–286. https://doi.org/10.1039/b300555k

27. Iatrou EI, Tsygankov V, Seryodkin I et al (2019) Monitoring of environmental persistent organic pollutants in hair samples collected from wild terrestrial mammals of Primorsky Krai, Russia. Environ Sci Pollut Res 26:7640–7650. https://doi.org/10.1007/s11356-019-04171-9

28. Tsygankov VY, Boyarova MD, Lukyanova ON (2016) Bioaccumulation of organochlorine pesticides (OCPs) in the northern fulmar (Fulmarus glacialis) from the Sea of Okhotsk. Mar Pollut Bull 110:82–85. https://doi.org/10.1016/j.marpolbul.2016.06.084

29. Tsygankov VY, Lukyanova ON, Boyarova MD (2018) Organochlorine pesticide accumulation in seabirds and marine mammals from the Northwest Pacific. Mar Pollut Bull 128:208–213. https://doi.org/10.1016/j.marpolbul.2018.01.027

30. Erdman L, WMO (2001) Atmospheric input of persistent organic pollutants to the Mediterranean sea. United Nations Environment Programme, Athens

31. Quémerais B, Lemieux C, Lum KR (1994) Concentrations and sources of PCBs and organochlorine pesticides in the St. Lawrence River (Canada) and its tributaries. Chemosphere 29:591–610. https://doi.org/10.1016/0045-6535(94)90446-4

32. Zeng EY, Yu CC, Tran K (1999) In situ measurements of chlorinated hydrocarbons in the water column off the Palos Verdes Peninsula, California. Environ Sci Technol 33:392–398. https://doi.org/10.1021/es980561e

33. Muir DCG, Grift NP, Lockhart WL et al (1995) Spatial trends and historical profiles of organochlorine pesticides in Arctic lake sediments. Sci Total Environ 160–161:447–457. https://doi.org/10.1016/0048-9697(95)04378-E

34. Harner T (1997) Organochlorine contamination of the Canadian Arctic, and speculation on future trends. Int J Environ Pollut 8:51–73. https://doi.org/10.1504/IJEP.1997.028158

35. Allen-Gil SM, Gubala CP, Wilson R et al (1997) Organochlorine pesticides and polychlorinated biphenyls (PCBs) in sediments and biota from four US Arctic lakes. Arch Environ Contam Toxicol 33:378–387. https://doi.org/10.1007/s002449900267

36. Jones KC, de Voogt P (1999) Persistent organic pollutants (POPs): state of the science. Environ Pollut 100:209–221. https://doi.org/10.1016/S0269-7491(99)00098-6

37. Burkow IC, Kallenborn R (2000) Sources and transport of persistent pollutants to the Arctic. Toxicol Lett 112–113:87–92. https://doi.org/10.1016/s0378-4274(99)00254-4

38. Crosby DC (1979) Transport and transformation of pesticides in the atmosphere. In: Proceedings of the I All-Union conference. Gidrometeoizdat, Moscow, pp 5–10

39. Tinsley IJ (1982) Behavior of chemical pollutants in the environment. Mir, Moscow

40. Isidorov VA (1999) Introduction to chemical ecotoxicology. Himizdat, St. Petersburg

41. Crosby DC (1983) Atmospheric reactions of pesticides. Pesticide Chem: Human Welfare Environ 3:327–332

42. Melnikov NN (1974) Chemistry and technology of pesticides. Khimiya, Moscow

43. Vrochinsky KK, Telitchenko MM, Merezhko AI (1980) Hydrobiological migration of pesticides, Monograph. Moscow State University, Moscow

44. Galiulin RV (1993) Biogeochemical approach to ecological regulation of persistent organochlorine compounds in agricultural landscapes. In: Ivanov MV, Bashkin VN, Snakin VV (eds) Biochemical foundations of environmental regulation. Nauka, Moscow, pp 49–64

45. Zelenin KN (2000) What is chemical ecotoxicology? Soros Educ J 6:32–36

46. Tanabe S (2007) Chapter 18 Contamination by persistent toxic substances in the Asia-Pacific region. In: Li A, Tanabe S, Jiang G et al (eds) Developments in environmental science. Elsevier, pp 773–817

47. Walker CH (1975) Variation in the intake and elimination of pollutants. In: Moriarty F (ed) Organochlorine insecticides: persistent organic pollutants. Academic Press, London, pp 73–131

48. Gold B, Leuschen T, Brunk G, Gingell R (1981) Metabolism of a DDT metabolite via a chloroepoxide. Chem Biol Interact 35:159–176. https://doi.org/10.1016/0009-2797(81)90140-X

49. Popova GV, Shamrova LD (1987) Accumulation of pesticides in the reproductive system of fish and their gonadotoxic effects. In: Experimental water toxicology, pp 191–201

50. Maslova OV (1981) Dependence of DDT accumulation on the content of lipids in the tissues of estuary fish. Hydrobiol J 17:75–77

51. Tanabe S, Tanaka H, Tatsukawa R (1984) Polychlorobiphenyls, DDT, and hexachlorocyclohexane isomers in the western North Pacific ecosystem. Arch Environ Contam Toxicol 13:731–738. https://doi.org/10.1007/BF01055937

52. Doelman P, Haanstra L, de Ruiter E, Slange J (1985) Rate of microbial degradation of high concentrations of α-hexachlorocyclohexane in soil under aerobic and anaerobic conditions. Chemosphere 14:565–570. https://doi.org/10.1016/0045-6535(85)90249-8
53. Virchenko EP, Borovikov TI (1985) Behavior of isomers in HCH in soils. IEM Proceedings. Gidrometeoizdat, Moscow, pp 18–23
54. The main toxic … (1989) The main toxic metabolites of pesticides widely used in the national economy. Obninsk
55. Gunther FA, Gunther JD (1979) Residue reviews. Springer, New York, New York, NY
56. Melnikov NN, Volkov AI, Korotkova OA (1977) Pesticides and the environment. Khimiya, Moscow
57. Israel YA, Tsyban AV (2009) Anthropogenic ecology of the ocean, Monograph. Nauka, Moscow
58. Braginsky LP, Komarovsky FY, Pischolka YK, Maslova OV (1980) Migration of persistent pesticides in freshwater ecosystems. In: Proceedings of the All-Union Conference. Gidrometeoizdat, Leningrad, Russia, pp 226–231
59. Braginsky LP (1981) Theoretical aspects of the problems of "norm and pathology" in aquatic ecotoxicology. In: Proceedings of the 3rd Soviet-American symposium. Nauka, Leningrad, Russia, pp 29–40
60. Tsygankov VY, Khristoforova NK, Lukyanova ON et al (2017) Selected organochlorines in human blood and urine in the south of the Russian Far East. Bull Environ Contam Toxicol 99:460–464. https://doi.org/10.1007/s00128-017-2152-0
61. Bespamyatov GP, Krotov YA (1985) Maximum permissible concentrations of chemical substances in the environment. Khimiya, Leningrad
62. SanPin 2.3.2.1078-01 (2001) SanPin 2.3.2.1078-01. Hygienic requirements of safety and nutritional value of food products
63. UNEP (2020) Stockholm convention on persistent organic pollutants
64. Fidler H (1998) Polychlorinated biphenyls. In: Proceedings of the subregional meeting on the identification and assessment of emissions of persistent organic pollutants, Moscow, pp 233–252
65. ATSDR (2000) Toxicological profile for polychlorinated biphenyls (PCBs). Department of Health and Human Services, Public Health Service, Atlanta, U.S
66. Giesy JP, Kannan K (1998) Dioxin-like and non-dioxin-like toxic effects of polychlorinated biphenyls (PCBs): implications for risk assessment. Crit Rev Toxicol 28:511–569. https://doi.org/10.1080/10408449891344263
67. Mizukami Y (1999) Exploratory ab initio MO calculations on the structures of polychlorinated biphenyls(PCBs): a possible way to make a coplanar PCB stable at coplanar conformation. J Mol Struct (Thoechem) 488:11–19. https://doi.org/10.1016/S0166-1280(98)00615-0
68. Van den Berg M, Birnbaum L, Bosveld AT et al (1998) Toxic equivalency factors (TEFs) for PCBs, PCDDs, PCDFs for humans and wildlife. Environ Health Perspect 106:775–792. https://doi.org/10.1289/ehp.98106775
69. Erickson MD (1997) Analytical chemistry of PCBs, 2nd edn. CRC/Lewis Publ, Boca Raton, Florida
70. Baker JE, Eisenreich SJ (1990) Concentrations and fluxes of polycyclic aromatic hydrocarbons and polychlorinated biphenyls across the air-water interface of Lake Superior. Environ Sci Technol 24:342–352. https://doi.org/10.1021/es00073a009
71. Achman DR, Brownawell BJ, Zhang L (1996) Exchange of polychlorinated biphenyls between sediment and water in the Hudson river estuary. Estuaries 19:950. https://doi.org/10.2307/1352310
72. Atkinson R (1995) Atmospheric chemistry of PCBs, PCDDs and PCDFs. In: Harrison RM, Hester RE (eds) Chlorinated organic micropollutants. Royal Society of Chemistry, Cambridge, pp 53–72
73. Barr JR, Oida T, Kimata K (1997) Photolysis of environmentally important PCBs. Organohalogen Compd 33:199–204

74. Abramowicz DA (1995) Aerobic and anaerobic PCB biodegradation in the environment. Environ Health Perspect 103(Suppl 5):97–99. https://doi.org/10.1289/ehp.95103s497
75. Williams WA, May RJ (1997) Low-temperature microbial aerobic degradation of polychlorinated biphenyls in sediment. Environ Sci Technol 31:3491–3496. https://doi.org/10.1021/es970241f
76. Van Dort HM, Smullen LA, May RJ, Bedard DL (1997) Priming microbial *meta*-dechlorination of polychlorinated biphenyls that have persisted in Housatonic river sediments for decades. Environ Sci Technol 31:3300–3307. https://doi.org/10.1021/es970347a
77. Kania-Korwel I, Lehmler H-J (2016) Chiral polychlorinated biphenyls: absorption, metabolism and excretion—a review. Environ Sci Pollut Res 23:2042–2057. https://doi.org/10.1007/s11356-015-4150-2
78. Dunaeva MN, Pankratov DV, Surovyi AL et al (2022) Reconstruction of epizootic outbreak provoked the largescale death of Rhinoceros auklet on the coast of the Japan Sea in the Southern part of Primorsky Krai (July, 2021). Acta biomedica scientifica 7:90–97. https://doi.org/10.29413/ABS.2022-7.3.10
79. Tanabe S, Subramanian A (2006) Bioindicators of POPs: monitoring in developing countries. Kyoto University Press, Kyoto, Japan; Trans Pacific Press, Melbourne
80. Kukharchik TI, Kakareka SV, Tsytik PV (2003) Polychlorinated biphenyls in electrical equipment. Minsktipproekt, Minsk
81. Lee B-K (2010) Sources, distribution and toxicity of polyaromatic hydrocarbons (PAHs) in particulate matter. In: Villanyi V (ed) Air pollution. Sciyo
82. Clar E (1964) Polycyclic hydrocarbons. Springer, Berlin Heidelberg, Berlin, Heidelberg
83. Badger GM (1962) The chemical basis of carcinogenic activity. Thomas, Springfield, Ill
84. Dikun PP (1970) Fluorescent spectral study of sources and distribution of polycyclic aromatic hydrocarbons in the human environment. Abstract of the dissertation for the degree of doctor of technical sciences
85. Chernogaeva GM (2017) Review of the state and environmental pollution in the Russian Federation for 2016. Federal Service for Hydrometeorology and Environmental Monitoring (Roshydromet), Moscow
86. Milukaite AA, Shopauskas KK (1979) Benz(a)pyrene in atmospheric fallout in the Lithuanian SSR. In: Slepyan EI (ed) Plants and chemical carcinogens. Nauka, Leningrad, pp 194–195
87. Pavlova NA, Donina IL (1979) The value of the solubility of benzo(a)pyrene in water for its transfer from soil to plant. In: Slepyan EI (ed) Plants and chemical carcinogens. Nauka, Leningrad, pp 99–100
88. Ugrekhelidze DSh (1976) Metabolism of exogenous alkanes and aromatic hydrocarbons in plants. Metsniereba, Tbilisi
89. Slepyan EI (1979) Transforming elements and compounds in the biosphere and plants. In: Slepyan EI (ed) Plants and chemical carcinogens. Nauka, Leningrad, pp 54–58
90. Shabad LM (1973) On the circulation of carcinogens in the environment. Meditsina, Moscow
91. Cooke M (1983) Formation, metabolism and measurement: 7th International Symposium. Battelle Press, Columbus
92. Dikun PP (1971) Carcinogenic properties of polynuclear aromatic hydrocarbons with condensed rings and their heterocyclic analogues. In: Lazarev NV (ed) Harmful substances in industry. Khimiya, Leningrad, pp 129–143
93. National Research Council (1972) Particulate polycyclic organic matter. The National Academies Press, Washington, DC
94. Dorbon M, Schmitter JM, Garrigues P et al (1984) Distribution of carbazole derivatives in petroleum. Org Geochem 7:111–120. https://doi.org/10.1016/0146-6380(84)90124-4
95. Zakharov IA (1981) Transforming elements and compounds in the biosphere and plants. In: Slepyan EI (ed) Problems of phytohygiene and environmental protection. Nauka, Leningrad, pp 194–200
96. Krylov A, Lopushaskaia E, Alexandrova A, Konopelko L (2012) Definition of polyaromatic hydrocarbons by the method of gas chromatography—mass-spectrometry with isotope dilution (GC/MS/IR). Analytics 6–16

97. GN 2.1.7.2041-06 (2006) GN 2.1.7.2041-06. Maximum permissible concentrations (MPC) of chemicals in the soil

98. MU 2.1.7.730-99 (1999) MU 2.1.7.730-99. Hygienic assessment of soil quality in populated areas. Guidelines

99. GN 2.1.5.1338-03 (2003) GN 2.1.5.1338-03. Maximum permissible concentrations (MPC) of pollutants in the atmospheric air of populated areas

100. GN 2.1.6.2309-07 (2007) GN 2.1.6.2309-07. Approximate safe exposure levels (SEL) of pollutants in the atmospheric air of populated areas

101. GN 2.2.5.1313-03 (2003) GN 2.2.5.1313-03. Chemical factors of the industrial environment. maximum permissible concentrations (MPC) of harmful substances in the air of the working area

102. GN 2.1.5.1315-03 (2003) GN 2.1.5.1315-03. Maximum permissible concentrations (MPC) of chemical substances in the water of water bodies of household and drinking and cultural and household water use

103. SanPin 1.2.2353-08 (2008) SanPin 1.2.2353-08. Carcinogenic factors and basic requirements for the prevention of carcinogenic danger

104. Kogut BM, Schulz E, Galaktionov AY, Titova NA (2006) Concentrations and composition of polycyclic aromatic hydrocarbons in the granulodensimetric fractions of soils in Moscow parks. Eurasian Soil Sc 39:1066–1073. https://doi.org/10.1134/S1064229306100048

105. Kapel'kina LP (2010) Contaminating substances in the megalopolis soils. The problems and the standardization paradoxes. In: Ecology of urban areas, pp 13–19

Chapter 2
Methods to Determine Persistent Organic Pollutants in Various Components of Ecosystems in the Far Eastern Region

Abstract The methods to determine persistent organic pollutants (POP) in various components of ecosystems in the Far Eastern region of Russia are considered. The techniques for the preparation of biological samples, chromatographic analysis, and POP estimations are described.

Keywords POPs · OCCs · OCPs · Sample preparation methods · Chromatography

The importance of persistent organic pollutants (POPs) due to their toxic effects has necessitated searching for sensitive and selective methods such as chromatography to determine these substances in many environmental and medical matrices (water, sediments, organs and tissue of animals, human blood, organs, adipose tissue, etc.). Organochlorine compounds (OCCs) are a group of lipophilic chemicals that comprise organochlorine pesticides (OCPs) and other POPs including polychlorinated biphenyls (PCBs). Exposure to OCCs results in the bioaccumulation of these substances in the human body (in adipose tissue, breast milk, etc.). The techniques used for sample preparation and determination of POPs are standardized and recommended for a wide range of plant- and animal-derived substances. Mass content of persistent organic pollutants (POPs) in biological samples is usually measured by gas chromatography coupled with an electron capture detector and a mass selective detector.

In this chapter, we consider the standard and modified techniques to prepare samples for gas chromatographic determination of POPs in biotic and abiotic components of ecosystems. The authors of the book use the methods described below in their research.

The results achieved by applying these methods are an increase in the efficiency and precision due to a more complete extraction of POPs chemically bound to lipids with hexane and also a reduction in the number of steps needed for the extraction and clarification of co-extrusive substances with concentrated sulfuric acid. Therefore, these methods allow identification of the lowest OCP concentrations in the organs and whole body. These methods were used to extract POPs from samples of biological liquids of human [1–4], marine mammals from the Bering Sea [5–7], seabirds from

© The Author(s), under exclusive license to Springer Nature Switzerland AG 2023
V. Tsygankov, *Persistent Organic Pollutants in the Ecosystems of the North Pacific*,
Earth and Environmental Sciences Library,
https://doi.org/10.1007/978-3-031-44896-6_2

the Sea of Okhotsk [6–8], Pacific salmon [6, 9–12], and some abiotic components of the environment [13].

2.1 Preparation of Samples from Abiotic Components of Ecosystems to Determine POPs

To preserve all the detectable components and properties of water, samples were taken and fixed with concentrated hydrochloric acid and methylene chloride. Samples of bottom sediments were collected with a bottom grab sampler (10 cm of the upper layer), packed in aluminum containers, and immediately frozen to − 18 °C. Further analysis was carried out by the officially approved methods with some modifications [14, 15].

The POP determination in water and bottom sediments was based on quantitative extraction of OCPs and PCBs from a sample with n-hexane, concentration of the extract through evaporation, and analysis by gas chromatography coupled with mass selective detector and/or electron capture detector.

Water. A water sample of 100 cm^3 for the analysis was placed in a 100 cm^3 graduated cylinder. Then the content of the cylinder was transferred to a 250 cm^3 separatory funnel, supplemented with 10 cm^3 of n-hexane, and then the separatory funnel was placed in a shaker at 60–80 shakes per min for 10 min. After stopping the shaker, the separatory funnel was left until the complete separation of phases (~15 min). The hexane extract was filtered through 10–15 g sodium sulfate (Na$_2$SO$_4$) pre-moistened with n-hexane (~2 cm^3) in a 50 cm^3 conical flask. The sodium sulfate was washed twice with 2 cm^3 portions of n-hexane; the wash was then added to the total extract.

The extract was concentrated through evaporation on a rotary evaporator or in a sand bath under air stream at 65 ± 5 °C. The extract was concentrated to a volume of ~3 cm^3 and transferred to a 10 cm^3 graduated test tube. The emptied flask was flushed with 1–2 cm^3 of n-hexane into the same tube, and the extract was concentrated through evaporation to a final volume of 1.0 cm^3. Then the extract was placed in a 2 cm^3 vial and tightly sealed. The resulting extract was analyzed on the same day.

Bottom sediments. A collected sample was poured onto a sheet of paper, freed from mechanical inclusions (undecayed roots, plant residues, stones, etc.), then ground, if necessary, in a porcelain mortar, and mixed. Afterwards, the sample was spread flat into an even layer on another sheet of paper and divided diagonally into four triangles (quartering method), of which two opposite ones were discarded, and the remaining two were used to make an averaged sample from which specimens were taken for chromatographic analysis and for measuring humidity.

A sample of 10.0 ± 0.5 g was placed in a 100 cm^3 conical flask with a ground glass stopper. The weighted sample was supplemented with 10 cm^3 of acetone and 10 cm^3 of n-hexane, closed with the stopper, and placed in an ultrasonic bath for 30 min. After the separation of layers (~5 min), the upper, liquid phase was decanted into a 250 cm^3 separatory funnel, the sample and the glassware were flushed with 5–10 cm^3 of n-hexane, the flush was added to the extract in the separatory funnel, which was also supplemented with approximately 20 cm^3 of distilled water in this funnel, and stirred for 10 min. The separatory funnel was placed in a clamp of a retort stand and left until the phases were completely separated (~10 min).

The lower, aqueous layer was discarded, and 6–8 cm^3 of concentrated sulfuric acid (H$_2$SO$_4$) was added. The mixture was gently agitated (but not shaken), with the stopper periodically opened, and left for 5–10 min for layer separation. Then the layer containing sulfuric acid (lower) was decanted, and the organic layer was supplemented with another 6–8 cm^3 of acid. The hexane layer was clarified with concentrated H$_2$SO$_4$ until the layer containing sulfuric acid became transparent.

The clarified extract was washed, first, with 10–15 cm^3 of distilled water, then with 8–10 cm^3 of a sodium bicarbonate (NaHCO$_3$) solution, and again with 10–15 cm^3 of distilled water until the washing water became neutral (the pH value of the washing water was determined using a common indicator paper). The clarified and washed n-hexane extract was filtered through 5–7 g sodium sulfate (Na$_2$SO$_4$) and collected in a 50 cm^3 flask. The sodium sulfate was washed twice with 2 cm^3 portions of n-hexane; the wash was then added to the total extract.

The clarified and dried extract was concentrated through evaporation on a rotary evaporator or in a sand bath under air stream at 65 $\pm$ 10 °C to a volume of ~3 cm^3 and then transferred to a 10 cm^3 calibrated tube.

The emptied flask was flushed with 1–2 cm^3 of n-hexane into the same tube, and the extract was concentrated through evaporation to a final volume of 1.0 cm^3. Then the extract was placed in a 1.5–2.0 cm^3 vial and tightly sealed. The resulting extracts were analyzed on the same day.

2.2 Preparation of Samples from Aquatic Organisms for POP Determination

The purpose of sample preparation is to extract lipids with acetone and hexane and then decompose fat components with concentrated sulfuric acid.

Frozen samples (– 20 °C) were delivered to the laboratory. A weighed tissue sample (10 g) was homogenized on a micro tissue homogenizer for 5 min in a mixture of 20 cm^3 acetone and 10 cm^3 hexane. Afterwards, the vessel with the homogenate was centrifuged for 15 min (at 3000 rpm) and the liquid part was transferred to a 250 cm^3 separatory funnel. A mixture of 20 cm^3 hexane and 2 cm^3 diethyl ether was added to the remaining portion of the biological material in the vessel, which was then homogenized for 5 min, centrifuged, and the liquid part was combined with the first portion. The remaining portion in the vessel was washed with a mixture of 10 cm^3 hexane and 1 cm^3 diethyl ether. The combined extracts were supplemented with 60 cm^3 of 0.9% sodium chloride solution and then shaken for 2–5 min. The hexane layer was separated, and the aqueous/acetone layer was extracted twice in 10 cm^3

portions each time. Hexane was evaporated on a rotary evaporator and the resulting fat portion was weighed. Then hexane was added again.

The hexane extract was clarified with concentrated sulfuric acid to obtain a colorless layer of sulfuric acid. The hexane layers were washed from the acid with a sodium bicarbonate solution and then with distilled water to a neutral pH, as was indicated by a common pH indicator. The washed extract was dried by filtering through anhydrous sodium sulfate. The clarified hexane extract was evaporated on a rotary evaporator. The resulting extract was separated by non-polar (for PCBs) and polar (for OCPs) solvents on a chromatographic column with a Florisil® sorbent [16–19].

2.3   Preparation of Human Biological Fluid Samples for POP Determination

Blood was collected into Vacuette tubes with sodium citrate (9 cm^3). Breast milk was collected into containers for biological material, frozen, and delivered to the laboratory.

A weighed blood sample was placed immediately in a flask with acetone. For example, 18 cm^3 of acetone was proportionally added to a 9 g blood sample, vigorously shaken and supplemented with 13.5 cm^3 hexane, placed on a shaker for 30 min, then left to subside for 15 min. The liquid phase of the homogenate was decanted into a separatory funnel through a simple filter paper (White Ribbon) pre-moistened with hexane. The content of the funnel was washed twice with 2–3 cm^3 portions of hexane. A 50 cm^3 portion of 1% KCl solution was introduced into the separatory funnel, shaken, and left until phase separation. After the separation, the aqueous/acetone layer was discarded. Afterwards, the extract was washed with concentrated sulfuric acid and separated on a column with a sorbent as described for biological samples of aquatic organisms.

A urine sample was filled with acetone, shaken, supplemented with hexane, extracted on a shaker for 30 min, transferred to a separatory funnel, and left to subside for 15 min. After the phase separation, the aqueous/acetone layer was discarded. Then, the extract was washed with concentrated sulfuric acid and separated on a column with a sorbent as described for biological samples of aquatic organisms.

To prepare breast milk samples, 2–5 cm^3 of breast milk, 1 cm^3 of 5% potassium oxalate, and 5 cm^3 ethanol were placed into a separatory funnel. The funnel was shaken for 1 min. After that, the mixture in the funnel was supplemented with 10 cm^3 diethyl ether and shaken for 1 min, then with 5 cm^3 of n-hexane and again shaken for 1 min. After the phase separation, the ether layer was decanted. The remaining layer in the funnel was filtered through anhydrous sodium sulfate moistened with hexane and placed on a rotary evaporator until the hexane was completely evaporated. After stabilizing the weight of the lipophilic extract, the flask was weighed to calculate the lipid weight. Afterwards, the extract was again supplemented with hexane, washed with concentrated sulfuric acid, and separated on a column with a sorbent as described above for samples of aquatic organisms [19, 20].

2.4 Preparation of Standard POP Solutions and Calibration of Equipment

To prepare standard OCP and PCB solutions, we used reference materials (Dr. Ehrenstorfer and AccuStandard) of α-HCH, β-HCH, γ-HCH, δ-HCH, p,p'-DDT, p,p'-DDD, p,p'-DDE, o,p'-DDT, o,p'-DDD, o,p'-DDE, and a mixture of PCB congeners (**28**, **52**, 155, **101**, 118, 143, **153**, **138**, **180**, and 207) with known metrological characteristics: a content of the main substance of 99.4–99.6% and a measurement error of 0.4%. For chromatography, we used working standard solutions of OCPs and PCBs with a concentration of 20 ng/mL, which were prepared by diluting the standard solutions with a specified volume of *n*-hexane. The Pesticides library was also used.

2.5 Chromatographic Analysis, QA/QC, and Estimation of POP Content

2.5.1 Abiotic Components of Ecosystems

Gas chromatographic analysis was performed on an Agilent 6890 Plus gas chromatograph coupled with a 5973N mass selective detector (Agilent Technologies, USA). A HP5-MS quartz capillary column was used (30 m × 0.25 mm × 0.25 μm); the carrier gas was helium; flow rate, 1 mL/min; injector temperature, 300 °C; the temperature protocol was as follows: 50 °C for 1 min, then increase at 20 °C/min to 300 °C; full scanning in a range of 60–350 m/z; injected sample size, 2 μL (in the splitless mode); solvent retention, 2 min. The results were interpreted using the AGILENT software and NIST libraries.

To assess the degree of extraction of the pesticides under study, a standard amount of DDT and γ-HCH (50 μL of solution in hexane at a concentration of 1 μg/cm³) was introduced into the model samples. Sample preparation and chromatographic analysis were carried out in accordance with the techniques described above. Chromatograms of the standard solutions of the compounds analyzed were used to determine the lower limit of measurements of OCP concentration. The mass of the component that gave a signal from the chromatograph detector threefold higher than the average level of the noise signal was assumed to be the lower limit of concentration measurements. To estimate the limits of OCP measurements, a series of model water and sediment samples were prepared, in which specified amounts of the pesticides under study were introduced. The samples were passed through all stages of sample preparation and chromatographic analysis. In this case, the limit of detection was assumed to be the amount of the substance introduced into the sample that gave a peak corresponding to the lower limit of concentrations measured [13].

2.5.2 Biotic Components of Ecosystems and Human Biological Material

Contents of OCPs and PCBs in biological samples were measured on a Shimadzu GC MS-QP 2010 Ultra gas chromatograph coupled with mass spectrometer equipped with an AOC-5000 autosampler [7]. For the study, an SLB-5 capillary column was used; the carrier gas was helium (at a flow rate of 1 mL/min); the injector and detector temperatures were 250 and 150 °C, respectively. The heating protocol was as follows: an increase to 100 °C for 4 min, heating to 310 °C at 7 °C/min, and maintaining the final temperature for 6 min. A 2 μL portion of the test mixture was injected in the splitless mode, with a 1-min period between sample injection and opening of the splitter. The substances in the gaseous phase were ionized in the electronic ionization mode. The selected ion monitoring (SIM) was developed according to the settings and detection limits of the instrument. Two ions (M + and [M + 2] +) were monitored for each chlorination level. To identify the compound under study, the retention time, mass, and relative abundance of confirming ions were used as the confirmation criteria. A relative percentage of uncertainty lower than ± 20% was assumed to be acceptable. Areas of peaks were measured using the GCMS Postrun Analysis software.

The results obtained were tested on a Shimadzu GC-2010 Plus gas chromatograph coupled with an electron capture detector. The capillary column was Shimadzu HiCap CBP5. The column temperature was 210 °C; the injector temperature, 250 °C; the detector temperature, 280 °C. The carrier gas was argon; the inlet pressure, 2 kg/cm^2; the flow splitter, 1:60; the rate of carrier gas flow through the column, 0.5 mL/min [7, 18].

The chromatographs were calibrated using the standard POP solutions (Fig. 2.1). Identification was performed on the basis of relative retention time. Quantification was carried out using a calibration curve based on the standard solutions of pesticides.

To assess quality of the methods performed, the standard addition method was used. Specified amounts of the compounds under study were added to the muscle

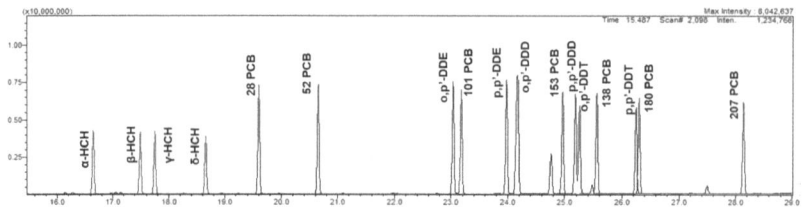

Fig. 2.1 Typical chromatogram of a standard solution of persistent organic pollutants (POP)

tissues of at least seven (from seven to 10) samples of biomaterial. The sample preparation and analysis of mixed samples were carried out by the techniques above. The results showed that the average reproducibility of analyte concentrations ranged from 94.6 to 103.7%, which indicates reliability of the data obtained, and also reproducibility and efficiency of the analytical methods. Detection limits were calculated as three standard deviations of 7–10 samples mixed with the standards. For the analytes that were not determined in mixed samples, detection limits were estimated as the amount of analyte in the sample that corresponded to the minimum concentration of the calibration reference solution. For the OCPs and PCBs under study, the detection limits were as follows (in ng/g): α-HCH, 0.2–0.3; β-HCH, 0.1–0.2; γ-HCH, 0.3–0.5; p,p'-DDT, 0.6–0.7; o,p'-DDT, 0.2–0.6; p,p'-DDD, 0.0–0.1; o,p'-DDD, 0.1–0.2; p,p'-DDE, 0.1–0.2; o,p'-DDE, 0.1–0.4; PCB 28, 0.5–0.6; PCB 52, 0.4–0.7; PCB 155, 0.1–0.5; PCB 101, 0.6–0.8; PCB 118, 0.7–0.8; PCB 143, 0.2–0.7; PCB 153, 0–0.1; PCB 138, 0.2–0.3; PCB 180, 0.5–0.6; and PCB 207, 0.7–0.8.

Statistical analysis of the results was performed using the IBM SPSS Statistics package for Mac OS and MS Windows. Statistical significance of data was tested using the two-way Kruskal–Wallis test at a significance level of $p \leq 0.05$, the nonparametric Mann–Whitney U test at a significance level of $p \leq 0.05$, and the Spearman's correlation coefficient. The results are presented as range of concentrations (min–max) and mean value ± standard deviation (Mean ± SD).

Acknowledgements The work was supported by the Ministry of Science and Higher Education of the Russian Federation, project no. FZNS-2023-0011.

References

1. Tsygankov VY, Boyarova MD, Kiku PF, Yarygina MV (2015) Hexachlorocyclohexane (HCH) in human blood in the south of the Russian Far East. Environ Sci Pollut Res 22:14379–14382. https://doi.org/10.1007/s11356-015-4951-3
2. Tsygankov VY, Khristoforova NK, Lukyanova ON et al (2017) Selected organochlorines in human blood and urine in the South of the Russian Far East. Bull Environ Contam Toxicol 99:460–464. https://doi.org/10.1007/s00128-017-2152-0
3. Tsygankov VY, Gumovskaya YP, Gumovskiy AN et al (2020) Bioaccumulation of POPs in human breast milk from south of the Russian Far East and exposure risk to breastfed infants. Environ Sci Pollut Res 27:5951–5957. https://doi.org/10.1007/s11356-019-07394-y
4. Tsygankov VY, Gumovskaya YP, Gumovskiy AN et al (2020) Organic chlorine compounds in breast milk of women in the south of the Russian Far East. Ekologiya Cheloveka (Human Ecology):12–18. https://doi.org/10.33396/1728-0869-2020-4-12-18
5. Tsygankov VY, Boyarova MD, Lukyanova ON (2015) Bioaccumulation of persistent organochlorine pesticides (OCPs) by gray whale and Pacific walrus from the western part of the Bering Sea. Mar Pollut Bull 99:235–239. https://doi.org/10.1016/j.marpolbul.2015.07.020
6. Tsygankov VY, Boyarova MD, Lukyanova ON, Khristoforova NK (2017) Bioindicators of organochlorine pesticides in the Sea of Okhotsk and the Western Bering Sea. Arch Environ Contam Toxicol 73:176–184. https://doi.org/10.1007/s00244-017-0380-2

7. Tsygankov VY, Lukyanova ON, Boyarova MD (2018) Organochlorine pesticide accumulation in seabirds and marine mammals from the Northwest Pacific. Mar Pollut Bull 128:208–213. https://doi.org/10.1016/j.marpolbul.2018.01.027
8. Tsygankov VY, Boyarova MD, Lukyanova ON (2016) Bioaccumulation of organochlorine pesticides (OCPs) in the northern fulmar (Fulmarus glacialis) from the Sea of Okhotsk. Mar Pollut Bull 110:82–85. https://doi.org/10.1016/j.marpolbul.2016.06.084
9. Lukyanova ON, Tsygankov VY, Boyarova MD, Khristoforova NK (2016) Bioaccumulation of HCHs and DDTs in organs of Pacific salmon (genus Oncorhynchus) from the Sea of Okhotsk and the Bering Sea. Chemosphere 157:174–180. https://doi.org/10.1016/j.chemosphere.2016.05.039
10. Tsygankov VY, Lukyanova ON, Boyarova MD et al (2019) Organochlorine pesticides in commercial Pacific salmon in the Russian Far Eastern seas: food safety and human health risk assessment. Mar Pollut Bull 140:503–508. https://doi.org/10.1016/j.marpolbul.2019.02.008
11. Tsygankov VY, Donets MM, Gumovskiy AN, Khristoforova NK (2022) Temporal trends of persistent organic pollutants biotransport by Pacific salmon in the Northwest Pacific (2008–2018). Mar Pollut Bull 185:114256. https://doi.org/10.1016/j.marpolbul.2022.114256
12. Donets MM, Tsygankov VY, Gumovskiy AN et al (2021) Organochlorine pesticides (OCPs) and polychlorinated biphenyls (PCBs) in Pacific salmon from the Kamchatka Peninsula and Sakhalin Island, Northwest Pacific. Marine Poll Bull 169:112498. https://doi.org/10.1016/j.marpolbul.2021.112498
13. Chernyaev AP, Rychkova EY, Kondrikov NB, Zyk EN (2017) Modern modification of the method for determination of organochlorine pesticides in organic vehicles. Izvestiya TINRO 188:244–250
14. PND F 14.1:2:3:4.204-04 (2004) PND F 14.1:2:3:4.204-04. Quantitative chemical analysis of water. Method of measuring the mass concentrations of organochlorogenic pesticides and polychlorinated biphenils in samples of drinking, natural and waste water by the method of gas chromatography
15. PND F 16.1:2.2:2.3:3.61-09 (2009) PND F 16.1:2.2:2.3:3.61-09. Quantitative chemical analysis of water. Method for measuring the mass fractions of organochlorine pesticides and polychlorinated biphenyls in samples of soils, bottom sediments, sewage sludge, production and consumption wastes by the gas chromatographic method with mass-selective detection
16. Klisenko MA, Kalinina AA, Novikova KF, Khokholkov GA (1992) Methods for determining the microquantities of pesticides in food products, feed and external environment. Kolos, Moscow
17. Tsygankov VY, Boyarova MD (2015) RU 2543360 C1. Method of sample preparation for gas-chromatographic determination of pesticides in biomaterial. 1–6
18. Tsygankov VY, Boyarova MD (2015) Sample preparation method for the determination of organochlorine pesticides in aquatic organisms by gas chromatography. Achievements Life Sci 9:65–68. https://doi.org/10.1016/j.als.2015.05.010
19. Gumovskaya YP, Gumovskiy AN, Tsygankov VY et al (2020) RU 2713661 C1. Method of sample preparation for gas-chromatographic determination of organochlorine compounds in biomaterial. 1–11
20. Gumovskaya YP, Gumovskiy AN, Tsygankov VY et al (2020) RU 2727589 C1. Method of sample preparation for gas-chromatographic determination of organochlorine compounds in biomaterial. 1–9

Chapter 3
Methods for Determining Total Hydrocarbons and Polycyclic Aromatic Hydrocarbons in Various Components of Ecosystems of the Far Eastern Region

Abstract Some methods to determine total hydrocarbon content (THC) and polycyclic aromatic hydrocarbons (PAH) in various components of ecosystems in the Far Eastern region of Russia are considered in this chapter. The general protocol of analysis proposed includes freeze-drying of sample, extraction of target components through ultrasonication, saponification of lipids, and sorbent-based purification with subsequent analysis by the HPLC and IR spectrometry methods. Metrological characteristics of the method are calculated.

Keywords THC · PAH · Sample preparation techniques · Chromatography

3.1 Introduction

Total hydrocarbons (THC), including polycyclic aromatic hydrocarbons (PAH), are widely distributed in the environment. A major part of these compounds are always synthesized and formed in nature. However, anthropogenic hydrocarbons having negative effects also enter the environment in substantial amounts.

Hydrocarbon pollutants can build up in various environmental components: in water, bottom sediments, and tissues of living organisms. Many PAHs are mutagenic and carcinogenic. Thus, measuring THC and PAH levels in various matrices is a relevant and highly demanded approach [1–4]. Being components of industrial emissions, PAHs enter the air, sometimes from soil or groundwater. PAHs can be sorbed on microparticles suspended in the air, migrate in the atmosphere with air currents, and settle in the form of dry or wet (rain, dew, etc.) atmospheric precipitates. Upon entering lakes and rivers, they settle onto the bottom and are deposited in bottom sediments. Some of them penetrate the soil layer and enter groundwater.

For some of PAHs, safety standards have been introduced to regulate their levels in environmental components and food. In the Russian Federation, there are hygienic standards for benzo(*a*)pyrene [5, 6], which until recently has been considered an indicator of pollution by the PAH group, in some environments (including food). Therefore, the major attention in Russia, as earlier in the USSR, is paid to the

© The Author(s), under exclusive license to Springer Nature Switzerland AG 2023
V. Tsygankov, *Persistent Organic Pollutants in the Ecosystems of the North Pacific*,
Earth and Environmental Sciences Library,
https://doi.org/10.1007/978-3-031-44896-6_3

methodological provision of benzo(*a*)pyrene determination. In modern standards abroad, 16 priority PAHs have been selected (naphthalene, acenaphthene, fluorene, acenaphthylene, phenanthrene, anthracene, fluoranthene, pyrene, chrysene, benz(*a*)anthracene, benzo(*b*)fluoranthene, benzo(*k*)fluoranthene, benzo(*a*)pyrene, dibenz(*a,h*)anthracene, indeno(1,2,3-*cd*)pyrene, and benzo(*g,h,i*)perylene), which are to be determined in environmental components [7–9].

In this chapter, we present a modification of the method for determining PAHs in abiotic and biotic components of nature and estimating the current PAH content.

3.2 Preparation of Samples from Abiotic Components of Ecosystems for PAH Determination

Samples of bottom sediments were collected with a grab sampler (from the upper 10 cm), packed in aluminum containers, and immediately frozen to – 18 °C. After being delivered to the laboratory, the samples were dried in a freeze-drier at – 51 °C and under a residual pressure of 1 Pa. The dried bottom sediments were classified by the sieve method, with a particle size fraction of no larger than 0.25 mm selected for analysis.

Methylene chloride was used as an extracting agent. To build a calibration graph, solutions of standard PAH (Standard Reference Material) with weight concentration as a certified characteristic were prepared. A prepared working solution of standard PAH was stored in a freezer at – 14 °C for no more than 7 days.

After the classification of bottom sediments, a weighed sample of 5 g was placed in a 125-cm^3 laboratory flask. The target components were extracted by solid–liquid extraction according to a previously developed technique [10]. To do this, a dried sample of bottom sediments was supplemented with 20 cm^3 methylene chloride and extracted in an ultrasonic bath for 15 min. The extract was then filtered through a folded filter paper; the residue was washed with 10 cm^3 dichloromethane. The combined extract was evaporated dry on a rotary evaporator under vacuum in a water bath (T = 30 °C). The dry residue was dissolved in 1 cm^3 acetonitrile.

3.3 Preparation of Samples of Aquatic Organisms for Determining THCs and PAHs

In the laboratory, specimens of aquatic organisms were thawed at room temperature and dissected, with muscle tissue cut out. Then the samples were dried in a drying cabinet at 50 °C. As samples of model environment components, we used pollock muscle tissue preliminarily freeze-dried. The dried samples were crushed to a homogeneous state on a homogenizer and sieved through a 0.5-mm mesh sieve.

A weighed sample of pre-crushed and dried tissues ($\approx$2 g) was placed in a laboratory flask, and the target components were extracted with 20 cm^3 dichloromethane twice in an ultrasonic bath during 15 min; after each extraction, the extract was filtered through a specially prepared filter.[1] The resulting extract was evaporated dry in a round-bottom flask on a rotary evaporator (T = 50 °C) under vacuum (P = 0.01 MPa); the oil residue was saponified with 2 cm^3 2% sodium hydroxide solution in ethyl alcohol for 24 h.

The dry residue was re-dissolved in distilled water, supplemented with fine-crystal sodium chloride (to break down emulsion), and extracted with 20 cm^3 dichloromethane. The organic layer was separated using a separatory funnel. The resulting extract was supplemented with 5 g anhydrous sodium sulfate to remove traces of water. Then the extract was clarified on a prepared cartridge.[2] The resulting extract was evaporated dry in a round-bottom flask on a rotary evaporator (T = 50 °C) under vacuum (P = 0.01 MPa). The dry residue was dissolved in 1 cm^3 carbon tetrachloride. A 0.5 cm^3 portion of the extract was collected with a pipette for subsequent measurement of petroleum hydrocarbons by IR spectroscopy.

The residue was evaporated dry, blown off in a nitrogen stream (to completely remove traces of carbon tetrachloride), and dissolved in 0.5 cm^3 acetonitrile for subsequent analysis of PAHs by the reversed-phase high-performance liquid chromatography (HPLC) method.

3.4 Preparation of Standard THC and PAH Solutions and Calibration of Equipment

To build a calibration curve, we used TP-22 turbine oil dissolved in carbon tetrachloride as a standard. A 0.1-g portion of TP-22 oil was weighed on an analytical balance and dissolved in 10 cm^3 carbon tetrachloride (basic solution): C_0 = 10 mg/mL (10 000 µg/mL). Calibration solutions were prepared by sequential dilution. Then these solutions were placed in 10-mm-long quartz cuvettes and the absorbance of the solutions was measured in a range of 2500–3500 cm^{-1} relative to the initial solvent, carbon tetrachloride. Before measuring each solution, the cuvettes were rinsed with the pure solvent.

In our study, we used standard solutions of PAH compounds in acetonitrile with weight concentration as a certified characteristic. The spectral characteristics of the substances under study are presented in Table 3.1.

---

[1] *Filter preparation.* The filter housing is made of polytetrafluoroethylene (PTFE): the lower part is a funnel lined with a layer of filter paper (white tape) and fat-free cotton wool, overlaid with a PTFE mesh. Then the upper cylindrical part of the filter with a screw thread is screwed into the funnel, tightening and sealing the lower layers.

[2] *Cartridge preparation.* The cartridge consists of a polymeric base in the form of a syringe (d = 1 cm, V = 5 cm^3) with a glass porous membrane installed in the lower part. A layer of silica gel (100–200 mesh) weighing 1.0 g and a layer of anhydrous sodium sulfate 2.0 g are placed on top of the membrane. The layers are tightly packed and overlaid with a layer of fat-free cotton wool.

Table 3.1 Spectral characteristics of compounds

Name	Acronym	Fluorescence	
		$\lambda_{excit.}$, nm	$\lambda_{emiss.}$, nm
naphthalene	N	220	330
biphenyl	BF	248	311
2-methylnaphthalene	MN	222	331
fluorene	FL	261	310
acenaphthene	ANT	226	334
phenanthrene	FN	250	364
anthracene	AN	251	402
fluoranthene	FLR	236	473
pyrene	P	240	397
benz(a)anthracene	BAA	287	398
chrysene	CR	266	379
benzo(b)fluoranthene	BBF	255	435
benzo(k)fluoranthene	BKF	306	405
benzo(a)pyrene	BAP	296	404
dibenz(a,h)anthracene	DAHA	296	397
benzo(g,h,i)perylene	BGHIP	299	414

Calibration graphs for certain PAH representatives were built in the LCSolution software using at least five concentrations. The linearity of the obtained relationships in the selected range was also close to 1 ($R^2 > 0.95$).

3.5   Spectral and Chromatographic Analysis, QA/QC

The THC content was determined as the absorbance intensity in the IR region of the spectrum at $2926 \, cm^{-1}$. THC concentrations were determined by the calibration graph method in the ranges 5–100 and 50–1000 mg/dm^3. An IR spectrum of a standard THC solution is shown in Fig. 3.1.

The PAH content was determined by the reverse-phased HPLC with a fluorescent detector as retention time.

Gradient elution was performed by the following protocol:

- 0–7 min—50% water: 50% acetonitrile
- 7–15 min—increase in acetonitrile level to 100%
- 15–30 min—100% acetonitrile
- 30–35 min—decrease in acetonitrile level to 50%
- 35–45 min—50% water: 50% acetonitrile.

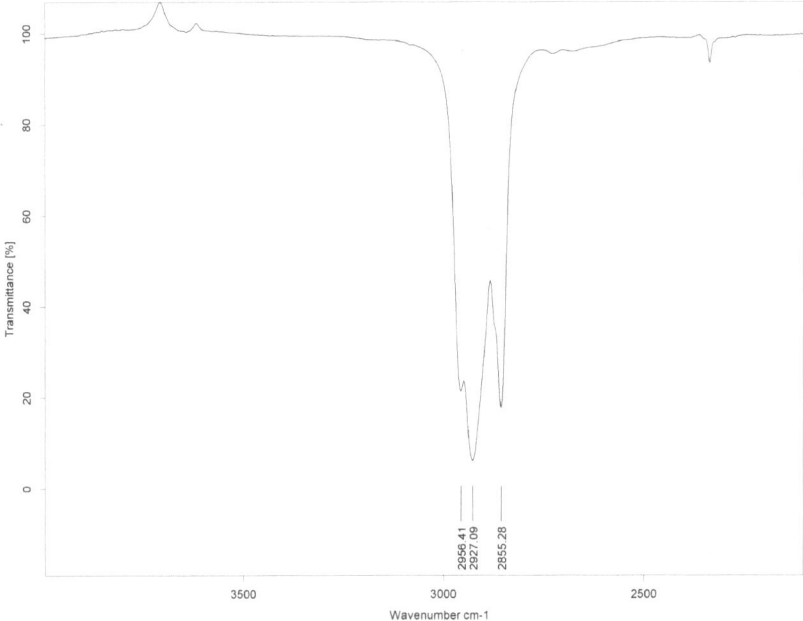

Fig. 3.1 IR spectrum of a standard THC solution

Analytical measurements of THCs were carried out on a Shimadzu IRAffinity-1S Fourier transform infrared spectrophotometer (Japan).

The PAH analysis was carried out on the following equipment:

1. Shimadzu LC-20AD liquid chromatograph (Japan) equipped with a RF-10AXL fluorescence detector and a Waters PAH C_{18} column (25 cm × 0.46 cm, 5 μm). An acetonitrile/water mixture was used as eluent in the gradient elution mode. The eluent flow rate was 1 cm^3/min. Sample was taken with a SIL-20A autosampler. Prior to being injected into the chromatography column, the eluent was degassed with a DGU-20A$_5$ degasser.

2. Shimadzu LC-10AD VP liquid chromatograph (Japan) equipped with a RF-10AXL fluorescence detector and a Discovery C_{18} column (25 cm × 0.46 cm, 5 μm). An acetonitrile/water mixture (90: 10) was used as eluent. The eluent flow rate was 0.5 cm^3/min. Sample was taken with a SIL-10AD VP autosampler. Prior to being injected into the chromatography column, the eluent was degassed with a DGU-14A degasser.

A chromatogram of a standard PAH solution is shown in Fig. 3.2.

The lower limit of PAH concentrations determined was set as a 3: 1 signal-to-noise ratio, and that of THC concentrations as a confidently distinguishable signal at the selected wavelength (Table 3.2). Various model samples were prepared to estimate the degree of THC extraction. Each sample of 1 cm^3 TP-22 solution at a concentration of 1000 μg/mL was injected using an autosampler.

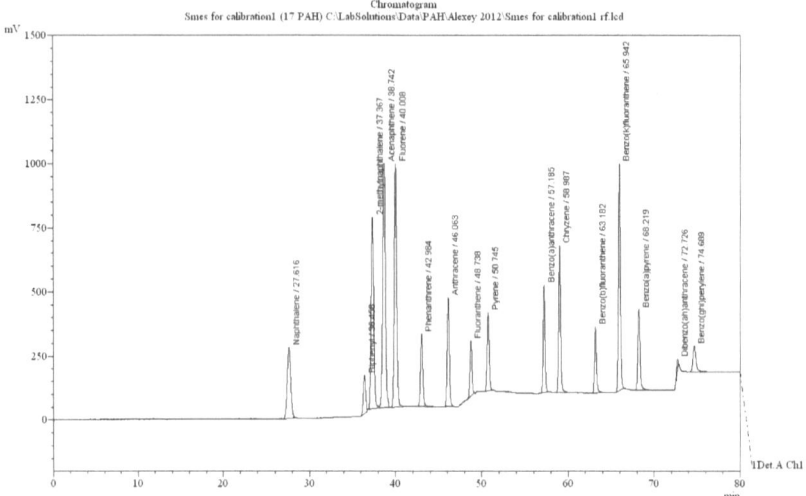

Fig. 3.2 Chromatogram of a standard PAH solution

Table 3.2 Values of lower limit of detected concentrations (DCLL) for certain THCs and PAHs

Compound	DCLL, ng
N	0.53
BF	1.6
MN	0.23
ANT	0.11
FL	0.16
FN	0.53
AN	0.4
FLR	3.2
P	0.4
BAA	0.4
CR	0.23
BBF	0.53
BKF	0.08
BAP	0.8
DAHA	0.8
BGHIP	0.8
Total hydrocarbons	1000

To determine the degree of PAH extraction, we prepared a mixture of 16 standard PAHs. The standard solutions were sampled with microliter syringes (Hamilton, USA). The model samples (freeze-dried pollock muscle tissue, m = 2 g) were injected as a 52 μL mixture containing a specified amount of certain PAHs. The obtained samples were subjected to the above-described sample preparation and clarification protocols with modifications in the defatting techniques. The values of the coefficient and degree of extraction for THCs and PAHs are presented in Table 3.3.

Statistical processing included the evaluation of the results obtained and the analysis of error deviation from mean value. The values evaluated were characterized by samples from five dimensions. To test the accuracy of the results obtained, we used the Student's t-test, which is an approach to objective statistical evaluation of analysis results. We assumed that the mathematical expectation of the sample tested has an a priori normal or close to normal distribution, and the sample representing it has a small size. Calculations were carried out using descriptive statistics available in the Microsoft Excel 2016 data analysis package.

Table 3.3 Values of the coefficient and degree of extraction for certain THCs and PAHs

Compound	Degree of PAH extraction, %	Values of coefficient k
N	16 ± 4	6.25
BF	21 ± 3	4.76
MN	29 ± 3	3.45
ANT	27 ± 5	3.70
FL	42 ± 10	2.38
FN	18 ± 1	5.56
AN	19 ± 1	5.26
FLR	50 ± 1	2.00
P	27 ± 2	3.70
BAA	54 ± 7	1.85
CR	94 ± 1	1.06
BBF	73 ± 2	1.38
BKF	62 ± 7	1.61
BAP	63 ± 3	1.59
DAHA	98 ± 1	1.02
BGHIP	74 ± 0.5	1.35
Total hydrocarbons	37 ± 1.9	2.70

3.6   Conclusion

All existing methods for determining THCs and PAHs imply their separate extraction and determination. The present modified method allows combined extraction and determination of components. In this case, the duration of the analysis and costs are significantly reduced.

Using this method, we managed to achieve high degrees of extraction and reproducibility of experiment for most polycyclic aromatic hydrocarbons, except for naphthalene and biphenyl. Although the latter are classified as persistent organic pollutants, these are much more reactive and easily degradable compounds than other representatives of PAHs due to their chemical structure (a carbon atom at the β-position and a phenyl bridge). The introduced hydrocarbons are well extracted, and their losses during further manipulations are minimal. Nevertheless, the defatting stage is required to implement the protocol of combined extraction, chromatographic separation of PAHs, and IR spectrometry of THCs.

Acknowledgements The work was supported by the Ministry of Science and Higher Education of the Russian Federation, project no. FZNS-2023-0011.

References

1. Shurubor EI (2000) Polycyclic aromatic hydrocarbons in the soil-plant system of an oil field (Kama River Region of Perm Oblast). Eurasian Soil Sci 33:1329–1333
2. Lung S-CC, Liu C-H (2015) Fast analysis of 29 polycyclic aromatic hydrocarbons (PAHs) and nitro-PAHs with ultra-high performance liquid chromatography-atmospheric pressure photoionization-tandem mass spectrometry. Sci Rep 5:12992. https://doi.org/10.1038/srep12992
3. Ju W, Lu C, Liu C et al (2020) Rapid identification of atmospheric gaseous pollutants using fourier-transform infrared spectroscopy combined with independent component analysis. J Spectrosc 2020:1–14. https://doi.org/10.1155/2020/8920732
4. Onopiuk A, Kołodziejczak K, Marcinkowska-Lesiak M, Poltorak A (2022) Determination of polycyclic aromatic hydrocarbons using different extraction methods and HPLC-FLD detection in smoked and grilled meat products. Food Chem 373:131506. https://doi.org/10.1016/j.foodchem.2021.131506
5. SanPin 2.3.2.1078-01 (2001) SanPin 2.3.2.1078-01. Hygienic requirements of safety and nutritional value of food products
6. GN 2.2.5.1313-03 (2003) GN 2.2.5.1313-03. Chemical factors of the industrial environment. Maximum permissible concentrations (MPC) of harmful substances in the air of the working area
7. Jira W, Ziegenhals K, Speer K (2008) Gas chromatography-mass spectrometry (GC-MS) method for the determination of 16 European priority polycyclic aromatic hydrocarbons in smoked meat products and edible oils. Food Addit Contaminants: Part A 25:704–713. https://doi.org/10.1080/02652030701697769
8. Yuan X-X, Jiang Y, Yang C-X et al (2017) Determination of 16 kinds of polycyclic aromatic hydrocarbons in atmospheric fine particles by accelerated solvent extraction coupled with high performance liquid chromatography. Chin J Anal Chem 45:1641–1647. https://doi.org/10.1016/S1872-2040(17)61047-8

9. Jeffery J, Carradus M, Songin K et al (2018) Optimized method for determination of 16 FDA polycyclic aromatic hydrocarbons (PAHs) in mainstream cigarette smoke by gas chromatography–mass spectrometry. Chem Cent J 12:27. https://doi.org/10.1186/s13065-018-0397-2

10. Chernyaev AP, Rychkova EY, Kondrikov NB, Zyk EN (2017) Modern modification of the method for determination of organochlorine pesticides in organic vehicles. Izvestiya TINRO 188:244–250

Chapter 4
Current Levels of Organochlorine Pesticides in Abiotic Components of Ecosystems in the Sea of Japan Basin and Khanka Lake

Abstract Levels of organochlorine pesticides (OCP)—isomers of HCH and DDT and its metabolites—were measured in abiotic components of freshwater and marine ecosystems of the Sea of Japan and Lake Khanka, Sea of Okhotsk basin. An assessment of the duration of presence and entry of toxicants in these regions is done. The coefficients indicating time periods since entry of OCPs into ecosystems are estimated.

Keywords HCHs · DDTs · Bottom sediments · Water · Sea of Japan · Khanka lake

4.1 Introduction

Pesticides are a particular group of chemical compounds intentionally introduced into the environment in order to destroy certain living organisms. Besides the effects desired by humans (crop protection and pest control), these compounds pollute the biota and abiotic components of ecosystems, reducing sustainability and biodiversity of the latter. Furthermore, many pesticides can negatively affect human health, causing poisoning or increasing the risk of cancer [1].

Pesticides are widely used in agriculture to control pests of cereals, legumes, technical crops, fruit trees and vineyards, vegetable and field crops.

Organochlorine pesticides (OCPs) are recognized among the most hazardous compounds due to their high toxicity, biological activity, environmental persistence, and bioaccumulation and biomagnification potentials. Products of their degradation or transformation, which are more stable than the original pesticides, also pose danger to human health. The well-known DDT (*p,p*-dichlorodiphenyltrichloroethane) and chlorine derivatives of dioxin are the examples of such OCPs. Certain members of this class of substances are the most potent toxins, dozens of thousands of times more toxic than potassium cyanide [2, 3].

Organochlorine pesticides mainly enter the human body with food, the air breathed, and via skin contact. The major ways of OCP removal are through the kidneys and gastrointestinal tract. Organisms' sensitivity to organochlorine

© The Author(s), under exclusive license to Springer Nature Switzerland AG 2023   67
V. Tsygankov, *Persistent Organic Pollutants in the Ecosystems of the North Pacific*,
Earth and Environmental Sciences Library,
https://doi.org/10.1007/978-3-031-44896-6_4

compounds is individual-, species-, and age-dependent [4]. OCPs have a *polytropic effect*, are parenchymal toxins affecting the central nervous system, liver, kidneys, heart muscle, gastric and intestinal mucosa, and organs of internal secretion (mainly adrenal glands, thyroid gland, and ovaries). Pronounced morphological changes in bodies of homeotherms caused by OCP poisoning range from minor circulatory disorders and reversible dystrophic changes to focal necrosis, which are related with the organism, dose, duration of exposure and other factors [5].

Effectiveness of pesticides depends on their capability of entering a target organism, be transported inside it to the site of action and suppress vital processes, on the amount of substance applied (measured by a dose in mg per whole organism or its unit weight or concentration of the active agent in the working solution), and on the time of exposure. The measure of pesticide activity is LD_{50} or AC_{50} (the lethal dose or average concentration of solution causing mortality of 50% of organisms, respectively). In practice, the measure is the rate of consumption of the active agent per unit area (weight or volume), at which the desired protective effect is achieved [3].

Ecological safety of pesticides is related to their selectivity, as well as to their greater or lesser persistence which is the capability of staying for a certain time in the environment without losing the biological activity. Persistence of the same pesticide can vary significantly between different environmental components and different climatic conditions.

Many pesticides are toxic to humans and other homeotherms. On the basis of acute toxicity with oral administration (measured mostly on rats), the following classification for pesticides was accepted in the USSR: extremely hazardous, LD_{50} up to 15 mg/kg; highly hazardous, 15–150 mg/kg; moderately hazardous, 150–5000 mg/kg, low hazardous, above 5000 mg/kg [3].

An insecticide notoriously known under the abbreviation **DDT** (**dichlorodiphenyltrichloroethane**) has drawn the greatest attention of scientists and the public. The expectation for promising effects and safety of this compound quickly changed from hope and enthusiastic feedbacks to bitter disappointment and accusations. The pesticides have been called by scientists a miracle of modern chemistry and, simultaneously, an unforgivable mistake [3, 6].

DDT is highly resistant to degradation: critical temperatures, enzymes, or light did not have any marked effect on this compound and degradation process. As a result, after entering into the environment, DDT is involved in the food chain where it builds up to significant quantities, first, in plants, then in animals, and eventually in the human body [3, 6].

HCH (hexachlorocyclohexane, hexachlorane) is a mixture of several stereoisomers. A total of eight isomers of this substance have been obtained in pure forms, of which only the γ-isomer (lindane) has pronounced insecticidal properties. HCH is categorized as a toxic compound with skin-resorptive effect. It also exhibits marked cumulative properties. It causes hyperemia of the skin, swelling, vesicles and pustules, and irritation of the eye's conjunctiva. HCH is retained for a long time in the organs and tissues of the body (especially in adipose tissue), excreted through the kidneys and colon, enters the milk of breast-feeding women, etc. Application of lindane is allowed in some countries. It is highly toxic to mammals (LD_{50}, 25–200 mg/kg) [1].

OCPs have high chemical resistance to various environmental factors and belong to the group of highly and ultra-persistent pesticides. Having these properties, they build up in aquatic organisms and are transferred along food chains, increasing by about an order of magnitude per each subsequent link. However, not all organochlorine compounds have the same persistence and cumulative properties. In the hydrosphere and aquatic organisms, they gradually degrade to form metabolites. Therefore, OCP residues and their metabolites in bodies of aquatic organisms are constantly detected in areas of intensive agriculture, which should be taken into account in poisoning diagnostics [7].

Besides organochlorine pesticides, similar compounds such as polychlorinated biphenyls (PCBs) and terphenyls (PCTs) used in various industries are also found in freshwater and marine bodies of water, as well as in aquatic species. In their physical and chemical properties and physiological effects on organisms, and also in methods of analysis, these are very close to organochlorine pesticides. Therefore, differentiation of these groups of chlorinated hydrocarbons is required [7].

4.2 Pollution of the Seas and Oceans by Organochlorine Compounds

Pollution of marine ecosystems by OCPs such as DDT and HCH has become their integral characteristic and has acquired a global dimension. This is explained by a number of factors: the continued use of OCPs, their extremely high persistence, the potential to accumulate in living organisms, as well as large-scale atmospheric, hydrodynamic transport and high migration potential within ecosystems [5].

Although, at first glance, OCP levels in marine waters seem insubstantial, these compounds can negatively affect ecosystems of the world's oceans due to their bioconcentration and bioaccumulation in marine organisms.

DDT and its metabolites are regularly detected at ultra-low concentrations in the near-water layer of the atmosphere above the world's oceans. Over the most polluted part of the oceans, waters of the North Atlantic, the concentration of the aerosol fraction was in the range of 0.002–0.020 ng/m^3 in the 1970s–1980s [5].

OCPs are concentrated in the surface microlayer (SML) of water and above the so-called pycnocline (usually from dozens to 150–200 m). A substantial portion of these compounds is involved in biochemical cycles, being present in the particulate suspended matter, bottom sediments, and in marine organisms. DDT is characterized by the greatest degree of accumulation; HCH isomers, mostly the γ-form, were detected in some of zooplankton and fish samples. Thus, the potential danger of pollution of the seas and oceans on a global and regional scale is obvious [5].

4.3 Transboundary Transport

Although OCPs have a low saturated vapor pressure, these can evaporate from the soil and water surface into the air. In particular, significant amounts of these compounds enter the atmosphere after being sprayed from agricultural aircrafts and, thus, can be transported thousands of kilometers by air currents.

OCPs are well adsorbed by soil organic matter or bottom silt and can be transported by surface waters. The degree of OCP adsorption increases in the series sand–bottom sediments–soil.

A very important characteristic of transport of OCPs in the environment is their concentration in atmospheric precipitation. Contamination of water bodies is mainly caused by pollutants in surface runoff, as well as by their precipitation from the atmosphere. Upon entering water bodies, these compounds are actively redistributed between the water and bottom sediments. One substantial proportion of OCPs is intensively absorbed by aquatic organisms and metabolized over time; the other is adsorbed on particles suspended in the water column and sediments on the bottom, where it stays for dozens or even hundreds of years. The greatest OCP concentrations in surface waters are observed during flood seasons. The highest levels of these pollutants are recorded from Africa, Latin America, China, and India, which is a result of the active use of pesticides in these regions [8].

4.4 Monitoring of Surface Waters and Bottom Sediments

Although the world's oceans hold a major place in the overall balance of the hydrosphere, fresh water plays an important role for the biosphere in general. For this reason, special attention is paid to the study of water bodies as regards contamination with xenobiotics persistent in the environment.

Natural water is an open system that exchanges substances and energy with other environments (atmosphere, water objects, and bottom sediments) and with its biological components. Furthermore, natural water contains plenty of suspended particles and microbubbles. Many OCPs contaminate groundwater, where their concentration can reach a few ng/L.

The most important source of organochlorine pesticides entering irrigated fields can be irrigation water. Bottom sediments of irrigation fields are also characterized by a high OCP content, which is associated with the removal of pesticides with surface runoff and their further deposition in bottom sediments. A consequence of this is the transfer of pesticides from bottom sediments to water. High OCP levels are found in water bodies that are more polluted due to the reuse of water for irrigation. In the total OCP balance, the proportion of metabolites is significantly higher than that in original pesticides [9].

To date, many pollutants have been observed to be distributed over long distances. Environmental pollution caused by migration of pollutants between natural environments has a complex pattern. As the experience of ecological research shows, all elements of the biosphere are exposed to anthropogenic impact regardless of sources: surface and groundwater, atmosphere, soil ecosystems, plants, etc.

4.5 Physical and Geographical Characteristics of the Sea of Japan and the Sea of Okhotsk

The Sea of Japan is a vast deep-sea body of water separated from the ocean and the Sea of Okhotsk by the Sakhalin Island Arc. Its area is estimated at 1,008,000 to 1,062,000 km^2; the water volume, from 1,360,000 to 1,700,000 km^3; the maximum depth, from 3669 to 3695 m [10].

The shores of the northwestern Sea of Japan are mostly high and steep. The vast Muravyov-Amursky Peninsula juts deep into Peter the Great Bay, dividing it into two large bays, Amur and Ussuri.

The coastline of the Sea of Japan is slightly indented in general. This influences the pattern of the water circulation and the biological features of aquatic organisms [10].

The bays of Amur, Ussuri, and Nakhodka are receivers of normatively treated and contaminated wastewater from the cities of Vladivostok, Nakhodka, Artem, Bolshoy Kamen, and Fokino, an also from the Shkotovsky, Khasansky, Nadezhdinsky municipal districts.

One of the major sources of pollutants' emission into the marine environment is terrigenous runoffs, entering mostly rivers, freshwater lakes, reservoirs, and further the seas. One of the largest freshwater ecosystems of Primorsky Krai is Lake Khanka (Sea of Okhotsk basin).

Lake Khanka is located in the middle of the Khanka Lowland, at the Russia's border with Heilongjiang Province, the People's Republic of China. The northern part of the lake belongs to China. The lake has a pear-like shape expanding in the northern part. The area of the water surface is unstable, varying depending on climatic events. The inflow to the lake is, on average, 1.94 km^3 per year; the outflow from the lake, about 1.85 km^3. Thus, the sources of pollutants' emissions can be located both in Russia and in China. Rice and soybeans are actively cultivated on the banks of the lake. There are also many industrial enterprises operated in the coastal zone. Furthermore, Lake Khanka is a large recreational zone. In this regard, the waters of the lake receive huge amounts of pollutants of various classes: toxic elements, organochlorine compounds (pesticides), and hydrocarbons of various origins [11].

4.6 OCPs in the Northwest Pacific

The coastal zone of Peter the Great Bay, Sea of Japan, is one of the most densely populated areas in the Russian Far East. The economic activity leads to the high anthropogenic pressure on the waters of the bay and coves along the coastline. The major pollutants of marine waters are industrial (power supply facilities, shipbuilding, chemical and coal industries, mechanical engineering and metalworking, and also commercial, military, fishing and small-tonnage fleet) and municipal (discharges from residential areas) wastewater, river and stormwater runoff, and also solid waste and garbage discharges into the sea. Rivers make a significant contribution to the pollution of the coastal zone of the bay. About two hundred water users of Primorsky Krai discharge wastewater into surface bodies of water via more than five hundred organized outlets [12].

In the Russian Federation, regulatory standards for the level of organochlorine pesticides in the water of fishery areas were established by Order No. 552 of the Ministry of Agriculture of the Russian Federation dated December 13, 2016. The maximum permissible concentration (MPC) in the water of water bodies is 1 ng/L for DDT, 10 ng/L for HCH, and 10 ng/L for hexachlorobenzene (HCB).

For the purpose of convenience, the information is divided into blocks: the Sea of Okhotsk basin (the drainage basin of Lake Khanka) and the Sea of Japan basin (Peter the Great Bay and the drainage basins of its second-order bays: Amur, Ussuri, Nakhodka, and Possyet) (Fig. 4.1). Also, only p,p'-forms of DDT are considered in this study.

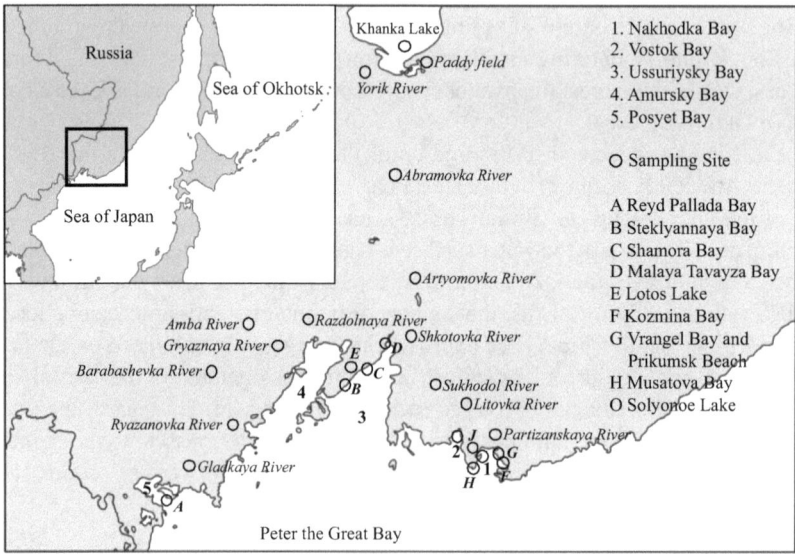

Fig. 4.1 Sampling stations in Lake Khanka and the Sea of Japan

4.6.1 Lake Khanka (Sea of Okhotsk Basin)

Pollution of Lake Khanka is a serious problem resulting from the impact that the industrial activity exerts on the ecological condition of the region. According to the Primorsky Environmental Pollution Monitoring Center, the water of Lake Khanka is characterized as *very polluted*. This body of water can no longer assimilate pollutants that enter with household wastewater and surface runoff. The methods of irrigation systems applied to paddy fields for rice cultivation near the lake cause serious damage to the environment. The removal of the fertile layer of soil from these fields (into drainage canals and further into the lake) negatively affects the environment and the flora and fauna due to the use of a variety of pesticides and agrochemicals on the Chinese side, as well as due to the non-observance of conditions of storage and use of pesticides in the Russian territory.

The level of DDT in the water from Lake Khanka ranged from 24.9 to 49.6 ng/L; DDD, from 20.2 to 34.1 ng/L; hexachlorobenzene (HCB), from 6.1 to 25 ng/L; γ-HCH, from 3.0 to 14.9 ng/L; α-HCH, from 9.5 to 17.8 ng/L; and β-HCH, from 15.7 to 64.9 ng/L (Fig. 4.2). In terms of duration of exposure, HCH was used for a relatively long time, as evidenced by the ratio of the γ-isomer to the total of α- and β-isomers. At all the stations studied, the ratio was 0.2, which indicates a long period since the entry of the active isomer, lindane. The DDT/(DDE + DDD) ratio was, vice versa, above 2.5, which undoubtedly confirms the active application of DDT as an insecticide currently in agriculture, despite the ratification of the Stockholm Convention on Persistent Organic Pollutants.

The Astrakhanka River, flowing near the same-name village and emptying in Lake Khanka, proved to be most polluted watercourse. The total level of organochlorine pesticides was 1034 ng/L. HCH exceeded MPC 38-fold; DDT, 315-fold; and HCB,

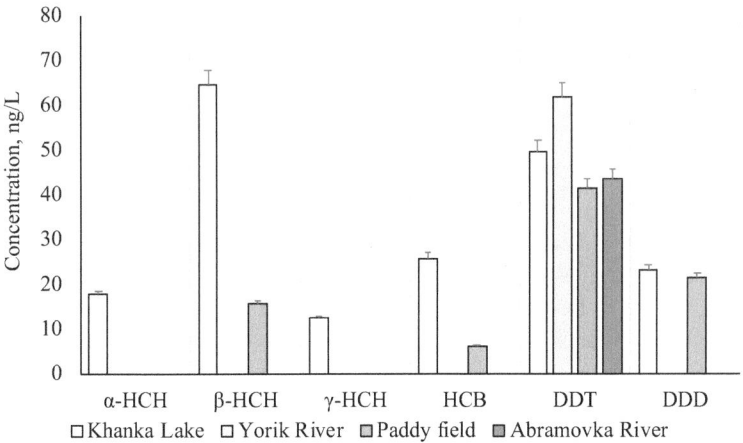

Fig. 4.2 Levels of OCPs in freshwater bodies of water in the Lake Khanka basin, ng/L

32-fold. The entry of DDT into the Astrakhanka River is considered relatively old, but that of HCH fresh.

The Yerik is a small spawning river that empties into Lake Khanka near the village of Kamen-Rybolov. In the areas adjacent to the river banks, there are zones of high agricultural activities, as everywhere in the Khanka Lowland. A visual inspection showed that water from the agricultural areas freely enters this watercourse. The level of non-transformed DDT was 61.9 ng/L. No DDT metabolites were found, which is direct evidence of the illegal use of this banned insecticide for agrochemical purposes in Primorsky Krai. All the HCH isomers, whose entry was assumed to have occurred long ago (α, 10.5; β, 8.4; and γ, 5 ng/g), DDT metabolites (DDT, 17.2; DDE, 7.2; and DDD, 16 ng/g), and HCB were found in the bottom sediments of this river. The total OCP content was 67.7 ng/g, with the ratio of transformed isomers and metabolites indicating a chronic contamination.

A similar situation was observed in the ecosystem of the Abramovka River. The DDT level was slightly lower, 43.7 ng/L; DDT metabolites were also not detected. Although HCH and HCB were not found in the water, their concentrations in bottom sediments were 11.7 and 4.5 ng/g, respectively. DDD concentrations were comparable to those of DDT. The total content of pesticides in bottom sediments was 46.7 ng/g.

4.6.2 Sea of Japan

The major source of organochlorine pesticides from the Russian Federation in the waters of the Sea of Japan is Peter the Great Bay (PGB) and, therefore, we paid attention primarily to the water areas within the bay that are exposed to severe anthropogenic pressure.

The largest second-order bays of Peter the Great Bay with large industrial facilities located on their shores are Amur Bay, Ussuri Bay, Nakhodka Bay, and Possyet Bay.

Ussuri Bay is part of Peter the Great Bay. Its western shore is occupied by the eastern districts of Vladivostok, and the city's municipal solid waste landfill, reclaimed in 2011, is also located there. The Artemovka River with a large tributary, the Knevichanka River, draining the city of Artem and its suburbs, empties into the northern part of the bay, Muravinaya Cove. The Shkotovka River empties into the eastern part of the bay, where wastewater from the village of Shkotovo is discharged. Further south, on the eastern coast, there is Sukhodol Cove with the same-name river that brings pollution from the villages of Smolyaninovo and Romanovka. The town of Bolshoy Kamen, known for its enterprise OAO Dal'nevostochnyi Zavod "Zvezda" (Zvezda Far Eastern Plant), which is a shipbuilding and ship repair center, is located on the shores of the bays of Bolshoy Kamen and Andreev [13].

The following watercourses of the area are exposed to the major anthropogenic pressure as a result of discharge of contaminated wastewater: the rivers of Artemovka, Shkotovka, Sukhodol, and Ryazanovka. The Razdolnaya River with its tributaries receives polluted and insufficiently treated wastewater from the city of Ussuriysk and the Oktyabrsky municipal district; the Partizanskaya River, from the town of Partizansk and its municipal district [12]. In the drainage basin of the Razdolnaya River, MUP Vodokanal (a water supplying enterprise), AO Primorsky Sakhar (a sugar producing enterprise), AO Dalsoya (a soy-processing enterprise), a cardboard factory, and other water-using enterprises take clean river water for production and then discharge insufficiently purified wastewater. It is not always discharged directly into the river or its tributaries: quite often this is done on the land surface and then pollutants enter the river with stormwater runoff [14].

The calculation results showed (Fig. 4.3) that there was no fresh entry of γ-HCH into the waters of the rivers of Artemovka, Shkotovka, and Sukhodol during the study period, and the levels of the α-isomer were 22.8, 37.0, and 13.3 ng/L, respectively. In contrast, DDT was detected in the water samples from all the rivers, with its content ranging within 64.0–79.6 ng/L. HCB levels ranged from 12.8 to 27.0 ng/L.

DDT is known to be quite stable in the environment, and its transformation to take a long time. The detected concentrations of DDD, a DDT metabolite, in the waters of these rivers were quite high and reached 31.4 ng/L (in the Shkotovka River). Thus, on the one hand, the contamination was acute (with DDT concentrations exceeding the MPC 31-fold) and, on the other hand, chronic, because the level of the deeply transformed metabolite DDD was comparable to that of DDT. Based on the quantitative proportions of concentrations of HCH isomers, we may conclude that lindane has almost ceased to be used in farmland areas within the catchment basins

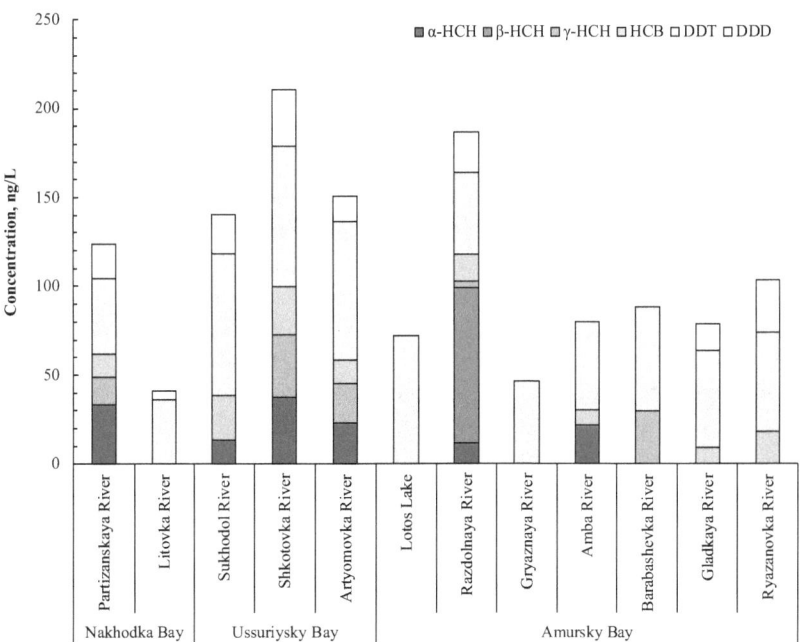

Fig. 4.3 OCP contents in freshwater bodies of water in the Sea of Japan, ng/L

of these rivers, while the compounds formed as a result of HCH isomerization over time still enter in significant quantities.

The situation as regards bottom sediments of the rivers of Ussuri Bay was less critical, although sufficiently high OCP concentrations were found in sediments from the rivers of Shkotovka, Artemovka, and Partizanskaya. However, there the total content did not exceed 94.7 ng/g, and the entry of DDT and HCH was assumed to have occurred long ago. The source of pollution of bottom sediments in the study water areas is undoubtedly the high agricultural activity on the banks of rivers, as well as activities of large power-supplying facilities and fish processing industry.

The major emitters of organochlorine pesticides in the waters of Amur Bay are wastewater discharges from Vladivostok and Ussuriysk. The city of Ussuriysk makes its contribution via the Razdolnaya River discharge, and the contribution of Vladivostok is discharges of insufficiently purified or untreated waters within the city: from the Verkhne-Portovy, Pervaya Rechka, Vtoraya Rechka, and De-Friz wastewater treatment facilities.

The level of organochlorine pesticides in the apex of Amur Bay (the estuarine zone of the Razdolnaya River) was 186.8 ng/L and exceeded the MPC 160-fold for certain components. In bottom sediments from the Razdolnaya River estuary, the total OCP content was 32.6 ng/g. A calculation of coefficients of time period since the entry showed that DDT still continues to enter the waters of the bay (with terrigenous runoff from farmlands). HCH contamination can be assumed as relatively old, but the total level of its isomers indicates chronic contamination. In the middle part of the bay, the content of chlorinated pesticides was somewhat decreased, but remained substantial even in the open part. Thus, in the waters of the western part of the island territories (islands of Reineke and Rikord), the total OCP content exceeded the value recorded from the bay apex and amounted to more than 25 ng/L. Such a high concentration is probably associated with anchorage of ships in the roadstead, illegal bunkering, and discharge of bilge water in this area.

Hexachlorobenzene (HCB) was found in the waters of all the studied areas of Amur Bay. Its level was up to 27 ng/L and exceeded the MPC 2.7-fold.

One of the largest and most developed bays of Russky Island is Novik Bay, which juts deeply into the northern part of the western shore of the island. The contents of γ-HCH ranged from 7.2 to 10.5 ng/L; α-HCH, from 4.1 to 8.2 ng/L. HCB was found only in the eastern part of the bay at 6.5 ng/L; the metabolite DDD was also detected there at 6.4 ng/L. DDT, DDE, and β-HCH were not found at the rest of the stations, and their levels were below the limits of detectable concentrations. The total of organochlorine pesticides ranged from 11.2 to 27.1 ng/L (Fig. 4.4).

The calculation results showed a relatively *fresh* entry of γ-HCH at all of the stations sampled. The entry of DDT turned out to have occurred a long time ago at all stations. The metabolite DDD exceeded the MPC 6.4-fold.

Thus, the coastal zone of the eastern part of the bay proved to the most polluted area. The presence of pollution here is most likely associated with the recreational pressure, high shipping activity of small-tonnage boats, construction of facilities of the Far Eastern Federal University campus on the shore of Lesnik Cove, and also the unauthorized discharge of wastewater onto the land, which gradually gets into Novik

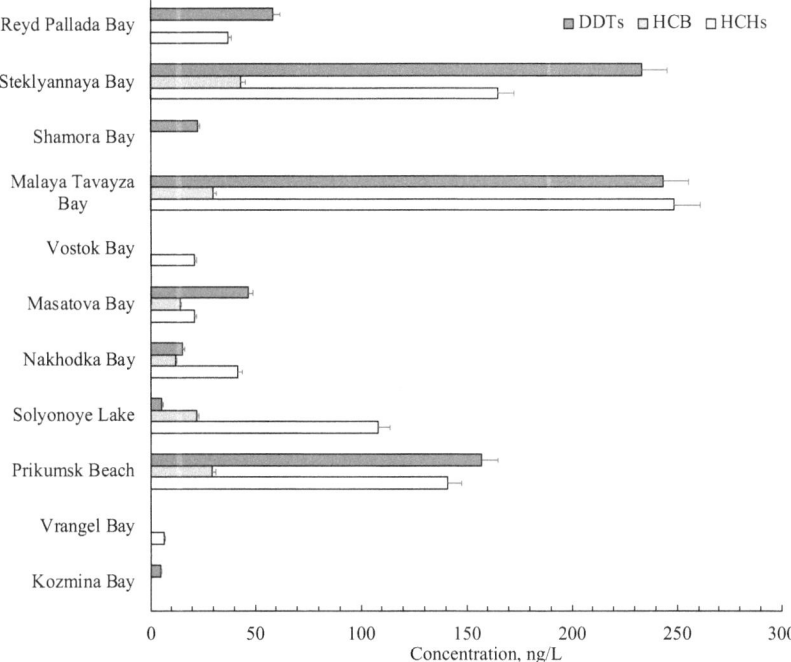

Fig. 4.4 OCP contents in coastal waters of the Sea of Japan

Bay. According to estimates, the biowaste of the Russky Island population makes up at least 10,000 m³ a day, or 4 million m³ a year. These volumes of dirty fresh water can significantly change the ecological condition of the narrow and shallow Novik Bay. Thus, the need to assess the condition of the waters of Novik Bay, and also to conduct regular monitoring of the condition of the environment and biota which would allow timely identification of emerging trends in the ecosystem of the bay, will certainly become acute in a few years [15].

As mentioned above, the major *supplier* of OCPs to Amur Bay is terrigenous runoff, but the extremely negative impact of seaports, railroad transport, and the actively developing network of highways (bridges, road junctions, and high-speed motorways running in the coastal part of the bay) should also be taken into account.

Unlike seawater, the organochlorine pesticide content of bottom sediments is much less subject to fluctuations and can provide more reliable evidence of level and degree of pollution.

α- and β-HCH were recorded from sediments sampled at all stations (5–17 ng/g). Lindane was found only in sediments from the open part of the bay (40.4–121.1 ng/g). Moreover, the ratio indicating a long-ago entry was significantly higher than unity, which is generally uncharacteristic of the ecosystems under study. The origin of lindane pollution in a part of Amur Bay, characterized by a sufficiently active hydro-chemical regime, can have only one explanation. Repair enterprises operated in the village of Slavyanka are located on the shores of the southwestern Amur Bay. These are equipped with wastewater treatment facilities with mechanical and biological methods of purification of industrial and domestic wastewater from enterprises and residential area. Sediments remaining during the purification process are removed from the treatment facilities and are not further disposed. The treated wastewater is used for technological purposes and discharged into the coastal waters of Amur Bay [16]. According to the data of the basin department of Rosvodresurs for Primorsky Krai (State's statistical reports on water use, document *2TP-vodkhoz*), in Possyet Bay in 2013, wastewater was discharged by six enterprises: OOO Zarubinskaya Baza Flota (Zarubino fleet base), OOO Morskoi Port (marine port), heat-supplying district Khasansky KGUP Primteploenergiya (Primorsky heat/energy supplying enterprise), OAO Slavyanka, trade port Possyet, and OOO Akvatekhnologii (Aquatechnologies). The largest amount of wastewater is discharged by water-using municipal utilities (heat-supplying district Khasansky and a branch of Ussuriysk's OAO Slavyanka), which constitutes approximately 75% of the total discharge. Wastewater discharged into Possyet is categorized as *without purification* and *insufficiently purified* [16].

In bottom sediments from the rivers of the Amur Bay basin, the total level of organochlorine pesticides ranged from 4.9 to 185.0 ng/g. Sediments from the rivers of Razdolnaya and Barabashevka were the most polluted. The maximum OCP level in bottom sediments was recorded from the Poima River, 185.0 ng/g. However, the entry of HCH and DDT was assumed to have occurred long ago. One of the possible OCP emitters is Pacific salmon that accumulate pesticides in tissues at feeding grounds and transport them to spawning grounds [17]. On the basis of level of OCP pollution, rivers can be arranged into the following sequence (ng/L): Narva and Poima (below the detection limits) < Gryaznaya (46.5) < Gladkaya (78.9) < Amba (79.8) < Barabashevka (88.5) < Ryazanovka (103.5) < Razdolnaya (186.8). When regarding bottom sediments as evidence of chronic pollution, the sequence of the rivers becomes different (ng/g): Ryazanovka (5.7) < Gryaznaya (16.8) < Razdolnaya (32.6) < Narva (49.1) < Barabashevka (55.3) < Poima (185.0).

The Tumen River, in its greatest extent, is the water border between China and North Korea; in the lower reaches and estuarine part, it is the border between North Korea and Russia. Almost 70% of the river's drainage basin is formed within the territory of China and almost 30% in North Korea. The share of Russia in the drainage formation is less than 1.0%. The annual discharge of the Tumen River is 5.7 km^3.

The ecological status of the river can be regarded as catastrophic. The detected OCP concentration was 5636 ng/L. HCHs exceeded the MPC 4470-fold, DDT 65-fold, and HCB 85-fold. It was shown that almost all pesticides underwent serious transformation: significant concentrations of α- and β-HCH (766.3 and 2080 ng/L, respectively) and metabolites of DDT (DDE, 165.1 ng/L; DDD, 80.3 ng/L), which may indicate a reduction in the use of these pesticides in China and North Korea. However, the current extremely high pollution of the drainage basin of the Tumen River in China and North Korea will, for a long time, negatively affect the ecological condition of not only the waters of the river, but also the waters of Peter the Great Bay where this river empties and emits highly toxic OCP transformation products. On the side of the Russian Federation, the Tumen River is not navigable, and there are no large human settlements on its banks. Thus, the conclusion that the major part of pollution comes from the right tributary of the river, located in the territory of China, where there has been an intensive industrial development in recent years, seems quite plausible.

In the water samples taken from Lake Lotos, located a few hundred meters from the Tumen River, the OCP content was almost 100-fold lower than in the waters from the Tumen. Only DDT was recorded, which indicates acute contamination (water sampling was carried out during heavy and stormy rains and non-transformed DDT got in a sample with surface runoff).

In the village of Vityaz, which is one of the favorite tourist attractions located in the southwestern Peter the Great Bay, the situation is relatively safe as regards OCP pollution. The total OCP content in the water was at a level of 50 ng/L. The coastal waters off Cape Shults are the cleanest part of Amur Bay, where the content of pesticides was below the detection limit.

Thus, on the basis of the level of OCP pollution, Amur Bay is categorized as waters exposed to the strongest anthropogenic pressure. The data obtained indicate the chronic contamination of this water area.

Nakhodka Bay is one of the most developed water areas in Primorsky Krai. It harbors ports and industrial enterprises located on its coast, and also commercial fishery areas for harvesting biological resources located within the bay. The eastern shore is exposed to a heavy recreational pressure. A source of pollution is also the Partizanskaya River, which empties into the bay. It drains wastewater from the town of Partizansk and agricultural farmlands nearby [18].

Currently, the ecological condition of Kozmin Bay, located in the easternmost part of Peter the Great Bay, is of particular interest and requires continuous monitoring. This is explained by the active industrial and economic development of the region and, as a consequence, with the increasing anthropogenic pressure on the waters of both the entire Nakhodka Bay and specifically Kozmin Bay, located in the its southeastern part.

Lake Vtoroye turned out to be one of the most polluted water areas of Nakhodka Bay studied. The total level of pesticides there made up 3171 ng/L. As calculated data shows, the entry of pesticides, both DDT and HCH, occurred long ago. γ-HCH exceeded of the MPC for fishery water areas more than 300-fold, DDT 120-fold, and HCB more than 60-fold. The poor ecological condition of this area is certainly associated with activities of the industrial enterprises of Nakhodka, located on the shores of the lake, as well as with the high terrigenous runoff.

The Partizanskaya River is, apparently, one of the major suppliers of OCPs to the waters of Nakhodka Bay. In the summer, during typhoons and heavy and stormy rains, the level of γ-HCH was 15.6 ng/L; the α-isomer was detected at 33.4 ng/L; HCB, 13.1 ng/L; DDT, 42.6 ng/L; and DDD, 18.8 ng/L. The total OCP content amounted to 123.4 ng/L, which was comparable to the influx of pesticides with the rivers of Artemovka, Shkotovka, and Sukhodol. The entry of DDT was assumed to be *fresh* and that of HCH to have occurred a long time ago. α-HCH, detected in sediments from the Partizanskaya River, suggests chronic contamination by this pesticide (5.8 ng/g). The finding of DDT (9.7 ng/g), on the contrary, indicates its recent entry (the metabolites DDD and DDE were not detected).

In the relatively small bodies of water in the Nakhodka Bay drainage basin, the total pesticide content was, on average, not higher than 100 ng/L, with the exception of abnormally high concentrations in the coastal waters off the Prikumsk berth (Vrangel Bay). On the basis of water pollution level, the studied areas of Nakhodka Bay were arranged into the following sequence (ng/L): Cape Kozmin (below detection limit) < Kozmin Bay (4.9) < Vrangel Bay, Vostochny Port (6.6) < Cape Petrovsky (18.1) < Lake Solenoye (27.4) < Musatov Bay (82.2) < Cape Shefner (82.3) < channel of Lake Solenoye (135.9) < Prikumsk berth (506.1).

As regards pollution of bottom sediments, the sequence was somewhat different, with the most polluted area being off Cape Shefner (138.7 ng/g) and the cleanest area being the Vostochny port (17.2 ng/g). This is also quite understandable, because the water area of Nakhodka Bay has been exploited much longer and more actively than Vrangel Bay. On the shores of the latter, a fuel-filling facility is operated, the active construction of facilities continues, and the pollution has not yet acquired the category of chronic, as can be evidenced by the OCP transformation ratios higher than unity.

4.6.3 Transboundary Transport in the Northwestern Pacific Ocean

From the viewpoint of global transport of pollution, including organochlorine pesticides, the Russian Federation should not be attributed to the main OCP-emitting countries. Although a major part of Russia is located in Asia, most of country's territory has never been exposed to anthropogenic pressure, and its agricultural farmlands, where pesticides could be used, are substantially smaller than those of neighboring countries. However, due to the specific geophysical, hydrological and atmospheric factors, significant amounts of pollutants from industrialized Asian countries are transferred to the territory of Russia. A general pattern of possible routes of pollutant transport in the Asia–Pacific region is shown in Fig. 4.5.

The waters of the world's oceans are undoubtedly the major receiver and, as a consequence, a source of transboundary transport of persistent organic pollutants. This is explained, first, by the tremendous volume of the oceans and different hydrological and hydrochemical regimes of their waters, which allows even almost insoluble substances to be transported over long distances. Second, the world's oceans are receivers of terrigenous wastewater from lands occupied by large industrial enterprises, which are one of the most important emitters of POPs of anthropogenic origin that, one or another way, enter the waters of the oceans. On a rather complicated pathway from release of POPs to their dissolution in ocean waters, there occur a number of chemical transformations, including photochemical, microbial degradation, interactions with ecosystem components, etc., that are often difficult to describe

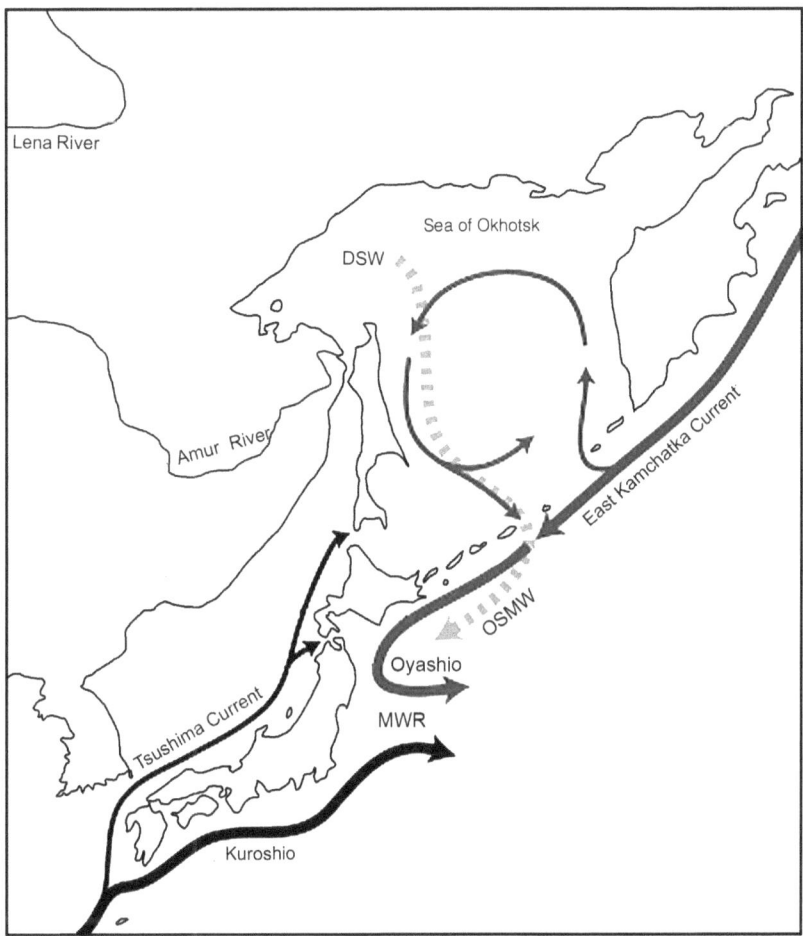

Fig. 4.5 Map of the currents in the northwestern Pacific Ocean. Cold (gray arrows) and warm (black arrows) surface currents, mixed water region (MWR) at the confluence of the Kuroshio and Oyashio. The dotted arrow schematically indicates the lowering of dense shelf water (DSW) and the outflow of the Okhotsk Sea Mode Water (OSMW) into the northwestern Pacific Ocean [19]

and simulate by the existing modeling methods. Due to the huge variety of these factors, it is sometimes challenging to predict the eventual fate of POPs. However, in recent years, the problem of transboundary transport of pollutants of various origins over large distances has been mentioned increasingly often. Thus, polychlorinated biphenyls, dioxins, organochlorine pesticides, polycyclic aromatic hydrocarbons, and alkylphenols are found in living organisms, ice, waters, and bottom sediments in the Arctic and Antarctic, i.e., where they have never been produced. Therefore, the issue of transboundary transport research is extremely important.

The East China and Yellow seas have periodic currents depending on the prevailing monsoon winds in them. In the East China Sea in winter, a current flows southward from the Taiwan Strait along the Asian mainland coast. Off the Vietnam coast, the current splits: one of the branches runs towards Borneo Island and, turning north near it, passes along the Philippine Islands, forming a counterclockwise gyre.

In winter, there are two northward currents flowing from the Taiwan Strait to the Yellow Sea. One is a cold current running north along the Asian mainland and tending to the Asian coast in the Taiwan Strait. There is a consequence of the existence here of a warm-water arm of the Kuroshio Current, which, thus, washes both the eastern and western shores of Taiwan. The second current of the Yellow Sea is formed by a branch of the Kuroshio, which deviates from it to the left and runs along the Korean Peninsula. Thus, a counterclockwise gyre is formed in the Yellow Sea.

In the Yellow Sea, during the northwest monsoon, there is a general northward movement of water along the Korean Peninsula, which is formed by a branch of the warm Kuroshio Current along with the current running southeast from the East China Sea through the Taiwan Strait.

In the Sea of Japan, there is a branch of the Kuroshio, referred to as Tsushima Current, that enters the sea from the south through the Korea Strait. In the Sea of Japan, this current deviates rightward and flows along the Japan coast. Most of the warm Tsushima water enters the ocean through the Tsugaru Strait (the strait between the Japan islands of Honshu and Hokkaido, connecting the Sea of Japan with the Pacific Ocean), and the rest of the current runs to the La Perouse Strait (between the islands of Hokkaido and Sakhalin) and through it, in part, flows to the Sea of Okhotsk. Cold waters, probably caused by upwelling from greater depths, are observed off the mainland coast in the Sea of Japan. Anyway, there is some movement of waters southward along the Russian coast and part of the Korean Peninsula. There is also evidence that a part of the Tsushima Current runs north along Sakhalin into the Tatar Strait.

In the Sea of Okhotsk, the current along the western Kamchatka coast flows north, and that along the mainland coast and Sakhalin runs south, forming a counterclockwise gyre.

There are no any clearly expressed and strong currents in the Bering Sea. The southward movement of cold water, observed along the eastern coast of the sea and along Kamchatka, continues further along the oceanic margin of the Kuril Islands and northern Japan islands, giving rise to the cold Oyashio Current, which also receives water through the Tsugaru Strait from a branch of the Tsushima Current.

The Bering Strait, as well as waters all along the American coast of this sea, is dominated by tidal currents. The current off the Asia coast is directed southward, and that off the American coast northward. No branches of the warm Kuroshio Current enter the Bering Sea, and, therefore, there cannot be such a branch in the Bering Strait.

Agricultural farmlands are located mainly on coastal plains and in river valleys, and, therefore, local rivers, receiving surface runoff from such lands, transport pesticide residues into estuaries and coastal zones of the seas. Pesticides coming with river water enter the sea in a dissolved state and are sorbed onto suspended particles.

The Zhujiang River in China drains vast fertile lands and potentially transports pesticide residues into the South China Sea. Then the river empties into the open part of the South China Sea and this water becomes involved by the current running north from the Taiwan Strait into the Yellow Sea. The Yellow Sea current, formed by a branch of the Kuroshio Current, runs along the Korean Peninsula. The Yellow Sea is an area of intensive shipping between the industrialized regions of China and South Korea. Moreover, due to the high shipping activity, anthropogenic disasters frequently occur here. Thus, the waters of the Yellow Sea are contaminated by pesticides. The branch of the Kuroshio Current of the Yellow Sea, referred to as the Tsushima Current, involves pollution coming from the South China Sea and enters the Sea of Japan through the Korea Strait along the Korean Peninsula coast. Most of the Tsushima Current enters the open part of the Pacific Ocean through the Tsugaru Strait, and a smaller part of the rest of the current enters the Sea of Okhotsk. The waters of the Sea of Okhotsk are cleaner compared to the waters of the Sea of Japan, and their pollution can be associated largely with transboundary transport by sea currents and air masses. Pesticides that have entered the Sea of Okhotsk with currents are brought by the Liman Current to the Primorsky Krai coast in Peter the Great Bay. Despite the polluted waters of Peter the Great, which is explained by the effect of some rivers common with China that empty directly into the bay, the transboundary transport of pollutants from the South-East Asia coast also greatly influences the general pollution of the bay.

Thus, we have attempted to describe the sources of OCP emissions in the northwestern Pacific Ocean, the pathways of pollutant transport, transformation, and migration with water masses from polluted waters of South-East Asia to the exclusive economic zone of the Russian Federation.

The ocean remains the final water body receiving persistent organic pollutants, and the bioaccumulation of pesticides up food chains in the ocean continues. Atmospheric precipitation and pollutants adsorbed on suspended particles and also toxic plastic degradation products accumulated by aquatic organisms can be concentrated at depths from 0 to 100 m [20].

The presence of pesticides in habitats of aquatic species remote from areas of industrial activities is a consequence of the global transport by winds and currents from the regions where these chemicals are applied (tropical and subtropical zones) to temperate latitudes. Furthermore, migrating fish have also been identified as a biotransport vector of organic pollutants from subtropical to subarctic ecosystems [21].

Acknowledgements The work was supported by the Ministry of Science and Higher Education of the Russian Federation, project no. FZNS-2023-0011.

References

1. Tsygankov VY (2019) Organochlorine pesticides in marine ecosystems of the Far Eastern Seas of Russia (2000–2017). Water Res 161:43–53. https://doi.org/10.1016/j.watres.2019.05.103
2. Maksimenko O (2003) Pesticides from the point of view of a chemist. In: Science and life, pp 56–57
3. Tsygankov VY (2020) Chapter 1. The dirty dozen of the Stockholm convention. Chemistry and toxicology of persistent organic pollutants (POPs): a review. In: Tsygankov VY (ed) Persistent organic pollutants (POPs) in the Far Eastern region: seas, organisms, human, monograph. Publishing House of Far Eastern Federal University, Vladivostok, Russia, pp 12–61
4. Franklin C, Worgan JP (2005) Occupational and residential exposure assessment for pesticides. J. Wiley, Chichester, West Sussex, England ; Hoboken, NJ
5. Fedorov LA, Yablokov AV (1999) Pesticides—a toxic blow to the biosphere and human. Nauka, Moscow, Russia
6. Tsygankov VY, Boyarova MD, Lukyanova ON (2015) Chemical and ecological aspects of persistent organic pollutants, textbook. Admiral Nevelskoy Maritime State University, Vladivostok, Russia
7. Hassan A, Gulzar S, Javid H, Nawchoo IA (2022) Pesticide toxicity and bacterial diseases in fishes. In: Bacterial fish diseases. Elsevier, pp 87–101
8. Galiulin RV, Galiulina RA (2011) Impact zones of persistent organochlorine compounds in the environment. Agrochemistry:83–89
9. Maistrenko VN, Klyuev NA (2012) Ecological and analytical monitoring of persistent organic pollutants. BINOM. Laboratoriya znaniy, Moscow
10. Shuntov VP (2001) Biology of the Far Eastern seas of Russia. TINRO-Center, Vladivostok
11. Nguyen DK, Malinin VN, Gordeeva SM (2016) The effect of water temperature on the formation of biological and fishery productivity of the South China Sea. In: Scientific notes of the Russian State Hydrometeorological University, pp 64–73
12. Korshenko AN (2016) Marine water pollution. 2015. Annual Report. Nauka, Moscow, Russia
13. Patrusheva OV, Chernova EN, Babicheva ON (2015) Petroleum hydrocarbons in the waters of the Ussuriysky Bay. In: New science: current state and ways of development, pp 5–10
14. Khmelnitsky VK (2005) Problems of utilization of industrial and household waste waters of the coastal territories of the Amursky and Ussuriysky Bays. In: Technical problems of the development of the World Ocean, pp 304–309
15. Khristoforova NK, Degteva YE, Berdasova KS et al (2016) Chemical and ecological state of the waters in the Novik Bay (Russky Island, Peter the Great Bay, Japan Sea). Izvestiya TINRO 186:135–144. https://doi.org/10.26428/1606-9919-2016-186-135-144
16. Nigmatulina LV, Chernyaev AP (2015) Pollution of the coastal waters in the Possyet Bay (Peter the Great Bay, Japan Sea) in conditions of current economic activity. Izvestiya TINRO 182:162–171. https://doi.org/10.26428/1606-9919-2015-182-162-171
17. Tsygankov VY, Lukyanova ON, Boyarova MD et al (2019) Organochlorine pesticides in commercial Pacific salmon in the Russian Far Eastern seas: food safety and human health risk assessment. Mar Pollut Bull 140:503–508. https://doi.org/10.1016/j.marpolbul.2019.02.008
18. Naumov YA (2015) Ecological state of the Nakhodka Bay. Izvestiya TINRO 122:524–537
19. Rella SF, Uchida M (2014) A southern ocean trigger for Northwest Pacific ventilation during the Holocene? Sci Rep 4:4046. https://doi.org/10.1038/srep04046
20. Lukyanova ON, Tsygankov VY, Boyarova MD, Khristoforova NK (2014) Pesticide biotransport by Pacific salmon in the northwestern Pacific Ocean. Dokl Biol Sci 456:188–190. https://doi.org/10.1134/S0012496614030089
21. Lukyanova ON, Tsygankov VY, Boyarova MD, Khristoforova NK (2015) Pacific salmon as a vector in the transfer of persistent organic pollutants in the Ocean. J Ichthyol 55:425–429. https://doi.org/10.1134/S0032945215030078

Chapter 5
Implementation of the Mussel Watch Program by Russian Scientists in the North Pacific (2000–2021)

Abstract Persistent organic pollutants (POP) are the most hazardous class of organic compounds in terms of the impact they exert on natural ecosystems and human health. POPs are characterized by high toxicity, bioaccumulation potential, and are subject to transboundary transport by migratory organisms and also with water and air masses. Despite the signing of the Stockholm Convention in 2001, a number of countries still retain the right to use DDT to control various pests and disease carriers. In the present study, we show the current levels of POPs in soft tissues of bivalves and, based on a comparison of data that we obtained and the results of earlier studies, carry out an analysis of time trends in the accumulation of toxicants using the example of Peter the Great Bay, Sea of Japan.

Keywords Bivalves · Mytilidae · POPs · OCPs · DDTs · HCHs · PCBs

5.1 Introduction

Currently, all natural ecosystems on Earth are exposed to anthropogenic pressure to one or another degree. The increasing pollution of the world's oceans has become one of the most serious environmental issues. Waters of the continental shelf are exposed to the greatest anthropogenic pressure associated with large amounts of terrigenous runoff, bilge water discharged, and high shipping activities [1].

Persistent organic pollutants (POPs) are considered the most poorly studied organic compounds among those harmful to living organisms. Many negative changes in human health and various functional disturbances that increasingly occur in natural ecosystems are associated with their adverse effects [2]. POPs have pronounced toxic properties, degrade extremely slowly, show a characteristic tendency to bioaccumulate, and are subject to transboundary transport by migratory organisms and also with water and air masses. They can be precipitated at a large distance from the source of pollution and build up in terrestrial, aquatic ecosystems, and fatty tissues of living organisms [3].

© The Author(s), under exclusive license to Springer Nature Switzerland AG 2023
V. Tsygankov, *Persistent Organic Pollutants in the Ecosystems of the North Pacific*,
Earth and Environmental Sciences Library,
https://doi.org/10.1007/978-3-031-44896-6_5

POPs are human-made chemicals used in agriculture to control various pests, in public health to control disease carriers, in production or technological processes (DDT and HCH), and are released during some production processes and combustion in the form of polychlorinated biphenyls (PCBs), dioxins, and furans [4].

Such POPs as dichlorodiphenyltrichloroethane (DDT) and hexachlorocyclo-hexane (HCH) had the widest application in the past [5, 6]. Now these pesticides are banned almost worldwide. However, despite the ban, some countries such as China and India still retain the right to produce and use DDT, which leads to contamination of not only users' but also their neighbors' territories.

To date, a large number of studies have been published that determined levels of chlorinated hydrocarbons in different regions of the world's oceans [7–10]. However, the problem of pollution by organochlorine compounds in the coastal waters of the Far Eastern seas of Russia has remained poorly understood [11–13] and, therefore, is highly relevant. In the present study, we aimed to analyze the temporal trends of POP accumulation (HCH isomers, DDT and its metabolites, and PCBs) in bivalve mollusks from the Sea of Japan by comparing the data that we obtained and the results of earlier studies.

5.2   Bivalves as Indicators of Water Body Pollution

Bioindication is a comprehensive assessment of the intensity and consequences of long-term pollution or other environmental effects based on the presence of indicator organisms, taxonomic structure of cenoses, disturbances in community functioning, or other deviations from the normal development of organisms [14, 15].

Bivalve (Lamellibranchia) mollusks are one of the groups of organisms most widely used for indication of coastal marine water pollution. This class includes members of benthic fauna closely associated with the substrate and widely distributed across the world. Bivalves are considered one of the richest groups in terms of abundance and biomass among other marine animal organisms. Furthermore, lamelli-branchs are filter-feeders, which contributes to the accumulation of various pollutants such as POPs and heavy metals in their tissues [16].

Their slow-moving or attached lifestyle, the potential to accumulate pollutants, and the relative ease of collecting material in large quantities make it possible to use bivalves as bioindicators [17].

Compared to fish and crustaceans, bivalves have a very low level of activity of enzyme systems capable of metabolizing POPs. Consequently, the concentration of toxicants in tissues of lamellibranchs more accurately indicates the degree of habitat pollution [18, 19].

Among the Bivalvia, the following mollusks of the family Mytilidae are often used to assess levels of pollutants, including organochlorine compounds, in a marine body of water: *Crenomytilus grayanus*, *Mytilus trossulus* and *Modiolus modiolus*. These mytilids are widely distributed in coastal waters of the seas and usually form large populations. In the Sea of Japan, they are found almost all along the coasts of Primorsky Krai, Sakhalin Island, and Japan [20].

Mussel aggregations are of applied importance: one mussel up to 6 cm in size is able to filter 60–70 L water a day, while a hectare of mussel bed can filter more than 5 km^3 of water within seven months [20].

Thus, bivalves are among the most accurate indicators of organic toxicant pollution of coastal waters due to their unique characteristics such as close associations with the substrate, sedentary lifestyle, and high abundance. Furthermore, regular monitoring of POP content in soft tissues of lamellibranchs will allow identification of the time trends in the accumulation of xenobiotics in natural ecosystems.

5.3 Mussel Watch in the Asia–Pacific Region

International Mussel Watch is a project implementing a program for monitoring the quality of coastal marine waters with bivalves used as bioindicators. The first phase of the project began in 1991–1993 and was carried out in South and Central America. The results of the studies revealed significant levels of organochlorine insecticides in the pre-industrial countries of the region.

The second phase was launched in 1994 and aimed at studying marine waters of the Asia–Pacific region, primarily the northwestern and western Pacific Ocean and tropical regions of the ASEAN countries. Highly toxic compounds (dioxins and furans) and organically bound metals (Cu and Sn) were also considered at the second phase of the project, in addition to traditional persistent organic pollutants such as PCBs, DDT and its metabolites, chlordane compounds, hexachlorocyclohexane (HCH) isomers, hexachlorobenzene (HCB), etc. The goals of the Asia–Pacific Mussel Watch project are to assess the current ecological status of marine waters in the Asia–Pacific region exposed to organic toxicants, study the time trends, the fate and consequences of POP accumulation, develop the programs to improve the condition of the marine environment in the region, and take countermeasures to protect coastal ecosystems [21].

Early surveys of the level of chlorinated hydrocarbons in marine waters of the Asia–Pacific region were conducted in the late 1990s and the early 2000s. An analysis of the POP content in the coastal zones of Thailand, the Philippines, and India [22] showed low levels of the organic toxicants in soft tissues of bivalves. However, the dominance of p,p'-DDT indicated the active use of this pesticide to protect public health and control pests of aquacultures. The highest HCH and DDT concentrations were recorded from coastal waters of India, while PCBs prevailed in the Philippines.

A monitoring study [23] of pollution, distribution, and possible sources of POPs in marine ecosystems of 12 countries of the Asia–Pacific region (Cambodia, China, Hong Kong, India, Indonesia, Japan, South Korea, Malaysia, the Philippines, Russian Far East, Singapore, and Vietnam), conducted from 1994 to 2001, showed the presence of chlorinated hydrocarbons in all collected specimens of bivalves. Significant organochlorine pesticide (OCP) concentrations were found in all samples. The pesticide levels in developing countries exceeded those in developed ones. The highest concentrations of DDT were recorded from China, Hong Kong, and Vietnam, and those of HCH isomers were higher in waters of India and China. Industrial countries are characterized by lower PCB levels compared to post-industrial ones. Such results suggest a close relationship between the PCB pollution of marine waters and the level of industrial activity in the study area.

The implementation of the Mussel Watch Program in the coastal waters of South Korea [24, 25] showed increasing environmental pollution by POPs in the period from 2001 to 2007. Although the production and use of DDT for agricultural purposes in South Korea has been banned since 1971, the study revealed high concentrations of this toxicant in soft tissues of collected bivalve individuals. Despite the uncertainty about the exact cause, the authors assumed that such results could be due to the resuspension of pollutants from bottom sediments. Nevertheless, DDE dominated the metabolites, which indicated the lack of fresh contamination of the water area by pesticides. The concentrations of HCH isomers were significantly lower compared to DDT and PCBs. According to the results, HCHs were distributed evenly along the coast, while the highest levels of DDT and its metabolites tended to areas with developed industrial zone and high population. The authors suggested that the difference in the distribution of pesticides could be related to the high volatility of HCHs that contributes to their transport over long distances by air masses. The use of PCB has also been banned in South Korea since the 1970s. Nevertheless, a study of 2007 [24] showed high levels of PCBs in industrial areas of the country. The highest concentrations were recorded from the bays of Ulsan and Masan, where large steel-casting and other industrial plants are operated.

To assess the degree of pollution of the marine waters along the northeastern coast of China, a monitoring study with bivalves as bioindicators was conducted in 2005 [26]. The results showed the presence of POPs dominated by DDT in all individuals. Although DDT metabolites prevailed, suggesting the degradation of the initial compound, high concentrations of original DDT (1000 ng/g dry weight) were recorded at 19% of the stations, which indicated the continuing entry of the toxicant into the environment. The β-isomer dominated HCHs. Despite the measures taken in China since the 1970s to control the entry of PCBs into marine waters, the study revealed relatively high concentrations of PCBs compared to those in waters off the southern coast of China and South Korea. The results indicated that the high PCB levels, as in studies by other authors, tended to areas with developed industries and high shipping activity. In general, the data obtained showed a decrease in the influx of POPs in the coastal waters off northeastern China.

The pollution of marine ecosystems by POPs, in particular OCPs, is one of the most urgent environmental problems for India. Since OCPs are still being used in India for crop pest control and malaria vector control, xenobiotics such as DDT and its metabolites and HCH isomers dominate the chemical pollutants recorded along the Indian coast. A study of presence, distribution, and possible sources of OCPs in the tropical coastal zone of India [27] revealed a significant difference between OCP concentrations on the eastern and western coasts, which could indicate differences in the use of pesticides in different parts of the country and their transport by several rivers into the marine environment. In some cases, the DDT levels exceeded the levels of minimum permissible exposure. In general, concentrations of metabolites varied between different regions of India, with the predominance of p,p'-DDT and p,p'-DDD. Furthermore, concentrations of DDT were higher than those of HCH, which the authors attributed to the greater solubility and biodegradability of HCH in water compared to DDT.

A study of the chlorinated hydrocarbon biomagnification in the food chains of the Yellow Sea [28] showed the presence of POPs in all collected bivalve individuals. DDT concentrations were one to two orders of magnitude higher than the levels of other pollutants. Since pesticides have actively been used in the agriculture of China, the water of the Yangtze River emptying into the Yellow Sea can be considered the major source of DDT pollution of the marine biota in the region. The recorded contents of HCH and PCB isomers were generally low and were consistent with the results of previous studies of marine organisms from coastal waters of China and South Korea.

A study of levels and distribution of POPs on the coast of Okinawa, Japan [29], revealed the presence of PCBs, DDT and its metabolites, and HCH isomers in almost all specimens of bivalves. The highest POP levels were recorded from samples taken on the southwestern coast near large cities. In particular, PCB pollution was mainly characteristic of the city of Naha. Earlier studies of toxicant levels in marine organisms from the coastal waters of Okinawa also showed higher concentrations of PCBs in specimens collected from the southwest of the island, which, presumably, could be related to PCBs used in electrical equipment and leaking into the environment. The DDT group was dominated by the metabolites p,p'-DDE and p,p'-DDD. However, p,p'-DDT was found in mollusks sampled near Onna Village, which indicated a local source of the pollutant entering the water.

A decadal study of POP distribution in deep-sea chemosynthetic bivalves [30] aimed to determine levels of organic toxicants and assess potential risks to deep-sea ecosystems. Bivalves were collected from two sites: near a densely populated area and relatively far from places of human activity. An analysis of samples showed a decrease in PCB concentrations over the 30-year study period and, therefore, the effectiveness of the environmental policy restrictions. However, the detected POP contamination of deep-sea ecosystems allowed the assumption about the global occurrence of toxicants in chemosynthetic organisms, in all the world's oceans. The results of the study showed the need for further investigation into levels of organic pollutants in chemosynthetic habitats in order to protect highly endemic species of aquatic organisms.

Surveys for the determination of POPs in soft tissues of bivalves from the Far Eastern seas of Russia have been conducted since 2000 [1, 12, 31, 32]. The results showed the presence of organic pollutants at low concentrations. The highest POP levels were recorded from the waters of Possyet Bay [1], which allows a conclusion about the substantial contribution that transboundary transport makes to the chlorinated hydrocarbon pollution of coastal waters in Primorsky Krai.

In general, OCPs, in particular DDT and its metabolites, are the pollutants most characteristic of the Asia–Pacific region. Their highest concentrations were recorded from China and India, where these toxicants are still applied in agriculture to control various pests and in public health to control disease carriers.

High PCB levels are also reported for Japan and China, which may be related with the developed industrial sector of the countries and high shipping activity. Several of the world's largest ports (Shanghai (China), Shenzhen (China), Ningbo-Zhoushan (China), and Yokohama (Japan)) are located in China and Japan, which may explain the pollution of coastal waters nearby. The results of surveys of POP levels in the marine environment of some of the Asia–Pacific countries are presented in Fig. 5.1.

Lower POP concentrations, compared to the other Asia–Pacific countries, were recorded from the territory of the Russian Federation. The absence of large industrial enterprises and progressive agricultural industry in Primorsky Krai explains the low levels of pollution of marine waters by organic pollutants in the region. Such results suggest that the transboundary transport of POPs by water and air masses makes the major contribution to the influx of fresh pollutants in the waters of Peter the Great Bay.

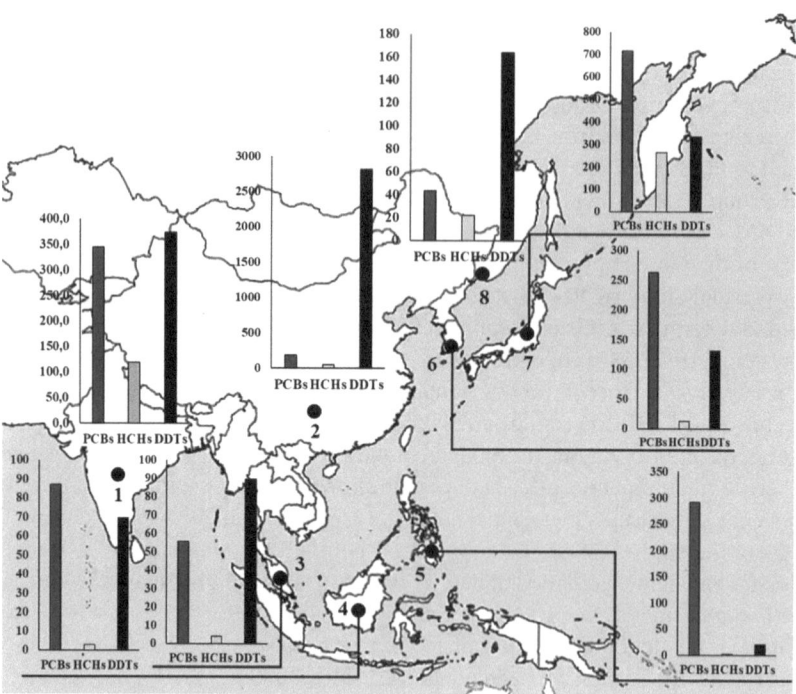

Fig. 5.1 Levels of persistent organic pollutants in several countries of the Asia–Pacific region: 1—India; 2—China; 3—Malaysia; 4—Indonesia; 5—Philippines; 6—South Korea; 7—Japan; 8—Russian Federation

5.4 Current POP Level in Soft Tissues of Bivalves from Peter the Great Bay, Sea of Japan

Soft tissues from a total of 94 individuals of three bivalve species of the family Mytilidae (*Crenomytilus grayanus*, *Modiolus modiolus*, and *Mytilus trossulus*), collected in different parts of Peter the Great Bay in the summer seasons of 2017 and 2018, were used in the present study as biological material to indicate pollution (Fig. 5.2).

The results of the analysis of soft tissues from bivalves in 2017 showed the presence of chlorinated hydrocarbons ($\Sigma OCP + \Sigma PCB$) within a range from 0.6 to 2769.7 ng/g lipid weight (l.w.) in all individuals. According to the data obtained, the highest POP concentrations were in *M. modiolus* collected from off Cape Krasny. The Bolshoy Pelis Island station, where POP levels ranged from 1.1 to 253.1 ng/g l.w., proved to be a water area least polluted by organic toxicants.

The concentrations of OCPs were higher than those of PCBs. DDT and its metabolites dominated the pesticides at all stations. The major metabolite was DDE, indicating the degradation of the initial compound and, as a consequence, the long-term exposure of the environment to the toxicant. The highest levels of OCPs, dominated by DDT, were recorded from soft tissues of bivalves sampled at Cape Krasny. The dominant metabolite was p,p'-DDE detected in a concentration range from 703.94 to 2769.7 ng/g l.w. The Partizanskaya River, draining the farmlands where pesticides were actively used for treating crops in the second half of the twentieth century, empties into Nakhodka Bay not far from Cape Krasny. This may be an explanation for the presence of these forms, indicating a long-ago entry of pollutants into the water, in soft tissues of mollusks.

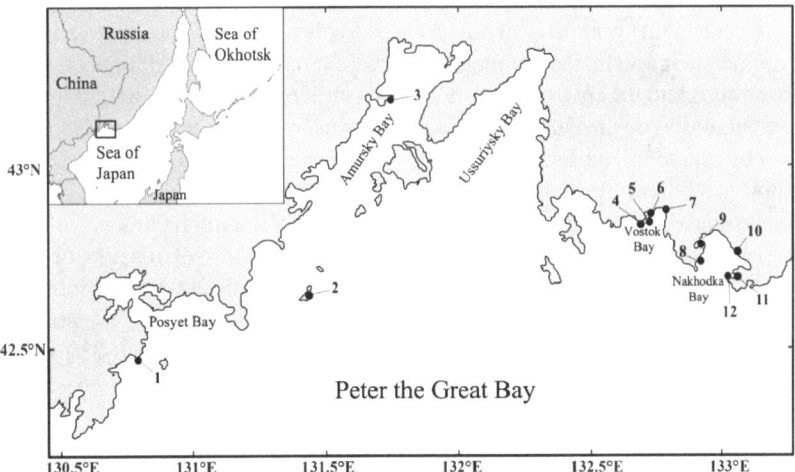

Fig. 5.2 Map of the sampling sites: 1—Vera Island; 2—Bolshoy Pelis Island; 3—Cape Chikhachev; 4—Strelbishche Cove; 5—Cape Tchaikovsky; 6—Sredniaya Cove; 7—Litovka Cove; 8—Musatov Cove; 9—Nakhodka Bay; 10—Cape Krasny; 11—Kozmin Bay; 12—Cape Kozmin

HCH isomers were present in almost all samples within a range from 0.6 to 553.1 ng/g l.w. The dominant isomer was β-HCH, the most stable form indicating that contamination of the water area occurred long ago. However, the ratio of α- and γ-HCH showed a predominance of the γ-isomer, which may be evidence of a recent entry of toxicants. Such results may be explained by the phenomenon of transboundary transport of POPs by water and air masses from territories where pesticides are still used in agriculture to control various pests and in public health to control disease carriers.

PCBs were detected in all samples within a range from 0.6 to 610 ng/g l.w. and were dominated by lower chlorinated congeners, 28 and 52, found in all soft tissue samples from bivalves. The presence of lower chlorinated congeners indicates fresh contamination of the water area. The highest concentrations were recorded from Kozmin Bay, within a range from 68.0 to 151.2 ng/g l.w. A source of PCBs entering the local waters may be the oil transloading facility operated since 2009 on the shore of the bay. Furthermore, Kozmin Bay was earlier used as an airfield for seaplanes and as a place for dismantling ships in the 1990s, which also resulted in the highest concentrations of pollutants among all the stations sampled. PCB 138, detected in a range from 3.7 to 52.1 ng/g l.w., dominated the higher chlorinated congeners. The results of the study are presented in Table 5.1.

5.5   Time Trends of POP Accumulation in Soft Tissues of Bivalves Exemplified by the Waters of Peter the Great Bay

Although the POP pollution of natural ecosystems has become a serious global environmental problem, the number of publications considering influx of chlorinated hydrocarbons into the environment in Russia is still low. To date, of particular interest have been studies on the Far Eastern Seas of Russia, including the Sea of Japan. Since POPs are characterized by high toxicity and resistance to natural factors, regular monitoring studies of toxicant levels in the water and commercial species of aquatic organisms have important human health and conservation implications.

An analysis of the organochlorine pesticide content in soft tissues of bivalve mollusks from Peter the Great Bay was carried out until 2002. A survey of the Amur Bay area in 1999 [31] revealed the presence of DDT and its metabolites in all *C. grayanus* individuals within a concentration range from 520 to 730 ng/g l.w. HCH isomers were also recorded from all samples in a range from 34 to 57 ng/g l.w. According to the data of a survey in Possyet Bay in 2002 [33], soft tissues of *C. grayanus* and *Mizuhopecten yessoensis* contained all HCH isomers that were dominated by β-HCH. The presence of DDT and its metabolites was recorded only from Gray's mussels, with the main metabolite being DDD. Since the results of the 2002 survey were expressed in terms of ng/g wet weight, we converted the data to be in ng/g l.w. (with an average fat content of 0.396%). A graphical representation

Table 5.1 Levels of persistent organic pollutants in soft tissues of bivalves of the family Mytilidae from Peter the Great Bay, 2017–2018, ng/g l.w.

Sampling site	N	Species	Fat, %	$\sum$DDT[4]	$\sum$HCH[5]	$\sum$PCB[6]
Nakhodka Bay	5	Bay mussel (*Mytilus trossulus*)	$\frac{0.29-0.41^{1}}{0.33\pm0.05^{2}}$	$\frac{<DL^{3}-47.5}{41.87\pm9.24}$	$\frac{<DL-27.2}{23.2\pm3.56}$	$\frac{<DL-21.9}{15.85\pm6.26}$
Cape Kozmin	10	Bay mussel (*Mytilus trossulus*)	$\frac{0.25-0.69}{0.27\pm0.16}$	$\frac{<DL-23.8}{18.25\pm7.85}$	$\frac{<DL-23.8}{20.90\pm4.10}$	$\frac{<DL-20.2}{37.34\pm13.72}$
Cape Chikhachev	10	Gray's mussel (*Crenomytilus grayanus*)	$\frac{0.03-3.16}{0.97\pm0.91}$	$\frac{<DL-35.7}{13.25\pm10.53}$	$\frac{<DL-553.1}{120.46\pm242.17}$	$\frac{<DL-610.5}{108.638\pm178.28}$
Sredniaya Cove	6	Northern horsemussel (*Modiolus modiolus*)	$\frac{0.20-1.33}{0.66\pm0.38}$	$\frac{<DL-161.9}{66.43\pm51.13}$	$\frac{<DL-30.0}{18.40\pm10.62}$	$\frac{<DL-230.2}{39.86\pm52.42}$
Kozmin Bay	6	Gray's mussel (*Crenomytilus grayanus*)	$\frac{0.15-0.39}{0.26\pm0.10}$	$\frac{<DL-89.0}{36.58\pm35.56}$	$\frac{<DL-29.9}{24.80\pm7.21}$	$\frac{<DL-151.2}{112.2\pm40.43}$
Cape Chaikovsky	5	Gray's mussel (*Crenomytilus grayanus*)	$\frac{0.44-0.65}{0.52\pm0.09}$	$\frac{<DL-665.0}{225.83\pm380.33}$	$<DL-12.5$	$\frac{<DL-54.9}{29.01\pm12.73}$
Cape Krasny	6	Northern horsemussel (*Modiolus modiolus*)	$\frac{0.32-0.91}{0.46\pm0.22}$	$\frac{<DL-2769.7}{769.18\pm974.02}$	$\frac{<DL-63.6}{31.07\pm16.16}$	$\frac{<DL-149.8}{121.12\pm50.71}$
Musatov Cove	6	Northern horsemussel (*Modiolus modiolus*)	$\frac{0.13-0.91}{0.46\pm0.25}$	$\frac{<DL-889.69}{102.52\pm248.29}$	$\frac{<DL-91.4}{37.85\pm27.86}$	$\frac{<DL-314.1}{105.23\pm97.20}$
Bolshoy Pelis Island	10	Gray's mussel (*Crenomytilus grayanus*)	$\frac{0.005-0.91}{0.009\pm0.003}$	$\frac{<DL-253.1}{74.57\pm66.76}$	$\frac{<DL-27.5}{6.38\pm6.16}$	$\frac{<DL-11.1}{9.18\pm7.72}$
Cape Vera	12	Gray's mussel (*Crenomytilus grayanus*)	$\frac{0.11-0.59}{0.35\pm0.13}$	$\frac{<DL-108.2}{39.73\pm45.68}$	$<DL$	$\frac{<DL-341.3}{80.74\pm81.27}$

(continued)

Table 5.1 (continued)

Sampling site	N	Species	Fat, %	$\sum$DDT[4]	$\sum$HCH[5]	$\sum$PCB[6]
Litovka Cove	10	Northern horsemussel (*Modiolus modiolus*)	0.003−0.009 / 0.005±0.002	<DL−453.4 / 169.33±123.93	<DL−179.3 / 29.59±45.27	<DL−43.1 / 19.10±13.92
Strelbishche Cove	10	Northern horsemussel (*Modiolus modiolus*)	0.004−0.50 / 0.015±0.014	<DL−1058.3 / 125.41±251.12	<DL−13.2 / 4.22±3.05	<DL−32.0 / 10.41±8.22

Note [1]Range, min–max (above line); [2]arythmetic mean ± standard deviation (below line); [3]below the detection limit; [4]total content of *p,p'*-DDT, *o,p'*-DDT, *p,p'*-DDE, *o,p'*-DDE, *p,p'*-DDD, and *o,p'*-DDD; [5]total content of α-HCH, β-HCH, γ-HCH, and δ-HCH; [6]total content of PCB 28, PCB 52, PCB 101, PCB 118, PCB 153, PCB 138, and PCB 180

of the temporal trends of OCP accumulation in soft tissues of bivalves is shown in Fig. 5.3.

The data obtained by comparing the levels of pesticides from 1999 to 2017 show a decrease in the concentrations of toxicants in the waters of Peter the Great Bay. In all the studies, degradation products of the initial compounds (DDE and DDD, β-HCH) in soft tissues of bivalves were observed to prevail. The results characterize a gradual reduction in the influx of pollutants to the environment both with the rivers receiving the surface runoff from agricultural fields of Primorsky Krai, where pesticides were previously used, and the through transboundary transport by water and air masses from territories where OCPs are still being used. Until the signing of the Stockholm Convention by a number of countries in 2001, the production and use of OCPs was not

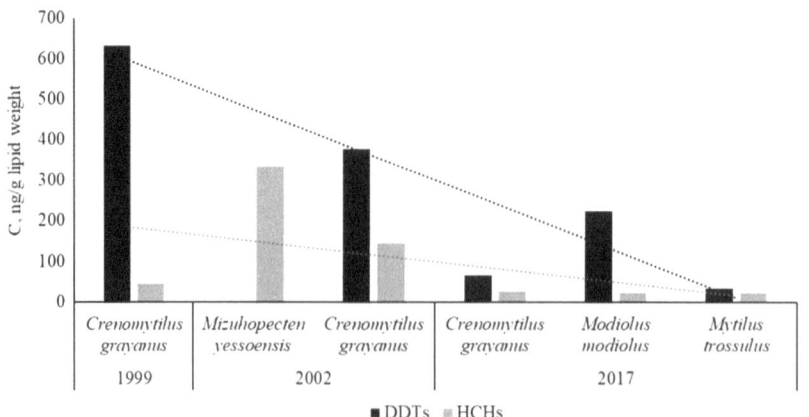

Fig. 5.3 POP levels in soft tissues of bivalves from Peter the Great Bay, Sea of Japan, in 1999, 2002, and 2017

limited, which may explain the higher levels of DDT documented in 1999 compared to 2002 and 2017.

An analysis of the PCB level in soft tissues of bivalves from Peter the Great Bay was conducted in 1999 [31]. The results of the study revealed the presence of PCBs in all *C. grayanus* individuals within a range from 2700 to 3700 ng/g l.w. The data that we obtained during our work in 2017 show a decrease in PCB concentrations. Nevertheless, despite the lack of developed industrial sector in Primorsky Krai, lower chlorinated congeners that indicate recent contamination prevail in soft tissues of bivalves. Such results may be associated with the impossibility to fully control disposal and entry of PCB into the environment, because, according to the Stockholm Convention, PCBs are categorized as unintentionally produced compounds.

5.6 Conclusion

Thus, the results obtained indicate the continuing pollution of the Peter the Great Bay waters by POPs. According to the data, there is a predominance of the xenobiotics' forms indicating that the entry of toxicants into the environment occurred long ago. Furthermore, a comparative analysis of the levels, distribution, and possible sources of POPs in the waters of Peter the Great Bay has shown lower concentrations of organic toxicants compared to those recorded from the neighboring countries of the Asia–Pacific region.

The refusal to use OCPs and the measures aimed at preventing unintentional PCB production in an increasing number of countries have led to a significant reduction in POP levels in soft tissues of bivalves from the bay over 18 years. Nevertheless, the monitoring studies of the influx and transformation of chlorinated hydrocarbons in coastal zones of the Far Eastern seas of Russia should be continued.

Acknowledgements The work was supported by the Ministry of Science and Higher Education of the Russian Federation, project no. FZNS-2023-0011.

References

1. Boyarova MD (2008) Current levels of organochlorine pesticides in aquatic organisms of Peter the Great Bay (Sea of Japan) and Khanka Lake. Abstract of the dissertation for the degree of candidate of biological sciences, Pacific State Economic University
2. Rovinsky FY, Voronova LD, Afanasev MI et al (1990) Background monitoring of ground ecosystems contamination by organochlorine compounds. Gidrometeoizdat, Leningrad
3. Mamontova EA, Tarasova EN, Mamontov AA et al (2012) Persistent organic pollutants in the atmospheric air of some territories of Siberia and the Far East of Russia. In: Geography and natural resources, pp 40–47
4. El-Shahawi MS, Hamza A, Bashammakh AS, Al-Saggaf WT (2010) An overview on the accumulation, distribution, transformations, toxicity and analytical methods for the monitoring

of persistent organic pollutants. Talanta 80:1587–1597. https://doi.org/10.1016/j.talanta.2009. 09.055

5. Tsygankov VY, Boyarova MD, Kiku PF, Yarygina MV (2015) Hexachlorocyclohexane (HCH) in human blood in the south of the Russian Far East. Environ Sci Pollut Res 22:14379–14382. https://doi.org/10.1007/s11356-015-4951-3

6. Tsygankov VY, Khristoforova NK, Lukyanova ON et al (2017) Selected organochlorines in human blood and urine in the south of the Russian Far East. Bull Environ Contam Toxicol 99:460–464. https://doi.org/10.1007/s00128-017-2152-0

7. Kljaković-Gašpić Z, Herceg-Romanić S, Kožul D, Veža J (2010) Biomonitoring of organochlorine compounds and trace metals along the Eastern Adriatic coast (Croatia) using Mytilus galloprovincialis. Mar Pollut Bull 60:1879–1889. https://doi.org/10.1016/j.marpolbul.2010. 07.019

8. Vieweg I, Hop H, Brey T et al (2012) Persistent organic pollutants in four bivalve species from Svalbard waters. Environ Poll (Barking, Essex: 1987) 161:134–142. https://doi.org/10.1016/j. envpol.2011.10.018

9. Ali N, Ali LN, Eqani SAMAS et al (2015) Organohalogenated contaminants in sediments and bivalves from the Northern Arabian Gulf. Ecotoxicol Environ Saf 122:432–439. https://doi. org/10.1016/j.ecoenv.2015.09.013

10. Korotkova LI, Sevostyanova MV, Votinova TV, Barabashin TO (2018) Accumulation of organochlorine pesticides and polychlorinated biphenyls in the organs of the Azov sea commercial fish species. Probl Fisheries 19:522–533

11. Khristoforova NK, Latkovskaya EM (1998) Organochlorine compounds in the bays of the North-East of Sakhalin. Vestnik of the Far East Branch of the Russian Academy of Sciences:34–45

12. Li A (2007) Persistent organic pollutants in asia: sources, distributions, transport and fate (Developments in environmental science; 7). Elsevier Science Limited

13. Lukyanova ON, Boyarova MD, Tsygankov VY (2016) Persistent organic pollutants in molluscs and fishes of the Sea of Japan and the sea of Okhotsk. In: Marine biological research: achievements and prospects. A.O. Kovalevsky Institute of Biology of the Southern Seas of RAS, Sevastopol, Russia, pp 136–139

14. Tsygankov VY, Boyarova MD, Lukyanova ON, Khristoforova NK (2017) Bioindicators of organochlorine pesticides in the Sea of Okhotsk and the Western Bering Sea. Arch Environ Contam Toxicol 73:176–184. https://doi.org/10.1007/s00244-017-0380-2

15. Khristoforova NK, Gnetetskiy AV (2022) The contents of heavy metals in long-lived Mytilids from Ussuriisky Bay. Russ J Mar Biol 48:26–32. https://doi.org/10.1134/S1063074022010060

16. Waykar B, Deshmukh G (2012) Evaluation of bivalves as bioindicators of metal pollution in freshwater. Bull Environ Contam Toxicol 88:48–53. https://doi.org/10.1007/s00128-011-0447-0

17. Harbo RM (2007) Shells & shellfish of the Pacific Northwest: a field guide. Harbour Pub., Madiera Park, B.C.

18. Lukyanova ON, Korchagin VP (2017) Integral biochemical index of the state of aquatic organisms under polluted conditions. Biol Bull 44:203–209. https://doi.org/10.1134/S10623590170 2011X

19. Walker CH (2019) Organic pollutants: an ecotoxicological perspective, 2nd edn. CRC Press, Boca Raton

20. Mikhaltsova OS, Galysheva YA (2014) Population and biological features of the settlements of Crenomytilus grayanus (Bivalvia: Mytilidae) in the Kievka Bay, Japan Sea. Izvestiya TINRO 177:125–138. https://doi.org/10.26428/1606-9919-2014-177-125-138

21. Tanabe S (2000) Asia-Pacific mussel watch progress report. Mar Pollut Bull 40:651. https:// doi.org/10.1016/S0025-326X(00)00019-9

22. Tanabe S, Prudente MS, Kan-atireklap S, Subramanian A (2000) Mussel watch: marine pollution monitoring of butyltins and organochlorines in coastal waters of Thailand, Philippines and India. Ocean Coast Manag 43:819–839. https://doi.org/10.1016/S0964-5691(00)00060-0

23. Monirith I, Ueno D, Takahashi S et al (2003) Asia-Pacific mussel watch: monitoring contamination of persistent organochlorine compounds in coastal waters of Asian countries. Mar Pollut Bull 46:281–300. https://doi.org/10.1016/S0025-326X(02)00400-9

24. Ramu K, Kajiwara N, Isobe T et al (2007) Spatial distribution and accumulation of brominated flame retardants, polychlorinated biphenyls and organochlorine pesticides in blue mussels (Mytilus edulis) from coastal waters of Korea. Environ Poll (Barking, Essex: 1987) 148:562–569. https://doi.org/10.1016/j.envpol.2006.11.034

25. Choi HG, Moon HB, Choi M et al (2010) Mussel watch program for organic contaminants along the Korean coast, 2001–2007. Environ Monit Assess 169:473–485. https://doi.org/10.1007/s10661-009-1190-4

26. Jin Y, Hong SH, Li D et al (2008) Distribution of persistent organic pollutants in bivalves from the northeast coast of China. Mar Pollut Bull 57:775–781. https://doi.org/10.1016/j.marpolbul.2008.04.045

27. Sarkar SK, Bhattacharya BD, Bhattacharya A et al (2008) Occurrence, distribution and possible sources of organochlorine pesticide residues in tropical coastal environment of India: an overview. Environ Int 34:1062–1071. https://doi.org/10.1016/j.envint.2008.02.010

28. Byun G-H, Moon H-B, Choi J-H et al (2013) Biomagnification of persistent chlorinated and brominated contaminants in food web components of the Yellow Sea. Mar Pollut Bull 73:210–219. https://doi.org/10.1016/j.marpolbul.2013.05.017

29. Mukai Y, Goto A, Tashiro Y et al (2020) Coastal biomonitoring survey on persistent organic pollutants using oysters (Saccostrea mordax) from Okinawa, Japan: Geographical distribution and polystyrene foam as a potential source of hexabromocyclododecanes. Sci Tot Environ 739:140049. https://doi.org/10.1016/j.scitotenv.2020.140049

30. Ikuta T, Nakajima R, Tsuchiya M et al (2021) Interdecadal distribution of persistent organic pollutants in deep-sea chemosynthetic bivalves. Front Mar Sci 8:751848. https://doi.org/10.3389/fmars.2021.751848

31. Tkalin AV, Lishavskaya TS, Hills JW (1997) Organochlorine pesticides in mussels and bottom sediments from Peter the Great Bay near Vladivostok. Ocean Res 19:115–119

32. Tsygankov VY (2019) Organochlorine pesticides in marine ecosystems of the Far Eastern Seas of Russia (2000–2017). Water Res 161:43–53. https://doi.org/10.1016/j.watres.2019.05.103

33. Boyarova MD, Lukyanova ON (2006) Chlorinated hydrocarbons in marine organisms of Posyet Bay (Japan Sea). Izvestiya TINRO 145:271–278

Chapter 6
Polycyclic Aromatic Hydrocarbons in Bottom Sediments and Fish from the Northwestern Pacific

Abstract Distribution and origin of polycyclic aromatic hydrocarbons were studied in bottom sediments collected from Peter the Great Bay, Sea of Japan, and in tissues of fish from the Sea of Okhotsk and Bering Sea. The results have shown that polyarenes in the bottom sediments from Peter the Great Bay are predominantly of petrogenic origin. In terms of the PAH level in bottom sediments, Amur and Ussuri Bays have proven to be the most polluted of the water bodies under study; the least polluted area is the open part of Peter the Great Bay. The highest levels of toxicants in aquatic organisms from the Sea of Okhotsk and Bering Sea have been recorded for Alaska pollock; the lowest level, for flounder.

Keywords PAHs · THCs · Aquatic organisms · Sea of Japan · Sea of Okhotsk · Bering Sea

6.1 Introduction

Total hydrocarbons (THCs), including polycyclic aromatic hydrocarbons (PAHs), are widely distributed in the environment. Quite a large portion of these compounds are synthesized and formed in nature, while anthropogenic hydrocarbons having negative effects also enter the environment in substantial amounts.

PAHs are organic compounds of the benzene series, differing in the number of aryl rings and specifics of their bonding [1]. The simplest representatives of this group are naphthalene, anthracene, and phenanthrene. PAHs are of both natural and technogenic origin. Their major sources are divided into pyrogenic (formed through combustion of fossil fuels, organic substances, and waste) and petrogenic (emerging in case of oil and petroleum products spills) [2, 3].

These pollutants can enter the human body with food ingested and air breathed. Due to their high lipophilicity, PAHs can build up in living organisms, cause gene mutation, disturbance of immune and metabolic functions, negatively affect enzyme activity, cause death of red blood cells, be accumulated in the liver, and contribute to the cancer prevalence [4].

© The Author(s), under exclusive license to Springer Nature Switzerland AG 2023
V. Tsygankov, *Persistent Organic Pollutants in the Ecosystems of the North Pacific*,
Earth and Environmental Sciences Library,
https://doi.org/10.1007/978-3-031-44896-6_6

In aquatic systems, PAHs, characterized by low mobility, high hydrophobicity, and persistence, tend to build up in bottom sediments containing fulvic and humic acids similar to toxicants in structure. As a result, benthic organisms that constitute food supply for fish, marine mammals, and other organisms suffer most of all [5]. PAHs are capable of biomagnification, which poses a threat to organisms at the top of the trophic pyramid. In recent decades, the world's scientific community has recognized that persistent organic pollutants (POPs), including PAHs, arc subject to transboundary transport, which means their emergence and accumulation in areas previously considered clean, e.g., Antarctica, the Arctic, Greenland, etc.

Due to their wide distribution, PAHs are considered priority environmental pollutants, both on the European Union list and on the lists of the U.S. Environmental Protection Agency (EPA) and Russia. For some of PAH representatives, special regulations have been introduced to limit their levels in environment components and food products. In the Russian Federations, safety standards have been established for benzo(a)pyrene (which until recently was considered an indicator of contamination by the PAH group) in some of environment components (including food). For this reason, mainly this compound is studied in Russia. Abroad, 16 priority PAHs are currently regulated: naphthalene, acenaphthene, fluorene, acenaphthylene, phenanthrene, anthracene, fluoranthene, pyrene, chrysene, benz(a)anthracene, benzo(b)fluoranthene, benzo(k)fluoranthene, benzo(a)pyrene, dibenz(a,h)anthracene, indeno(1,2,3-cd)pyrene, and benzo(ghi)perylene.

In this regard, our study aimed to determine concentrations of various PAHs in bottom sediments and fish from the Northwest Pacific. Bottom sediments were collected in Peter the Great Bay, Sea of Japan (Fig. 1a), a zone of particular risk due to the high level of anthropogenic pollution. A total of 53 stations were analyzed in the spring–summer period. Fish were caught in the Bering Sea and Sea of Okhotsk (Fig. 1b) that are the major fishery zones of Russia. A total of 84 samples from the Bering Sea and 81 samples from the Sea of Okhotsk (15 stations) were studied.

6.2   Influx of Pollutants into Peter the Great Bay, Sea of Japan

Coastal waters of Peter the Great Bay, Sea of Japan, are subject to severe anthropogenic pressure caused by a significant discharge of household and industrial waste. First, these are the water bodies directly adjacent to the port of Vladivostok: Amur and Ussuri Bays, which include the Golden Horn, Diomid, Patrokl, Uliss Bays, and also the Eastern Bosphorus Strait. The maximum amount of untreated wastewater, about 53% of the total inflow to Peter the Great Bay, is discharged into Golden Horn Bay and the Eastern Bosphorus Strait [6]. The waters under study are exposed to hydrocarbons of various origins that cause significant pollution. Thus, environmental monitoring of the condition of the waters in the Bay is an indispensable link in the management and conservation of aquatic biological resources.

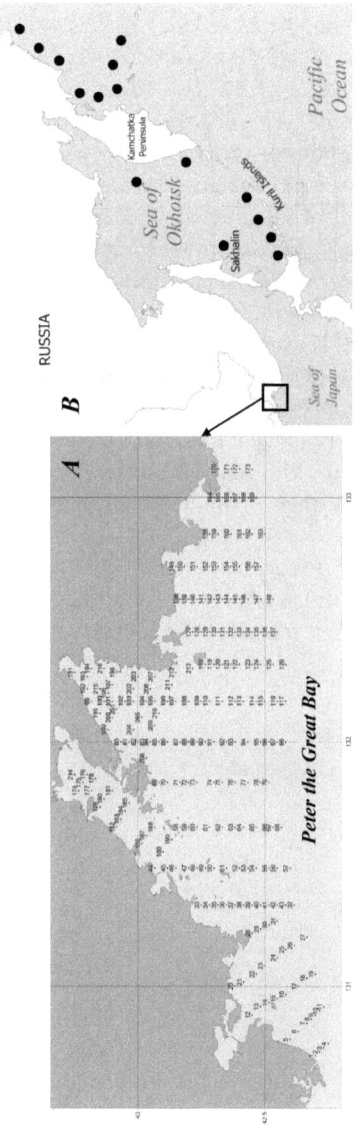

Fig. 6.1 Stations for sampling bottom sediments (**a**) and fish (**b**) in the Far Eastern Seas

The volume of wastewater entering the Golden Horn, Diomid, Uliss, Patrokl Bays, and the Eastern Bosphorus Strait was estimated on the basis of data from the Russian Federation statistics for the for 26 major enterprises operating in the bay area [6].

The total volume of wastewater flowing into the study areas of Peter the Great Bay, according to experts' estimates, amounts to 230,621,430 m^3. The largest volume of wastewater is discharged into Golden Horn Bay, 226,168,560 m^3/yr (98.1%); the Eastern Bosphorus Strait, 3,342,400 (1.4%); Uliss Bay, 865,120 (0.4%); and Diomid Bay, 245,350 m^3/yr (0.1% of the total volume of wastewater). The highest weight percentages of pollutants were recorded for suspended (33.4%) and organic substances (30.6%). Hydrocarbons of various origins, including polycyclic aromatic ones (PAH), were among dominant organic substances.

The major sources of PAHs in estuarine and marine sediments are their biogenic (bitumen, coal deposits, grass, forest fires, and fossil fuels) and anthropogenic forms (leakage of various oils, petroleum product spills, exhaust gases, sewage, and wood burning). Depending on origin, these compounds are divided into pyrogenic and petrogenic. Pyrogenic PAHs usually include high-molecular-weight compounds (four or more aromatic rings); petrogenic compounds have 2–3 rings [1]. An exact pattern of the origin of contamination can be identified using indicator ratios of PAH contents. A similar technique is actively used worldwide (Table 6.1) [7].

In bottom sediments from the study waters, two- and three-benzene-ring compounds slightly prevailed: their combined concentration averaged at 55–60% of the total PAH concentration.

The maximum permissible concentrations (MPC) for polycyclic aromatic compounds in bottom sediments have not been established in the Russian Federation. The exception is benzo(a)pyrene, for which the MPC level has been determined to be

Table 6.1 PAH origin indices

Ratio	Value	Origin
LMWC/HMWC	> 1	Petrogenic
	< 1	Pyrogenic
PHE/ANT	> 10	Petrogenic
	< 10	Pyrogenic
FLA/PYR	> 1	Pyrogenic
	< 1	Petrogenic
FLA/(FLA + PYR)	> 0.4	Petrogenic
	0.4–0.5	Pyrogenic
	< 0.5	Pyrogenic
BaA/228	> 0.35	Pyrogenic
	0.2–0.35	Pyrogenic

Note LMWC, low-molecular-weight PAHs; HMWC, high-molecular-weight PAHs; PHE, phenanthrene; ANT, anthracene; FLA, fluoranthene; PYR, pyrene; BaA, bezn(a)anthracene; 228, benz(a)anthracene + chrysene

0.02 mg/kg [8]. As a result of our study, we revealed the excess of the benzo(*a*)pyrene MPC in bottom sediments from each of the bays. The most polluted areas were as follows:

Amur Bay: transects with stations (stns.) nos. 179–181 and 182–185; also, an increased concentration was recorded from the eastern part of the area off the Peschany Peninsula and in the middle part of the bay; in the western part of the area off Popov Island and the northern part of the area off Pakhtusov Island.

Ussuri Bay: the southern area off Askold Island (stn. 118), near Cape Sedlovidny (stn. 198), in Emar Cove (stn. 195), and in Desantnaya Cove (stn. 199). In general, the least polluted waters were the open-water part of Peter the Great Bay.

According to the classification of levels of bottom sediment pollution by PAHs [9], a PAH concentration within a range of 0–100 ng/g is categorized as a low level of pollution; 100–1000 ng/g, moderate; 1000–5000 ng/g, high; and greater than 5000 ng/g, very high. Thus, Amur Bay was the most polluted body of water. Moderate pollution was found at the following stations: in the upper part (apex) of Amur Bay, 334.70 ng/g; Skrebtsov Island, 164.80 ng/g; in the middle of the northern part of the bay, 140.75 ng/g; Kirpichny Zavod Cove, 534.80 ng/g; Sportivnaya Harbor, 595.90 ng/g; transect to the Peschany Peninsula, 455.60 ng/g; Tokarevskaya Koshka Cove, 435.00 ng/g; the middle transect, 390.40 ng/g; and Popov Island, 133.30 ng/g. A relatively low level of pollution was observed in Ussuri Bay and in the open part of Peter the Great Bay, with average PAH concentration values of 6.16 ng/g and 10.86 ng/g, respectively.

Carcinogenic PAHs include benz(*a*)anthracene, chrysene, benzo(*b*)fluoranthene, benzo(*k*)fluoranthene, and benzo(*a*)pyrene. The latter is the most toxic compound (with its toxicity conventionally assumed to be a unity). High levels of carcinogenic PAHs > 50% were found in bottom sediments collected at stations near the Razdolnaya River estuary, south of the Peschany Peninsula, in the middle part of Amur Bay, in the middle of the transect between Popov Island and Perevoznaya Bay, in Emar Bay and near Cape Sedlovidny, off the western part of Popov Island, in Possyet Bay, in the middle transect of Ussuri Bay, and in Desantnaya Cove.

A low level of carcinogenic PAHs (less than 10% of total) was found in bottom sediments collected at the following stations: Peschany Peninsula, open part of Vostok Bay, the central and open part of Amur Bay, in Boisman Cove, in most of Ussuri Bay, in the open part of Peter the Great Bay.

In Peter the Great Bay, the ratio of fluoranthene to fluoranthene + pyrene was found to be greater than 0.5 at most stations, which allows a conclusion about the pyrogenic origin of the detected polyarenes.

In the bottom sediments from Amur Bay, the concentrations of the total of identified PAHs ranged from 6.2 to 595.9 ng/g dry weight (d.w.). The concentrations of polyarenes in the bottom sediments from the eastern, near-shore part of the bay (the coastal zone of Vladivostok) were 3–fourfold higher than the PAH concentrations recorded from relatively clean sites (stns. 41 and 225) located in the open part of the bay (off the islands of Reineke and Rikord). The minimum concentrations of total hydrocarbons (THCs) were recorded at the same stations (< 20 mg/kg d.w.). Carcinogenic PAHs in the bottom sediments from the coastal zone of Vladivostok (stns. nos. 1, 3, 4, 10, 14, and 227) accounted for more than 60% of the total of identified compounds. The proportion of benzo(a)pyrene at most stations was not greater than 1% of total PAHs. The exception was stn. 3, where the relative proportion of benzo(a)pyrene concentration was higher than 3%. The rate of occurrence of benzo(a)pyrene in bottom sediments was 85%.

A dominance of *technogenic* compounds was observed in the bottom sediments from the middle and eastern parts of Amur Bay. Fluoranthene, chrysene, benzo(b)fluoranthene, and fluorene dominated the detected PAHs. The weight percentage of *heavy*, four- and six-benzene-ring, PAHs that are formed mainly through pyrolytic processes averaged at 48%. This was also confirmed by the index of pyrogenicity for all samples, which was at least 0.5.

The surveys of THC pollution of bottom sediments from Amur Bay showed that the most polluted waters were the coastal zone of Vladivostok and the middle part of the bay. The highest concentrations of polyarenes were found in sediments from stns. 10, 14, and 21 located near the city coast and exposed to maximum anthropogenic pressure.

Thus, in terms of the PAH level, Amur Bay is the most polluted body of water, Ussuri Bay is moderately polluted, and the open part of Peter the Great Bay is characterized by a low level of pollution.

6.3   PAHs in Fish from the Sea of Okhotsk and Bering Sea

In the Russian Federation, the level of total hydrocarbons (THC) and polycyclic aromatic hydrocarbons (PAH) in aquatic organisms is not regulated. The exception is benzo(a)pyrene in smoked food products, where its concentration should not exceed 5000 ng/kg.

Of all the fish specimens analyzed, the maximum THC level was recorded from Alaska pollock, 362 ng/kg; the minimum level, from flounder, 0.6 ng/kg. Alaska pollock also showed the highest average THC concentration (121 ng/kg), which was twice as high as that in saffron cod (the *cleanest* fish).

To assess migration of THCs along a food chain, it is necessary to identify the feeding specialization of the organisms under study. By its feeding habits, adult Alaska pollock is a predator preying mainly on small fish (capelin and rainbow smelt) and squid. Saffron cod feeds on various worms, zooplankton, and eggs and juveniles of other fishes. Bivalve mollusks and polychaetes constitute the major part of flounder's diet. Thus, the assumption about the accumulation of THCs up food chains is fully cogent: Alaska pollock ingests the most THC-contaminated food (adult well-fed fish) and accumulate these compounds in its tissues. Saffron cod prefers smaller planktonic organisms that have not had enough time to accumulate organic pollutants in a short period of their life history. Flounder preys mainly on invertebrate organisms that, as shown above, almost do not absorb THCs (due to the low lipid content of their tissues). A comparison of PAH concentrations in the analyzed fishes is shown in Fig. 6.2.

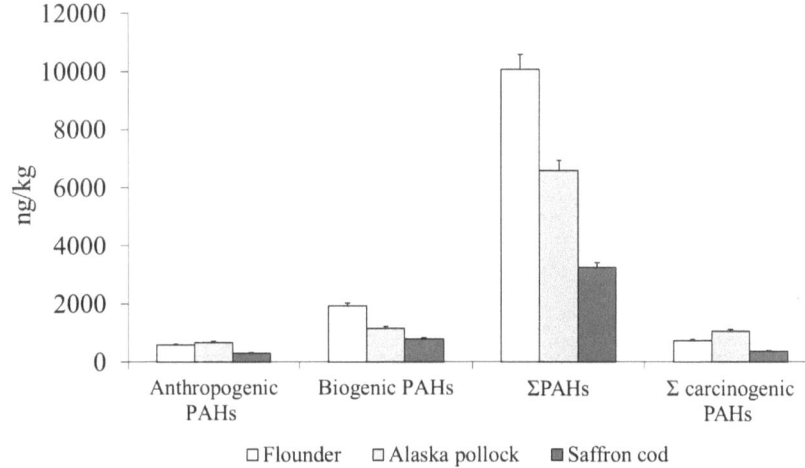

Fig. 6.2 PAH levels in tissues of fishes from the Bering Sea and Sea of Okhotsk

The highest total PAH level was observed in flounder, followed by Alaska pollock and saffron cod. It was most likely related with both the trophic level of each fish species and the conditions of its habitat. Flounder, as a demersal fish, is more exposed to PAHs from the environment that actively build up in bottom sediments. Furthermore, the accumulation of such lipophilic compounds is also influenced by the amount of lipids in tissues of organisms.

As the study of the origin of PAHs in fish tissues showed, the index of technogenicity (or anthropogenicity) was exceeded only in the saffron cod tissues, which probably indicates a specific accumulation of certain compounds in the organs of this fish (Fig. 6.3).

The concentrations of carcinogenic PAHs were approximately equal between all the specimens analyzed; however, the proportion of benzo(*a*)pyrene (having the greatest carcinogenic potential) was maximum in Alaska pollock tissues, which indicates a slightly greater hazard from eating this fish compared to the other species under study, even in case of lower concentrations. However, the maximum concentration of this toxicant was recorded from flounder, 735 ng/kg, as the fattiest organism among the others analyzed.

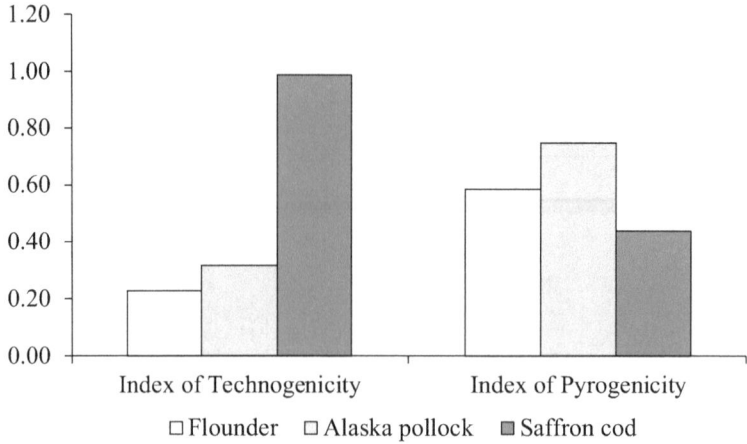

Fig. 6.3 Origin of PAHs in tissues of fishes from the Bering Sea and Sea of Okhotsk

6.4 Conclusion

Thus, the distribution of PAHs along the Peter the Great Bay coast is uneven and is determined by a number of associated factors. The major sources of these compounds are industrial enterprises and logistics hubs that discharge these compounds with sewage and bilge water and also as a result of fuel and lubricant spills. The highest PAH concentrations are found in bottom sediments from the upper parts of the bays, i.e. from waters with minimum hydrodynamics. Due to the lack of constant mixing, these compounds can be precipitated in significant quantities and at a higher rate, thus, forming potential *delayed-action bombs* that may *explode* during dredging operations or other industrial activities. Furthermore, increased PAH concentrations are found in areas with high shipping activities and transloading of coal and petroleum products, which indicates the effects of both shipping and improper handling of these combustible fluids. It is also worth noting that all the PAHs detected in Peter the Great Bay are of technogenic origin, which necessitates taking environmental protection measures.

As expected, the least polluted area has proven to be the open part of the bay with active hydrodynamics that provides redistribution of toxicants over a larger area of bottom sediments and reduces the local pressure on each of the stations analyzed.

The highest PAH concentrations among the fish species under study have been found in flounder, which is associated with its diet habits, demersal life, and the highest lipid content. Furthermore, this organism has shown the highest concentrations of carcinogenic PAHs (up to 755 ng/kg), which indicates the potential hazard of this fish family to humans when consumed as food. However, the greatest contribution of carcinogenic forms of this group to the total level of toxicants has been recorded for Alaska pollock, the second most PAH-contaminated species. This also suggests potential risks to human health, even in case of lower concentrations compared to those in flounder.

Thus, the scope of studies of polycyclic aromatic hydrocarbons in various environmental components (especially in areas of active shipping and logistics hubs on the coasts of the Russian Federation) needs to be extended, and maximum permissible concentrations of PAHs in fish and non-fish objects of marine fisheries to be established.

Acknowledgements The work was supported by the Ministry of Science and Higher Education of the Russian Federation, project no. FZNS-2023-0011.

References

1. Nemirovskaya IA, Lisitsyn AP (2004) Hydrocarbons in the ocean (snow-ice-water-bottom sediments). Nauchny mir, Moscow
2. Nelson-Smith A (1973) Oil pollution and marine ecology. Springer Science+Business Media, LLC, New York
3. Mai Q, Zeng EY et al (2003) Distribution of polycyclic aromatic hydrocarbons in the coastal region off Macao, China: assessment of input sources and transport pathways using compositional analysis. Environ Sci Technol 37:4855–4863. https://doi.org/10.1021/es034514k
4. Huang GP, Chen YJ, Lin T et al (2011) The distribution and ecological risk of polycyclic aromatic hydrocarbons of surface sediments in the intertidal zone of Bohai Bay, China. Zhongguo Huanjing Kexue/China Environ Sci 31:1856–1863
5. Zhang A, Zhao S, Wang L et al (2016) Polycyclic aromatic hydrocarbons (PAHs) in seawater and sediments from the northern Liaodong Bay, China. Mar Pollut Bull 113:592–599. https://doi.org/10.1016/j.marpolbul.2016.09.005
6. Chernyaev AP, Nigmatulina LV (2013) Quality monitoring of coastal waters in Peter the Great Bay (Japan Sea). Izvestiya TINRO 173:230–238
7. Khaustov AP, Redina MM (2014) Polycyclic aromatic hydrocarbons as geochemical markers of oil pollution of the environment. Oil Gas Exposition:92–97
8. Drugov YS, Rodin AA (2007) Contaminated soil and hazardous waste analysis: a practical guide. Binom, Moscow
9. Zhuravel EV, Khristoforova NK, Drozdovskaya OA, Tokarchuk TN (2012) Estimation the water state of Vostok Gulf (Peter the Great Bay, Japan Sea) on hydrochemical and microbiological parameters. Izvestia Samara Scien Center Russ Acad Sci 14:2325–2329

Chapter 7
Concentrations of Persistent Organic Pollutants in Benthic and Pelagic Fish from the Fishery Zones in the Far Eastern Seas of Russia

Abstract This chapter considers levels of polychlorinated biphenyl (PCB) congeners, hexachlorocyclohexane (HCH) isomers, dichlorodiphenyltrichloroethane (DDT) and its metabolites measured in the organs and tissues of Pacific salmon and flounder from the Far Eastern seas of Russia. Intraspecific, interspecific, and geographical differences in the accumulation of these hazardous toxicants by fish are described.

Keywords HCHs · DDTs · PCBs · Pacific salmon · Flounder · Far Eastern seas

In many countries, physicians recommend eating fish with high contents of ω-3 polyunsaturated fatty acids (PUFA) on a regular, weekly basis in order to decrease cholesterol and high blood pressure and also strengthen blood vessel walls. Fatty fish species, especially herring, mackerel, and salmon, are rich in such acids. Consuming these fishes and, accordingly, ω-3 fatty acids (eicosapentaenoic and docosahexaenoic acids) not only helps reduce the risk of cardiovascular diseases and endometrial cancer but also increases the level of micronutrients (primarily potassium and phosphorus that are extremely essential for human health), strengthens mental and cognitive functions, and also has a number of other positive effects [1–3]. Despite the obvious benefits and necessity to include these fatty fishes in the diet, their consumption may raise concern in some cases due to the bioaccumulation of environmental pollutants such as persistent organic pollutants (POP) in fish tissues [2, 4]. Up to 90% of these compounds enter the human body with food. The final *depot* for POPs in the environment is marine ecosystems and, therefore, these can be accumulated in various commercially important organisms that are harvested as seafood [5–7].

The Far Eastern seas (the Sea of Japan, Sea of Okhotsk, and Bering Sea) are the major fishery zones of the Russian Federation. Pacific salmon comprise the most common and, therefore, commercially very important group among the marine species harvested. The three main salmon species—pink, chum, and sockeye salmon—account for 90% of total catches. Pink salmon is the most abundant, smallest in size, and fastest growing species. In Russian waters, it is also the most commercially valuable of the Pacific salmons. Chum salmon ranks second after pink in abundance and is distributed wider than the other members of this genus [4].

Flounders harvested in the Far East are also among the most important target species of commercial fisheries, with their catch size making up 9.5% of the total fish catch from the region [8]. The wide distribution, the variety of species, and the low price on the market make them a particularly important item in the diet of local residents. One of the most popular flounder species is the Bering flounder (*Hippoglossoides robustus*) commonly distributed in the Sea of Okhotsk, Sea of Japan, and the Tatar Strait.

7.1 Pacific Salmon (Genus *Oncorhynchus*)

In our study, we selected the following species to be analyzed for POP content: pink (*Oncorhynchus gorbuscha*), chum (*O. keta*), masu (*O. masou*), Chinook (*O. tshawytscha*), and sockeye salmon (*O. nerka*). Fish samples were taken during the period 2010–2018 from various regions of the Far Eastern seas: the western Bering Sea (2010–2011); Sea of Okhotsk waters off the Kuril Islands (2012–2013); Lake Azabache (2017); the Bakhura River estuary (Sea of Okhotsk) (2017); the Poronay River estuary (Gulf of Patience, Sea of Okhotsk) (2017); and the Kamchatka River (eastern Kamchatka Peninsula) (2018). Muscles, liver, male gonads, and eggs were analyzed. In the fish caught in 2010–2013, only organochlorine pesticides (OCP) were measured. Of DDT and its metabolites, only *p,p'*-isomers were determined. In individuals caught in 2017–2018, PCB congeners (28, 52, 155, 101, 153, 118, 143, 138, 180, and 207) were also measured in addition to OCPs.

7.1.1 Intraspecific Differences in POP Accumulation Between Pacific Salmon from the Sea of Okhotsk and the Bering Sea

Pink salmon (*O. gorbuscha*). Samples were collected from the southern Sea of Okhotsk (waters off the Kuril Islands), Lake Azabache (Kamchatka Peninsula), and the Poronay River (Gulf of Patience, Sea of Okhotsk). POPs were found in all the samples analyzed (Tables 7.1 and 7.2).

In the pink salmon from the Sea of Okhotsk waters off the Kuril Islands, OCPs were detected in all samples within a range from 60.9 to 715.2 (with an average of 295.2 ± 203.6) ng/g lipid weight (l.w.). Their major part was represented by HCH isomers whose total concentrations ranged from 60.9 to 666.6 (277.0 ± 188.4) ng/g l.w. Among the HCH isomers, α-, β- and γ-HCH were found in ranges of 60.9–446.8, 4.4–171.0, and 5.2–65.6 ng/g l.w. with average values of 199.8 ± 130.1, 50.5

Table 7.1 Average OCP concentrations in the organs of pink salmon, ng/g l.w.

	α-HCH	β-HCH	γ-HCH	δ-HCH	o,p'-DDT	p,p'-DDT	o,p'-DDD	p,p'-DDD	o,p'-DDE	p,p'-DDE
Southern sea of Okhotsk (waters off the Kuril Islands)										
Muscles	101.0 ± 45.0	19.0 ± 9.0	12.0 ± 4.0	–[1]	–	<DL[2]	–	<DL	–	9.0 ± 6.0
Liver	191.0 ± 118.0	38.0 ± 18.0	54.0 ± 9.0	–	–	<DL	–	<DL	–	30.0 ± 21.0
Eggs	210.0 ± 96.0	41.0 ± 31.0	28.0 ± 19.0	–	–	<DL	–	<DL	–	10.0 ± 6.0
Gonads	405.0 ± 38.0	143.0 ± 26.0	52.0 ± 3.0	–	–	<DL	–	<DL	–	44.0 ± 5.0
Lake Azabache										
Muscles	<DL	18.4 ± 11.6	<DL	–	<DL	7.2 ± 2.6	3.2[3]	<DL	110.1	<DL
Liver	<DL	69.4 ± 70.3	<DL	–	<DL	<DL	<DL	<DL	<DL	<DL
Gonads	<DL	52.4 ± 34.3	<DL	–	<DL	<DL	61.5	3.3	<DL	<DL
Poronay River estuary, eastern Sakhalin coast										
Muscles	25.7 ± 25.3	5.1 ± 3.0	1.1 ± 1.2	1.8 ± 2.2	2.4 ± 0.9	4.0 ± 2.5	1.1 ± 0.6	3.1 ± 3.4	0.6 ± 0.5	0.7 ± 0.7
Liver	32.2 ± 29.9	8.1 ± 9.4	4.3 ± 4.4	6.1 ± 4.5	<DL	<DL	2.0 ± 0.4	5.0 ± 2.4	0.7 ± 0.2	<DL
Eggs	31.1 ± 45.6	11.1 ± 9.4	1.1 ± 1.4	3.2 ± 2.5	<DL	<DL	1.3	2.5 ± 2.0	0.2	1.0
Gonads	61.6 ± 41.8	3.2	15.0 ± 8.2	13.7 ± 6.8	<DL	<DL	<DL	<DL	3.0 ± 1.5	<DL

[1]Not studied; [2]below the detection limit (DL); [3]found in only a single specimen

Table 7.2 Average concentrations of PCB congeners in the organs of pink salmon, ng/g l.w.

	PCB 28	PCB 52	PCB 155	PCB 101	PCB 118	PCB 143	PCB 153	PCB 138
Lake Azabache								
Muscles	2.9 ± 1.8	2.6[1]	<DL[2]	9.7 ± 5.4	8.8 ± 5.3	4.2	7.6 ± 4.3	5.1 ± 3.8
Liver	3.9	<DL	<DL	4.3 ± 2.6	19.2 ± 25.3	16.2	1.7	10.0 ± 10.5
Gonads	43.2	24.3	<DL	74.3 ± 37.3	254.0 ± 180.2	<DL	174.6 ± 117.4	118.7
Poronay River estuary, eastern Sakhalin coast								
Muscles	1.9 ± 1.7	1.4 ± 1.1	0.8 ± 0.4	3.2 ± 1.6	3.6 ± 1.7	4.8 ± 2.7	4.2 ± 1.7	3.5 ± 1.6
Liver	8.5 ± 17.3	6.7 ± 4.6	0.9	2.1 ± 1.6	3.6 ± 0.2	<DL	3.6 ± 2.0	3.5 ± 0.4
Eggs	0.9 ± 0.5	5.1 ± 3.7	1.0 ± 0.2	3.1 ± 0.9	4.4 ± 1.1	9.0	5.1 ± 2.5	3.9 ± 1.4
Gonads	1.9	6.4 ± 4.3	<DL	2.2	<DL	<DL	<DL	<DL

[1]Found in only a single specimen; [2]below the detection limit (DL)

± 48.6, and 33.3 ± 20.9 ng/g l.w., respectively. The dominant isomer was α-HCH, which indicates a long-term circulation of the initial substance in the environment and the process of its degradation. Of the DDT metabolites, only DDE was found in all samples within a range from 0.8 to 49.6 (21.9 ± 18.2) ng/g l.w., which indicates both the lack of fresh contamination of the ecosystem and the degradation of the initial compound, DDT.

In the fish from the waters off the Kuril Islands, the minimum (for all organs) OCP concentrations ($p \leq 0.05$) were found in muscles, where the OCP values ranged within 89.3–222.8 (with an average of 141.0 ± 46.0) ng/g l.w. (Fig. 7.1). The total HCH concentration was 132.0 ± 49.0 ng/g l.w.; DDE, 9.0 ± 6.0 ng/g l.w. In the liver, the OCP levels ranged within 60.9–500.5 (279.0 ± 166.0) ng/g l.w. The total HCH concentration reached 259.0 ± 144.0 ng/g l.w.; DDE, 30.0 ± 21.0 ng/g l.w. In eggs and gonads of pink salmon from the Sea of Okhotsk, the range of OCP concentrations was from 131.5 to 399.4 and from 588.4 to 715.2 ng/g l.w. with average values of 285 ± 138 and 645 ± 65 ng/g l.w., respectively, which were the maximum values for all the organs under consideration. The concentrations of HCH isomers in eggs and gonads ranged within 279.0 ± 139.0 and 600.0 ± 60.0 ng/g l.w., respectively; DDE, 10.0 ± 6.0 and 44.0 ± 5.0 ng/g l.w., respectively.

In the pink salmon from *Lake Azabache*, the OCP range in all organs was 16.5–151.8 (with an average of 68.9 ± 49.7) ng/g l.w. HCH, represented only by the most stable β-form, constituted the major part of OCPs and had a level from 9.8 to 119.1 ng/g l.w. with an average of 43.9 ± 39.7 ng/g l.w. DDT and its metabolites were detected sporadically within a range from 6.7 to 120.2 (49.9 ± 54.1) ng/g l.w. The most common metabolite was *p,p′*-DDT whose concentrations in muscles varied from 4.9 to 10.1 (7.2 ± 2.6) ng/g l.w. *o,p′*-DDD was detected in two samples at concentrations of 3.2 and 61.5 ng/g l.w.

In muscles of the pink salmon from Lake Azabache, the range of OCP concentrations was from 16.5 to 151.8 (with an average of 63.4 ± 76.6) ng/g l.w. The average values of HCH were 18.4 ± 11.6 ng/g l.w.; DDT, 45.0 ± 65.1 ng/g l.w. (Fig. 7.2). In the liver, OCPs were represented by only β-HCH at 69.4 ± 70.3 ng/g l.w. In the

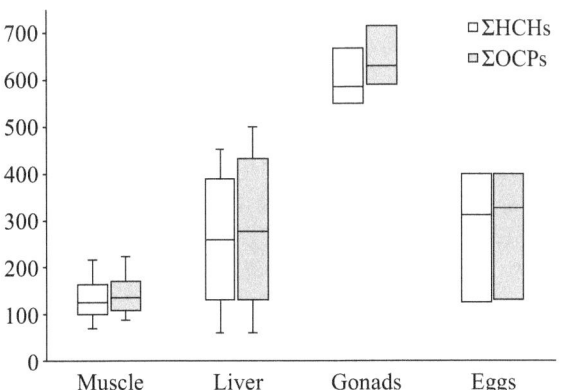

Fig. 7.1 Average total concentrations of HCH and OCPs (medians) in the organs of pink salmon from the southern Sea of Okhotsk (waters off the Kuril Islands), ng/g l.w.

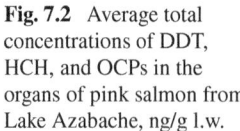

Fig. 7.2 Average total concentrations of DDT, HCH, and OCPs in the organs of pink salmon from Lake Azabache, ng/g l.w.

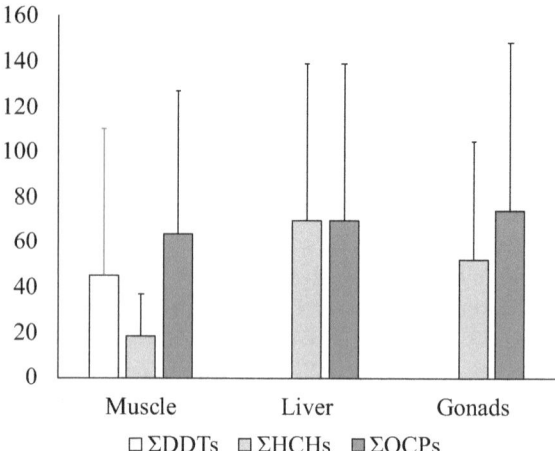

gonads, the OCP content varied from 57.4 to 83.9 (74.0 ± 14.4) ng/g l.w.; the β-HCH content was 52.4 ± 34.3 ng/g l.w. Of DDT and its metabolites, o,p'-DDD and p,p'-DDD were detected at concentrations of 61.5 and 3.3 ng/g l.w., respectively.

The total PCB concentrations in the organs of pink salmon from Lake Azabache varied from 9.6 to 739.7 (with an average of 175.4 ± 272.0) ng/g l.w. The most frequently detected pollutants were PCB 101 and PCB 118, whose levels amounted to 23.8 ± 34.5 and 74.0 ± 131.1 ng/g l.w., respectively. PCB 52 and PCB 143 were detected in two samples at the following concentrations: 2.6 and 24.3 ng/g l.w., respectively, in one sample and 4.2 and 16.2 ng/g l.w. in the other. PCB 155, 180, and 207 were below the detection limits in all samples.

In muscles, the total PCB concentrations ranged from 17.3 to 50.3 (with an average of 32.9 ± 16.6) ng/g l.w.; in the liver, from 9.6 to 89.2 (37.4 ± 44.9) ng/g l.w.; in the gonads, from 452.3 to 739.7 (596.0 ± 203.2) ng/g l.w. Mainly *heavy* PCBs were found (Fig. 7.3).

In the pink salmon from ***the Poronay River estuary (Gulf of Patience, Sea of Okhotsk)***, POPs were detected in all samples. The OCP concentrations ranged from 2.5 to 150.1 (with an average of 45.3 ± 41.5) ng/g l.w. The HCH levels were within a range of 1.3–147 (41.8 ± 41.5) ng/g l.w. and made up the major part of the detected OCPs. In addition to α-, β- and γ-HCH, the δ-isomer was also found in the samples. This isomer is penultimate in the sequence of pollutant's degradation (γ → α → δ → β). The dominance of α-HCH (35.2 ± 34.5 ng/g l.w.) indicates the degradation of the γ-form and the long period of circulation of the compound in the ecosystem. The levels of β-, γ-, and δ-HCH were 7.3 ± 7.4, 3.5 ± 5.1, and 5.0 ± 5.4 ng/g l.w., respectively. The total DDT concentrations in all samples ranged from 0.4 to 21.8 (4.4 ± 4.9) ng/g l.w. and significantly ($p \leq 0.05$) differed in the following row: eggs < gonads < muscles < liver (Fig. 7.4). The most detectable metabolites in all samples were o,p'-DDD, p,p'-DDD, and o,p'-DDE, which indicates the degradation of the initial compound, DDT.

Fig. 7.3 Average concentrations of the most detectable PCB congeners in the organs of pink salmon from Lake Azabache, ng/g l.w.

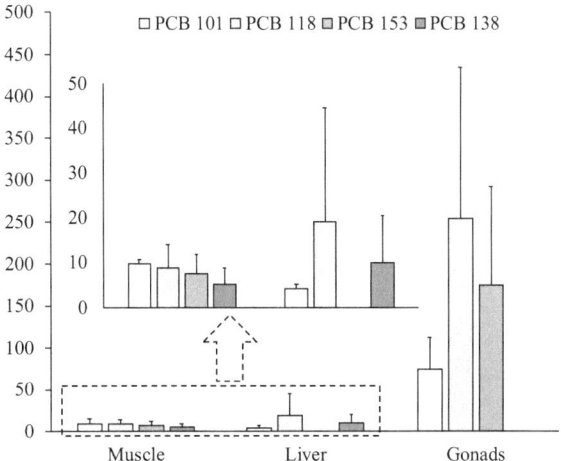

Fig. 7.4 Concentrations of ΣDDT, ΣHCH, and ΣOCP (medians) in the organs of pink salmon from the Poronay River estuary, ng/g l.w.

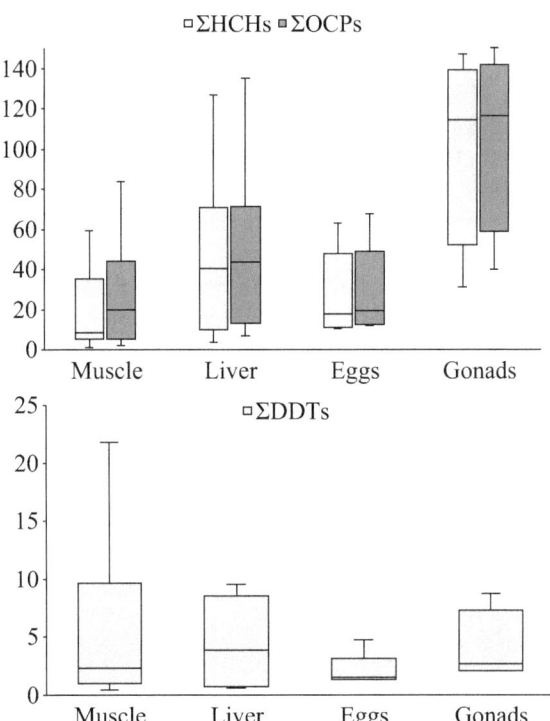

In muscles, OCP levels ranged from 2.5 to 83.9 (with an average value of 27.8 ± 28.1) ng/g l.w. The total HCH concentration was 23.1 ± 27.6 ng/g l.w.; DDT and its metabolites, 5.9 ± 7.4 ng/g l.w. In the liver, the OCP levels (6.7–135.3 ng/g l.w., with an average value of 48.3 ± 40.3 ng/g l.w.) were significantly ($p \leq 0.05$) higher than in muscles. The total HCH concentration was 45.2 ± 39.7 ng/g l.w.; DDT and its metabolites, 4.5 ± 3.7 ng/g l.w. In pink salmon eggs and gonads, the levels of OCPs were within 12.0–150.1 (42.2 ± 53.8) and 42.7–117.4 (85.8 ± 36.9) ng/g l.w., respectively. The total HCH concentration in eggs was 39.3 ± 53.3 ng/ g l.w.; in gonads, 83.6 ± 36.6 ng/g l.w. Of DDT and its metabolites, o,p'-DDT and p,p'-DDT were below the detection limits in all sex products. o,p'-DDD, o,p'-DDE, and p,p'-DDE in eggs were detected in single cases at concentrations of 1.3, 0.2, and 1.0 ng/g l.w., respectively. The level of p,p'-DDD was 2.5 ± 2.0 ng/g l.w. In gonads, only o,p'-DDE was detected at 3.0 ± 1.5 ng/g l.w.

The total PCB concentrations in the pink salmon from the Poronay River estuary ranged from 2.2 to 69.8 (with an average of 18.1 ± 14.2) ng/g l.w. Among all the samples, PCB 143 was the least detectable congener. PCB 180 and 207 were not found in all of the samples analyzed.

In muscles, the total PCB concentrations ranged from 9.0 to 33.5 (with an average value of 19.1 ± 8.0) ng/g l.w. (Fig. 7.5). PCB 28, 52, 155, 101, 118, 153, 138 were found in all the samples. In the liver, the total PCB concentrations ranged from 2.5 to 69.8 (17.9 ± 21.2) ng/g l.w. The most detectable congeners were PCB 28, 52, 101, 118, 153, and 138. In the sex products, the total PCB concentrations varied as follows: in eggs, from 13.3 to 41.5 (with an average of 24.8 ± 10.5) ng/g l.w.; in gonads, from 2.2 to 11.3 (5.8 ± 4.0) ng/g l.w. The most detectable congeners in eggs were 28, 52, 155, 101, 118, 153, and 138. In all the gonad samples, the concentrations of congeners 155, 118, 143, 153, and 138 were below the detection limits.

The concentrations of α- and γ-HCH were significantly ($p \leq 0.05$) higher in the pink salmon from the Sea of Okhotsk waters off the Kuril Islands than in those from the Poronay River estuary (Fig. 7.6).

The significant differences in concentrations may be explained by the fact that the fish collected during feeding in the Sea of Okhotsk had passed through the Great Pacific Garbage Patch (GPGP), a huge area of plastic debris floating on the surface

Fig. 7.5 Concentrations (median) of ΣPCB in the organs of pink salmon from the Poronay River, ng/g l.w.

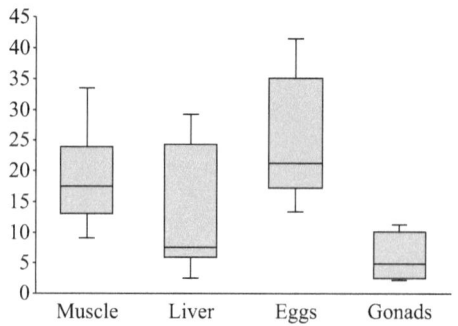

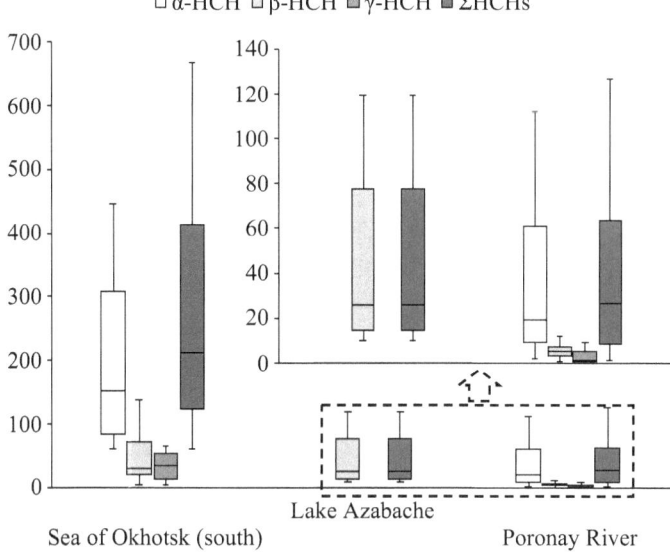

Fig. 7.6 Concentrations (median) of HCH in the organs of pink salmon from different areas of the Far Eastern seas, ng/g l.w.

of the Pacific Ocean. It is known that plastic particles can adsorb from pico- to nanograms of POPs on their surface [9]. Taking into account the life strategy of Pacific salmon (inhabiting predominantly the depth zone from 0 to 50 m, where most of the plastic is concentrated), the effect of the GPGP is the most plausible explanation for the increase in POP levels in the organs of pink salmon from the Sea of Okhotsk.

The β-HCH levels were significantly higher in the pink salmon from Lake Azabache than in those from the Sea of Okhotsk waters off the Kuril Islands and the Gulf of Patience. Lake Azabache is a landlocked body of water, having the water exchange with the sea mostly through the Kamchatka River which empties into the Kamchatka Gulf. There are no large agricultural farms in Kamchatka and, in particular, in the Kamchatka River valley. Thus, the atmospheric transport, biotransport by migrating species, and leachates from municipal solid waste landfills are assumed to be the major sources of OCPs and PCBs to enter the river and lake ecosystems. HCH in Lake Azabache was represented by only the β-isomer, which indicates a long-lasting contamination and the degradation of the original compound, γ-HCH, to the most persistent form.

The concentrations of DDT and its metabolites in fish were arranged in the following order: Poronay River estuary < southern Sea of Okhotsk < Lake Azabache ($p \leq 0.05$) (Fig. 7.7). This may be explained by more complex metabolic processes in the DDT degradation compared to those for HCH. In Lake Azabache, mainly DDT was detected, which indicates a recent entry of this toxicant into the ecosystem of this water body. Only DDE was detected in the Sea of Okhotsk waters off the Kuril

Islands, which indicates the degradation of the original compound. Almost all DDT metabolites were found in the Poronay River estuary. Despite the low concentrations, this may be evidence of a steady influx of these toxicants into the environment and the continuing degradation of intermediate compounds (DDD).

The PCB concentrations in the pink salmon from Lake Azabache were significantly higher ($p \leq 0.05$) than in those from the Poronay River estuary (Fig. 7.8). This may be due to the limited water exchange in the lake. In the case of pesticides, it reduces their possible entry with the river water. However, there are no local sources of these compounds near the lake. PCB, in turn, was widely applied across the former USSR. Moreover, PCB-containing transformers are still used in Russia. After entering the ecosystem of the lake, PCBs remain in it for a long time due to the weak evaporation at low temperatures and are accumulated in fish tissues. The low PCB concentrations in pink salmon from the Poronay River estuary may indicate their low concentrations in open waters of the world's oceans.

An analysis of variations in OCP concentrations in pink salmon from the Far Eastern seas of Russia has shown significantly lower ($p \leq 0.05$) total concentrations of HCH and DDT in 2017 compared to those in 2012. However, the decreases in HCH and DDT levels were not equivalent: the concentrations of the former compound decreased more rapidly than those of the latter.

Estimation of variations in PCB concentration over time was impossible, since the fish, in which PCB was measured, were collected simultaneously. As regards OCPs, their levels in pink salmon from the Far Eastern seas of Russia show a clear tendency to decrease over the past five years.

To identify the possible sources of OCPs and PCBs in fish, a correlation analysis was carried out between the detected compounds. The interrelation of HCH isomers is explained by a single source of entry into the ecosystem. The correlation between DDE and the HCH isomers also indicates a single source of contamination. Most

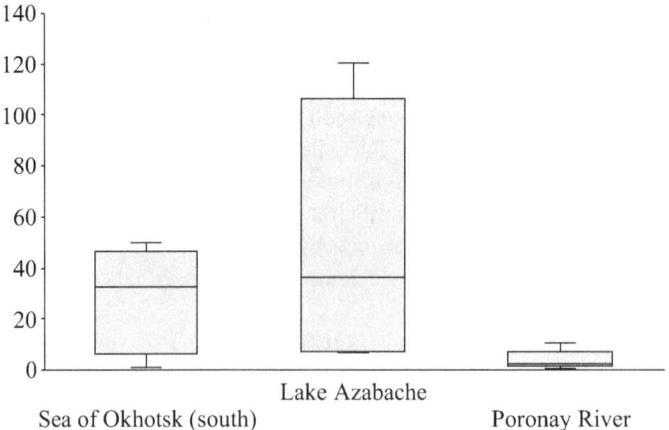

Fig. 7.7 Total DDT concentrations (medians) in pink salmon from the study areas, ng/g l.w.

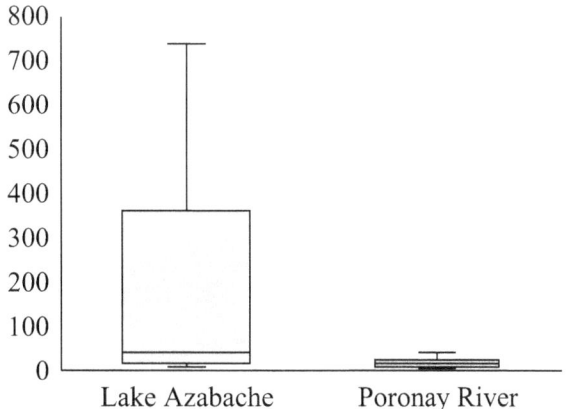

Fig. 7.8 Total PCB concentrations (medians) in pink salmon from Lake Azabache and the Poronay River estuary, ng/g l.w.

likely, atmospheric transport and sea currents are the major routes of pesticides' entry into the Sea of Okhotsk waters.

Among the pesticides detected in the fish from Lake Azabache (Table 7.3), a significant ($p \leq 0.01$) correlation was found between β-HCH and o,p'-DDE. The relationship between the two compounds may indicate similar degradation pathways, simultaneous emergence in organs, and/or a single source of entry. The former assumption is most likely, because, of all the DDT metabolites, only o,p'-DDE showed a high ($p \leq 0.01$) correlation with β-HCH. There was a very strong positive direct relationship between o,p'-DDD and p,p'-DDT ($p \leq 0.01$), which most likely indicates a single source of entry (atmospheric transport) into the ecosystem. o,p'-DDE showed a very pronounced negative correlation with p,p'-DDT, which is obviously evidence of the degradation of p,p'-DDT to o,p'-DDE in the fish body.

Among the PCBs, very high correlations were found between all the congeners at significance levels of $p \leq 0.01$ and $p \leq 0.05$, which indicates a single source of these compounds to enter the pink salmon body. The correlations between PCBs and OCPs ranged from medium to very high at levels of $p \leq 0.01$ and $p \leq 0.05$. These compounds are known to be used in different industries and enter ecosystems from a variety of sources. The most probable explanation for the results obtained is the effect of the municipal solid waste landfill sites located in the Kamchatka Peninsula that may contain simultaneously OCPs and PCBs. Leachates from such sites can be the major cause of the combined entry of pesticides and industrial waste chemicals into ecosystems and fish.

Similar correlations between the compounds were found in the pink salmon from the Poronay River, as well as from other areas (Table 7.4). The HCH isomers correlated with each other (positively) at a level of $p \leq 0.01$, and δ-HCH correlated with all the isomers. Among the DDT metabolites, a positive correlation ($p \leq 0.01$) was observed between the o,p'- and p,p'-isomers, which, as in other areas, indicates a similar source of POPs entering the environment.

The PCB congeners also showed a significant correlation, with the number of congeners, to which the compound correlates, increasing as the chlorination level

Table 7.3 Correlations between the organic pollutants in pink salmon from Lake Azabache

	β-HCH	p,p'-DDT	o,p'-DDD	o,p'-DDE	PCB 28	PCB 52	PCB 101	PCB 118	PCB 143	PCB 153	PCB 138
β-HCH	1.000	0.483	− 0.498	**1.000****	0.829	0.637	*0.690**	*0.749**	0.937	*0.804**	− 0.173
p,p'-DDT		1.000	**1.000****	− **1.000****	0.475	**1.000****	0.536	0.515	0.654	0.923	− 0.180
o,p'-DDD			1.000	0	0.993	**1.000****	0.989	*0.998**	0	*0.999**	*1.000**
o,p'-DDE				1.000	0	0	**1.000****	**1.000****	**1.000****	0	**1.000****
PCB 28					1.000	0.993	**0.996****	**0.997****	0	**0.999****	**0.997****
PCB 52						1.000	0.989	*0.998**	0	*0.999**	*1.000**
PCB 101							1.000	**0.984****	0.417	**0.986****	0.364
PCB 118								1.000	0.591	**0.995****	0.242
PCB 143									1.000	**1.000****	− 0.599
PCB 153										1.000	0.246
PCB 138											1.000

**Correlation is significant at a level of 0.01 (two-way) (bold emphasized values). *Correlation is significant at a level of 0.05 (two-way) (italic emphasized values)

Table 7.4 Correlations between the organic pollutants found in pink salmon from the Poronay river estuary

	HCH				DDT		DDD		DDE		PCB							
	α-	β-	γ-	δ-	o,p'-	p,p'-	o,p'-	p,p'-	o,p'-	p,p'-	28	52	155	101	118	143	153	138
α-¹	1.000	0.556**	0.645**	0.667**	−0.301	0.012	0.062	0.052	0.425	0.018	0.518**	0.18	−0.121	0.211	−0.01	−0.641*	0.126	−0.011
β-¹		1.000	0.441*	0.538**	−0.318	−0.258	−0.096	0.14	0.032	−0.155	0.502**	0.490**	0.078	0.445**	0.448*	−0.616*	0.454*	0.379
γ-¹			1.000	0.708**	−0.36	−0.195	0.379	0.498**	0.643**	0.09	0.288	0.379*	−0.114	0.098	0.178	−0.349	0.214	0.005
δ-¹				1.000	−0.393	−0.255	0.298	0.343	0.607**	−0.173	0.391*	0.525**	−0.025	0.218	0.343	−0.515	0.357	0.166
o,p'-²					1.000	0.843**	0.473	0.469	−0.165	0.424	0.226	−0.117	−0.088	−0.235	−0.477	−0.39	−0.243	−0.201
p,p'-²						1.000	0.294	0.057	−0.013	0.693**	0.271	−0.18	−0.348	−0.283	−0.505	−0.504	−0.261	−0.195
o,p'-³							1.000	0.662**	0.492	0.128	0.344	0.147	−0.301	−0.447	−0.481	0.062	−0.298	−0.552*
p,p'-³								1.000	0.364	−0.08	0.267	0.377*	0.132	0.302	0.344	−0.441	0.363	0.241
o,p'-⁴									1.000	−0.022	0.374	−0.133	−0.103	−0.025	−0.039	−0.27	0.262	−0.183
p,p'-⁴										1.000	−0.249	−0.214	−0.127	−0.254	−0.45	−0.446	−0.394	−0.372
28⁵											1.000	0.486**	0.084	0.493**	0.384	−0.175	0.587**	0.447*
52⁵												1.000	0.213	0.580**	0.682**	−0.239	0.676**	0.567**

(continued)

Table 7.4 (continued)

	HCH			DDT			DDD			DDE			PCB									
	α-	β-	γ-	δ-	o,p'-	p,p'-	o,p'-	p,p'-	o,p'-	p,p'-	28	52	155	101	118	143	153	138
155⁵													1.000	0.477*	0.557**	− 0.028	0.484*	0.432
101⁵														1.000	0.896**	− 0.349	0.889**	0.782**
118⁵															1.000	− 0.239	0.937**	0.877**
143⁵																1.000	− 0.165	− 0.166
153⁵																	1.000	0.904**
138⁵																		1.000

** Correlation is significant at a level of 0.01 (two-way) (bold emphasized values). * Correlation is significant at a level of 0.05 (two-way) (italic emphasized values).
^{1}HCH; ^{2}DDT; ^{3}DDD; ^{4}DDE; ^{5}PCB

rises. Of particular importance are the data on the negative correlations ($p \leq 0.05$) of PCB 143 with α- and β-HCH, as well as PCB 138 and PCB 101 with *o,p'*-DDD. The former may indicate the possible degradation of PCB 143 to HCH with the subsequent transition of one isomer to another, or different sources of the toxicants. The latter may be evidence of the possible degradation of *o,p'*-DDD to higher chlorinated PCBs 138 and 101 in the process of metabolic transformations.

Thus, the sources of the toxicants entering pink salmon bodies are similar for all the study areas and reflect both the global distribution of POPs and local pollution from the municipal solid waste landfill sites.

Chum salmon (*Oncorhynchus keta*). Chum salmon samples were collected from the southern Sea of Okhotsk (waters off the Kuril Islands) in 2013 and from the Kamchatka River (Kamchatka Peninsula) in 2018. Muscles, liver, eggs, and gonads were analyzed. POPs were found in all the samples studied (Tables 7.5 and 7.6).

In the chum salmon from the *Sea of Okhotsk waters off the Kuril Islands*, OCPs were detected in all the samples within a range from 56.0 to 4223.0 (with an average value of 841.7 ± 1209.9) ng/g l.w. The HCH isomers, with the total concentration ranging from 39.1 to 3850.3 (775.6 ± 1110.2) ng/g l.w., constituted the major part of OCPs. HCH was represented by the α-, β-, and γ-isomers. DDT was represented only by *p,p'*-DDE that ranged within 6.5–372.7 (85.0 ± 135.6) ng/g l.w.

In muscles of the chum salmon from the Sea of Okhotsk waters off the Kuril Islands, the OCP levels ranged from 78.8 to 174.1 (with an average of 125.4 ± 34.7) ng/g l.w. and were the lowest ($p \leq 0.05$) among the values for all the organs analyzed (Fig. 7.9).

The total HCH concentration amounted to 111.4 ± 40.9 ng/g l.w. DDT and its metabolites were represented only by *p,p'*-DDE at a concentration of 14.1 ± 6.6 ng/g l.w. In the liver, the OCP levels were within 56.0–294.0 (with an average value of 183.5 ± 84.2) ng/g l.w. The total concentrations of HCH and DDE were 166.0 ± 80.0 and 21.0 ± 9.0 ng/g l.w., respectively. In eggs of the chum salmon from the Sea of Okhotsk, the OCP levels ranged within 793.7–1825.4 (1472.0 ± 587.0) ng/g l.w. and were represented only by HCH isomers. In the gonads, OCPs ranged from 1485.3 to 4223.0 (2961.0 ± 1381.0) ng/g l.w., which were the highest values for all the organs analyzed. The total concentrations of HCH and DDE in the gonads were 2628.0 ± 1342.0 and 333.0 ± 40.0 ng/g l.w., respectively.

In the chum salmon from the *Kamchatka River*, the range of OCP concentrations in all organs was 1.0–20.7 (with an average of 6.2 ± 6.1) ng/g l.w. HCH, with its values from 1.0 to 14.7 (3.9 ± 4.3) ng/g l.w., made up most of OCPs. DDT and its metabolites were detected sporadically within a range from 0.3 to 9.2 (3.0 ± 3.4) ng/g l.w. The most detectable metabolite was *o,p'*-DDE whose level ranged from 0.3 to 2.5 (1.2 ± 0.8) ng/g l.w. *o,p'*-DDT and *o,p'*-DDD were detected in one liver sample at concentrations of 1.3 and 1.6 ng/g l.w., respectively.

In muscles of the chum salmon from the Kamchatka River, the OCP level ranged from 2.5 to 10.3 (with an average value of 5.9 ± 4.0) ng/g l.w. (Fig. 7.10). The average HCH concentration was 2.1 ± 1.7 ng/g l.w. The α- and β-isomers were found in one sample at 0.3 and 3.3 ng/g l.w., respectively. The average level of DDT and its metabolites in muscles was 3.8 ± 4.7 ng/g l.w.; the metabolites were represented

Table 7.5 Average OCP concentrations in the organs of chum salmon from the southern Sea of Okhotsk and the Kamchatka River, ng/g l.w.

	α-HCH	β-HCH	γ-HCH	o,p′-DDT	p,p′-DDT	o,p′-DDD	p,p′-DDD	o,p′-DDE	p,p′-DDE
Southern Sea of Okhotsk									
Muscles	63.0 ± 22.0	42.0 ± 27.0	19.5 ± 0.1	–[1]	<DL[2]	–	<DL	–	14.0 ± 7.0
Liver	92.0 ± 52.0	44.0 ± 26.0	45.0 ± 29.0	–	<DL	–	<DL	–	21.0 ± 9.0
Eggs	836.0 ± 309.0	338.0 ± 44.0	615.0 ± 193.0	–	<DL	–	<DL	–	<DL
Gonads	1098.0 ± 186.0	1591.0 ± 757.0	469.0 ± 140.0	–	<DL	–	<DL	–	333.0 ± 40.0
Kamchatka River									
Muscles	0.3[3]	3.3	0.9 ± 0.2	<DL	<DL	<DL	<DL	1.4 ± 0.6	7.2
Liver	0.9 ± 0.7	6.3 ± 6.3	1.2 ± 0.9	1.3	<DL	1.6	<DL	1.3 ± 1.1	1.1 ± 0.6
Eggs	0.1	3.6	0.5	<DL	<DL	<DL	<DL	0.3	<DL
Gonads	0.4 ± 0.1	3.1	0.9 ± 0.5	<DL	<DL	<DL	<DL	<DL	<DL

[1]Not studied; [2]below the detection limit (DL); [3]found in only a single specimen

Table 7.6 Average concentrations of PCB congeners in the organs of chum salmon from the Kamchatka River, ng/g l.w.

	PCB 28	PCB 52	PCB 155	PCB 101	PCB 118	PCB 143	PCB 153	PCB 138
Muscles	2.7 ± 1.0	5.4 ± 1.5	$< DL^1$	7.3 ± 1.6	10.7 ± 0.4	4.1 ± 0.8	5.7 ± 0.6	5.9 ± 1.1
Liver	1.3 ± 0.3	13.1 ± 7.3	$< DL$	2.8 ± 0.9	3.8 ± 1.2	3.3	2.3 ± 0.7	2.8 ± 2.2
Eggs	1.3^2	50.6	0.4	4.7	4.7	$< DL$	2.6	3.5
Gonads	2.9 ± 3.0	17.7 ± 21.2	2.0	2.4 ± 0.6	2.7 ± 0.8	$< DL$	2.2 ± 0.2	$< DL$

[1] Below the detection limit (DL); [2] found in only a single specimen

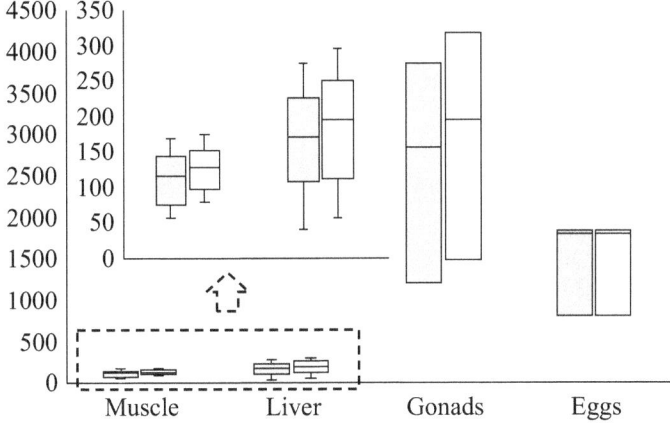

Fig. 7.9 Total OCP and HCH concentrations in the organs of chum salmon from the southern Sea of Okhotsk, ng/g l.w.

only by *o,p'*-DDE (1.4 ± 0.6 ng/g l.w.) and *p,p'*-DDE (7.2 ng/g l.w.). In the liver, the OCP concentrations ranged from 1.3 to 20.7 (9.4 ± 10.1) ng/g l.w. The average levels of HCH and DDT were 6.4 ± 7.3 and 3.0 ± 2.9 ng/g l.w., respectively. Of the DDT metabolites, *o,p'*-DDT (1.3 ng/g l.w.), *o,p'*-DDD (1.6 ng/g l.w.), *o,p'*-DDE (1.3 ± 1.1 ng/g l.w.), and *p,p'*-DDE (0.6 and 1.5 ng/g l.w.) were detected. In the gonads, OCPs were represented only by HCH within a range of concentrations from 1.0 to 4.7 (with an average value of 2.9 ± 2.6) ng/g l.w.

The total PCB concentrations in the organs of chum salmon from the Kamchatka River ranged from 9.5 to 67.6 (with an average value of 36.5 ± 17.1) ng/g l.w. PCB 28, 52, 101, 118, and 153 were found in all the samples analyzed at total concentrations of 2.1 ± 1.4, 15.7 ± 16.3, 4.4 ± 2.5, 6.0 ± 3.7, and 3.4 ± 1.8 ng/g l.w., respectively. The least detectable congener was PCB 155 detected in two samples at concentrations of 0.37 and 2.0 ng/g l.w. PCB 180 and 207 were below the detection limits in all the samples analyzed.

In muscles, the total PCB concentrations were within a range of 34.5–44.9 (with an average of 40.5 ± 5.4) ng/g l.w. Of all the congeners under study, only PCB

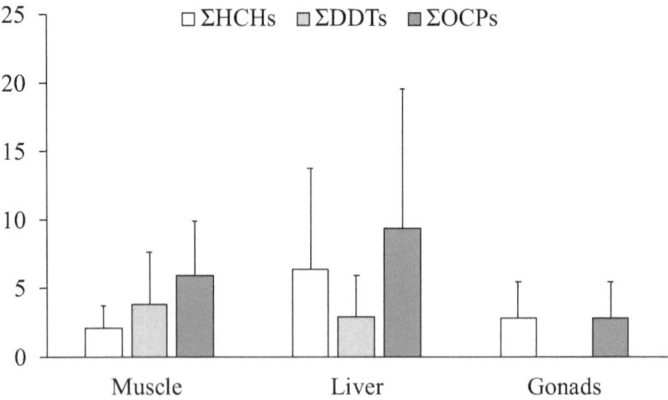

Fig. 7.10 Average total concentrations of DDT, HCH, and OCPs in the organs of chum salmon from the Kamchatka River, ng/g l.w.

155, 180, and 207 were not detected (Fig. 7.11). In the liver, PCB concentrations ranged from 18.2 to 35.6 (27.2 ± 8.7) ng/g l.w. PCB 143 was found in one sample at a concentration of 3.3 ng/g l.w. In the gonads, the total PCB concentration ranged from 9.5 to 48.1 (28.8 ± 27.3) ng/g l.w. PCB 155 was detected in one sample at a level of 2.0 ng/g l.w. PCB 138 was below the detection limit.

As a comparison of the results has shown, the concentrations of α-, β-, γ-HCH and DDE were significantly ($p \leq 0.05$) higher in the organs of the chum salmon from the Sea of Okhotsk waters off the Kuril Islands than in those from the Kamchatka River (Fig. 7.12). It was impossible to compare the levels of PCB accumulation between fish caught in different years, since PCBs were not measured in chum from the southern Sea of Okhotsk. In general, transport of water and air masses can be

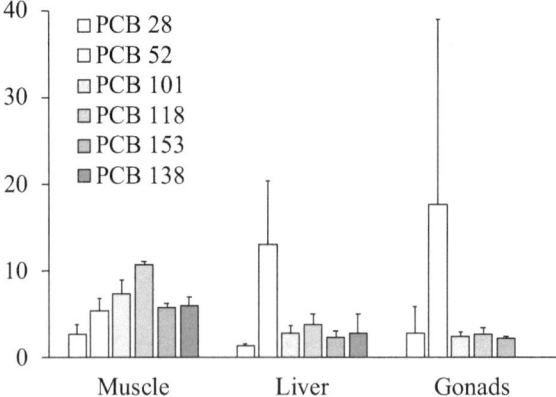

Fig. 7.11 Concentrations of the most detectable PCB congeners in the organs of chum salmon from the Kamchatka River, ng/g l.w.

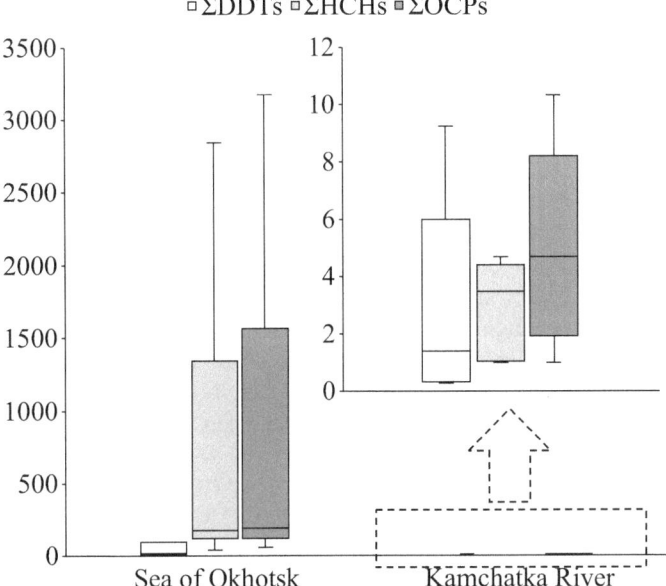

Fig. 7.12 Total concentrations of DDT, HCH, and OCPs in chum salmon caught in 2013 (Sea of Okhotsk) and 2018 (Kamchatka River), ng/g l.w.

considered the major source of OCPs in fish from the Sea of Okhotsk waters off the Kuril Islands.

The detected OCP concentrations in the Kamchatka River chum salmon may reflect both atmospheric transport of POPs and the effect of leachates from municipal solid waste landfills. This assumption is confirmed by the latest report from the Kamchatka Krai administration, which states that, in addition to the existing authorized landfills, local residents annually create many unauthorized ones, often containing dangerous substances. It is also worth noting that the Kamchatka River area is not exposed to chemicals from the Kozelsky landfill (a burial site for pesticides including OCPs), since the latter is located in the south of the Kamchatka Peninsula. PCBs, in turn, can enter the river with runoff from local landfill sites, or from spills of PCB-containing oils, or from leaking out-of-use transformers [10].

A consideration of temporal trends of variations in POP concentrations in the chum salmon organs has revealed a clear tendency of OCP amount to decrease from 2013 to 2018. This species showed a reduction in both HCH and DDT levels, and, unlike those in pink salmon, DDT concentrations decreased more rapidly than HCH, which may be explained by differences in the biology of these fishes (different life expectancies and feeding grounds).

According to the table of correlations between the compounds in the organs of the chum salmon from the Sea of Okhotsk waters off the Kuril Islands (Table 7.7), very high positive correlations were found between all the compounds under consideration at $p \leq 0.01$, which indicates a single source from which they enter the ecosystem and fish.

An analysis of the table of correlation coefficients between certain compounds in the fish from the Kamchatka River (Table 7.8) has revealed high positive correlations among pesticides as follows: between α- and γ-HCH ($p \leq 0.05$), between γ-HCH and o,p'-DDE ($p \leq 0.05$), and a very high negative correlation between β-HCH and p,p'-DDE ($p \leq 0.01$). The former two relationships reflect a single source of entry, while the latter one most likely indicates that the compounds entered separately.

PCB 28, 52, 101, 118, 153, and 138 did not show significant correlations with HCH and DDT. PCB 155 had a very high positive correlation ($p \leq 0.01$) with α- and γ-HCH and a very high negative correlation ($p \leq 0.01$) with the β-isomer. The former two may be evidence of a single source of the compounds, while the latter may indicate the possible degradation of the diphenyl base of PCB followed by the β-HCH formation. PCB 143 showed a very high negative correlation ($p \leq 0.01$) with α-HCH, which may be explained by similar processes.

PCB 28 had a very high positive correlation ($p \leq 0.01$) with PCB 155, which most likely indicates their single source. In turn, PCB 52 showed a very high negative correlation ($p \leq 0.01$) with PCB 155, which is most likely evidence that the latter degrades through the intermediate formation of congener 52. Higher chlorinated PCB 101, 118, and 153 had very strong negative correlations ($p \leq 0.01$) with congener 155, which indicates degradation and a decrease in chlorine in the biphenyl ring of PCB 155. However, PCB 101, 118, 153, 138, 118, and 153 showed a very strong positive relationship ($p \leq 0.01–0.05$) with each other, which also indicates their combined entry into the ecosystem.

Thus, OCPs and PCBs enter chum salmon in the Kamchatka River from various sources associated with atmospheric transport (in the case of OCPs) and local pollution (leachates from landfill sites and damaged transformers, in the case of PCBs).

Masu salmon (*Oncorhynchus masu*). Masu salmon samples were collected in 2017 from the Bakhura River estuary (Dolinsky District) on the eastern coast of Sakhalin Island.

POPs were found in all the samples analyzed (Tables 7.9 and 7.10). The OCP concentration in all the samples ranged from 4.1 to 479.7 (with an average value of 107.9 ± 147.1) ng/g l.w. HCH isomers, detected within a concentration range of

Table 7.7 Correlations between the organic pollutants in the organs of chum salmon from the Sea of Okhotsk

	α-HCH	β-HCH	γ-HCH	p,p'-DDE
α-HCH	1.000	0.844**	0.955**	0.972**
β-HCH		1.000	0.819**	0.944**
γ-HCH			1.000	0.947**
p,p'-DDE				1.000

[**]Correlation is significant at a level of 0.01 (two-way) (bold emphasized values)

Table 7.8 Correlations between organic pollutants found in chum salmon from the Kamchatka River

	HCH			DDE		PCB							
	α-	β-	γ-	o,p'-	p,p'-	28	52	155	101	118	143	153	138
α-HCH	1.000	0.935	0.760*	0.614	− 0.662	− 0.128	− 0.292	1.000**	− 0.155	− 0.036	− 1.000**	− 0.054	0.183
β-HCH		1.000	0.798	0.845	− 1.000**	− 0.207	− 0.178	− 1.000**	0.087	0.062	0	0.091	0.568
γ-HCH			1.000	0.815*	− 0.802	0.305	0.080	1.000**	0.056	0.031	0.262	0.013	0.238
o,p'-DDE				1.000	0.092	0.340	− 0.350	0	0.292	0.328	0.863	0.406	0.616
o,p'-DDE					1.000	0.934	− 0.976	0	0.908	0.910	0	0.918	0.650
PCB 28						1.000	0.173	1.000**	0.398	0.321	0.041	0.287	0.517
PCB 52							1.000	− 1.000**	− 0.158	− 0.344	− 0.365	− 0.394	− 0.277
PCB 155								1.000	− 1.000**	− 1.000**	0	− 1.000**	0
PCB 101									1.000	0.935**	0.350	0.870**	0.733
PCB 118										1.000	0.558	0.973**	0.836*
PCB 143											1.000	0.798	0.877
PCB 153												1.000	0.896**
PCB 138													1.000

**Correlation is significant at a level of 0.01 (two-way) (bold emphasized values). *Correlation is significant at a level of 0.05 (two-way) (italic emphasized values)

2.8–479.7 (104.3 ± 149.3) ng/g l.w., made up most of OCPs. The α-isomer was the dominant form of HCH. DDT and its metabolites were found sporadically, mainly in muscles. The total DDT concentrations ranged from 1.2 to 23.5 (7.3 ± 7.3) ng/g l.w. The most detectable metabolite was p,p'-DDD found at an average concentration of 4.4 ± 4.2 ng/g l.w.

In masu muscles, the OCP levels ranged from 4.1 to 35.4 (with an average value of 20.1 ± 11.0) ng/g l.w. (Fig. 7.13). The total HCH concentration was 11.8 ± 11.5 ng/g l.w. Among DDT and its metabolites, all the compounds under study were found. The total of DDT metabolites was at a level of 8.3 ± 8.6 ng/g l.w. The most detectable metabolite was p,p'-DDD at a concentration of 4.9 ± 4.7 ng/g l.w. In the liver, the OCP levels ranged from 32.2 to 141.7 (76.1 ± 45.2) ng/g l.w. The HCH concentration was 74.7 ± 46.4 ng/g l.w. In the gonads, the OCP concentrations ranged within 22.2–479.7 (305.5 ± 179.1) ng/g l.w.; in eggs, 16.5–17.0 (16.8 ± 0.4) ng/g l.w. OCPs in the gonads were represented only by HCH (305.5 ± 179.1); in eggs, by HCH and DDE (13.6 ± 1.8 and 3.2 ± 1.4 ng/g l.w., respectively).

The total PCB concentrations in the masu salmon organs ranged from 3.3 to 115.8 (with an average value of 25.3 ± 27.0) ng/g l.w. The most detectable congeners in all samples were PCB 52 and PCB 101 at concentrations of 10.3 ± 12.3 and 5.8 ± 7.8 ng/g l.w., respectively. PCB 143, 180, and 207 were below the detection limits.

The total PCB concentrations in muscles ranged from 9.7 to 45.3 (with an average value of 23.6 ± 11.7) ng/g l.w.; in the liver, from 3.3 to 115.8 (24.7 ± 48.8) ng/g l.w.; and in eggs, from 29.7 to 31.6 (30.7 ± 1.3) ng/g l.w. Most of the congeners were found in muscles (Fig. 7.14). In the liver, the most detectable congeners were PCB 52 (19.4 ± 19.0 ng/g l.w.) and PCB 101 (9.5 ± 12.8 ng/g l.w.). Of all the congeners, only PCB 52 (29.9 ng/g l.w.) was found in the gonads. In eggs, only PCB 143 was not detected.

A comparison of the POP levels between various organs showed that the concentrations of all HCH isomers in the gonads were significantly ($p \leq 0.05$) higher than in the liver, eggs, and muscles. DDT and its metabolites were found mainly in muscles. PCBs were also detected mainly in muscles.

Correlations between certain POPs are presented in Table 7.11.

Among pesticides, all the HCH isomers showed moderate and high positive correlations at a significance level from $p \leq 0.05$ to $p \leq 0.01$, which indicates a simultaneous entry of these pesticides into the body of masu salmon. The same relationship was observed between o,p'-DDD and p,p'-DDD ($p \leq 0.01$). p,p'-DDE had a very high correlation ($p \leq 0.01$) with α-HCH, which is probably due to the combined entry of these toxicants that leak from the facilities for the storage of banned pesticides currently existing in Sakhalin [11].

PCB 155 and 101 showed very high negative correlations ($p \leq 0.01$) with p,p'-DDE. This may be explained by the possible formation of these congeners from p,p'-DDE in the fish body. However, PCB 101 had moderate positive correlations ($p \leq 0.05$) with δ-HCH, PCB 52, and PCB 155, which may indicate not only their combined entry, but also the fact that these compounds appeared in the fish body simultaneously during the degradation of the initial POPs taken in from the environment. PCB 118 showed very high correlations ($p \leq 0.01$) with congeners 153

Table 7.9 Average OCP concentrations in the organs of masu salmon from the Bakhura River, ng/g l.w.

	α-HCH	β-HCH	γ-HCH	δ-HCH	o,p'-DDT	p,p'-DDT	o,p'-DDD	p,p'-DDD	o,p'-DDE	p,p'-DDE
Muscles	7.3 ± 8.1	2.9 ± 2.0	1.4 ± 1.5	2.6 ± 2.3	3.4 ± 0.6	3.9	1.4 ± 1.0	4.9 ± 4.7	3.9	1.2 ± 0.3
Liver	58.3 ± 43.4	5.6 ± 3.0	4.6 ± 3.5	8.3 ± 7.7	<DL[2]	<DL	<DL	<DL	8.3	<DL
Eggs	7.4 ± 1.1	4.3 ± 1.3	0.3[1]	1.8 ± 1.3	<DL	<DL	<DL	2.7 ± 0.6	<DL	1.1
Gonads	309.8 ± 149.6	27.7 ± 5.8	21.2 ± 7.2	70.9 ± 54.9	<DL	<DL	<DL	<DL	<DL	<DL

[1]Found in only a single specimen; [2]below the detection limit (DL)

Table 7.10 Average concentrations of PCB congeners in the organs of masu salmon from the Bakhura River, ng/g l.w.

	PCB 28	PCB 52	PCB 155	PCB 101	PCB 118	PCB 153	PCB 138
Muscles	4.6 ± 1.3	2.7 ± 1.5	1.2 ± 0.8	3.7 ± 2.1	4.1 ± 2.0	5.4 ± 2.0	3.6 ± 1.4
Liver	42.0[1]	19.4 ± 19.0	< DL[2]	9.5 ± 12.8	< DL	< DL	< DL
Eggs	2.31.1	13.8 ± 0.6	2.0	3.7 ± 0.1	3.1 ± 0.8	3.9 ± 0.9	3.1 ± 1.5
Gonads	< DL	29.9	< DL	< DL	< DL	< DL	< DL

[1]Found in only a single specimen; [2]below the detection limit (DL)

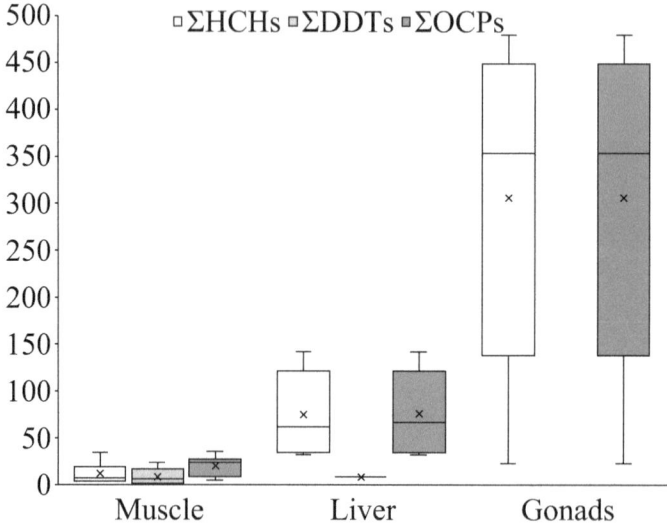

Fig. 7.13 Average total concentrations of DDT, HCH, and OCPs in the organs of masu salmon, ng/g l.w.

and 138, which most likely indicates their single source, as in the case of higher chlorinated PCBs in fish from other areas.

Thus, the major portion of DDT and PCBs entered the masu body probably during feeding. This is evidenced by the high concentrations of these compounds in fish muscles, which shows that the contamination occurred long ago.

Chinook salmon (*Oncorhynchus tshawytscha*). Samples were collected from the Bering Sea (Kamchatka Peninsula coast) (in 2010) and from the Kamchatka River (in 2018). In 2010, OCP concentrations were measured in muscles and liver; in 2018, OCP and PCB concentrations were measured in muscles, liver, and gonads. POPs were detected in all the samples analyzed (Tables 7.12 and 7.13).

In the Chinook salmon from the **Western Bering Sea**, OCP were found in all the samples within a range from 151.2 to 4219.0 (with an average value of 1277.7 ± 1026.4) ng/g l.w. HCH that ranged from 54.1 to 2168.2 (827.9 ± 664.5) ng/g l.w.

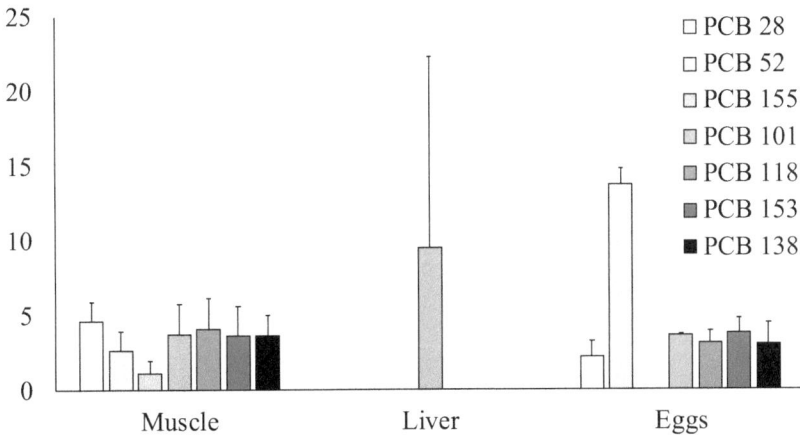

Fig. 7.14 Average concentrations of PCB congeners in the organs of masu salmon from the Bakhura River, ng/g l.w.

constituted the major part of OCPs. β-HCH was below the detection limits in all the organs analyzed. Of DDT and its metabolites, only p,p'-DDE was detected within a wide range, from 66.6 to 3022.1 (449.8 ± 596.8) ng/g l.w.

The OCP levels in Chinook muscles ranged from 265.0 to 2435.4 (with an average of 1240.9 ± 799.8) ng/g l.w.; in the liver, from 151.2 to 4219.0 (1304.7 ± 1192.7) ng/g l.w. (Fig. 7.15). The total HCH and DDT concentrations in muscles were 873.5 ± 687.4 and 367.4 ± 256.4 ng/g l.w., respectively; in the liver, 794.5 ± 669.4 and 510.2 ± 761.5 ng/g l.w., respectively. The average total OCP concentrations in muscles and liver did not differ statistically.

In the Chinook salmon from the ***Kamchatka River***, the OCP concentrations ranged from 7.7 to 70.8 (with an average value of 31.4 ± 23.1) ng/g l.w. HCH ranged within 6.4–70.8 (28.0 ± 26.3) ng/g l.w. and made up most of OCPs. The dominant isomer was β-HCH. DDT and its metabolites were found sporadically from 0.4 to 41.4 (12.9 ± 19.1) ng/g l.w. The most detectable metabolite was o,p'-DDE, 11.0 ± 20.1 ng/g l.w.

In muscles, the OCP levels ranged from 8.5 to 15.6 (with an average value of 12.0 ± 5.0) ng/g l.w. (Fig. 7.16). The HCH concentration was 6.9 ± 0.8 ng/g l.w. γ-HCH was detected in one sample at a concentration of 1.3 ng/g l.w. DDT and its metabolites were found sporadically at a level of 5.1 ± 4.2 ng/g l.w. The most detectable metabolites were o,p'-DDD and o,p'-DDE that amounted to 1.1 ± 0.1 and 1.3 ± 0.8 ng/g l.w., respectively. The OCP concentrations in the Chinook liver were significantly ($p \le 0.05$) higher than in muscles and ranged from 30.2 to 46.1 (39.1 ± 8.1) ng/g l.w. The total HCH level was 37.9 ± 11.5 ng/g l.w. Of DDT and its metabolites, only o,p'-DDE was detected in the liver at 20.8 ± 28.8 ng/g l.w. In the gonads, OCPs ranged from 7.7 to 70.8 (39.3 ± 44.6) ng/g l.w. and were represented by only HCH.

The total PCB concentrations in the organs of the Chinook salmon from the Kamchatka River ranged from 2.6 to 52.8 (with an average value of 26.1 ± 16.6)

Table 7.11 Correlations between organic pollutants in masu salmon from the Bakhura River

	HCH				DDD		DDE	PCB					
	α-	β-	γ-	δ-	o,p'-	p,p'-	p,p'-	52	155	101	118	153	138
α-HCH	1.000	0.589*	**0.862****	*0.644**	−0.500	0.054	**1.000****	0.232	−0.289	0.000	−0.608	−0.554	−0.359
β-HCH		1.000	*0.667**	**0.729****	0.400	0.126	−0.400	0.536	0.551	0.520	−0.142	−0.203	−0.017
γ-HCH			1.000	**0.794****	0.316	−0.018	0.000	−0.256	0.235	−0.032	−0.144	−0.027	−0.054
δ-HCH				1.000	1.000	−0.029	1.000	0.133	0.632	*0.709**	−0.116	−0.058	−0.029
o,p'-DDD					1.000	**1.000****	−0.500	0.400	0.800	0.800	0.800	0.316	0.400
p,p'-DDD						1.000	−0.200	0.370	0.551	0.611	0.517	0.323	0.402
p,p'-DDE							1.000	−0.600	**−1.000****	**−1.000****	−0.400	−0.775	−0.800
PCB 52								1.000	0.456	*0.715**	0.387	0.247	0.377
PCB 155									1.000	*0.899**	0.174	0.162	0.116
PCB 101										1.000	0.644	0.525	0.567
PCB 118											1.000	**0.919****	**0.895****
PCB 153												1	**0.932****
PCB 138													1

**Correlation is significant at a level of 0.01 (two-way) (bold emphasized values). *Correlation is significant at a level of 0.05 (two-way) (italic emphasized values)

Table 7.12 Average OCP concentrations in the organs of Chinook salmon, ng/g l.w.

	α-HCH	β-HCH	γ-HCH	o,p'-DDT	p,p'-DDT	o,p'-DDD	p,p'-DDD	o,p'-DDE	p,p'-DDE
Western Bering Sea									
Muscles	481.8 ± 426.3	<DL[1]	391.7 ± 309.1	–[2]	<DL	–	<DL	–	367.4 ± 256.4
Liver	356.7 ± 268.7	<DL	437.8 ± 443.3	–	<DL	–	<DL	–	510.2 ± 761.5
Kamchatka River									
Muscles	0.7 ± 0.5	5.6 ± 1.3	1.3[3]	1.2	<DL	1.1 ± 0.1	4.0	1.3 ± 0.8	0.4
Liver	2.3 ± 2.0	31.2 ± 5.9	4.5 ± 3.7	<DL	<DL	<DL	<DL	20.8 ± 28.8	<DL
Gonads	4.9 ± 5.3	29.3 ± 32.1	10.1	<DL	<DL	<DL	<DL	<DL	<DL

[1]Below the detection limit (DL); [2]not measured; [3]found in only a single specimen

Table 7.13 Average concentrations of PCB congeners in the organs of Chinook salmon from the Kamchatka River, ng/g l.w.

	PCB 28	PCB 52	PCB 155	PCB 101	PCB 118	PCB 153	PCB 138
Muscles	4.7 ± 2.0	5.3 ± 0.8	1.0 ± 0.1	5.0 ± 4.5	7.5 ± 6.6	6.0 ± 5.0	8.9 ± 2.4
Liver	10.5 ± 2.4	8.0 ± 4.5	< DL[1]	6.3 ± 3.7	3.1 ± 0.7	4.0 ± 1.1	5.7 ± 4.4
Gonads	2.5 ± 0.2	6.7 [2]	< DL	2.3	3.2	< DL	< DL

[1]Below the detection limit (DL); [2]found in only a single specimen

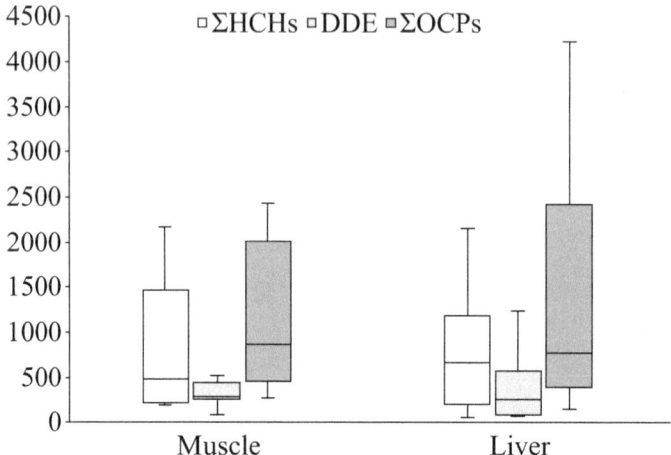

Fig. 7.15 Average concentrations of HCH, DDE, and OCPs in the organs of Chinook salmon from the western Bering Sea, ng/g l.w.

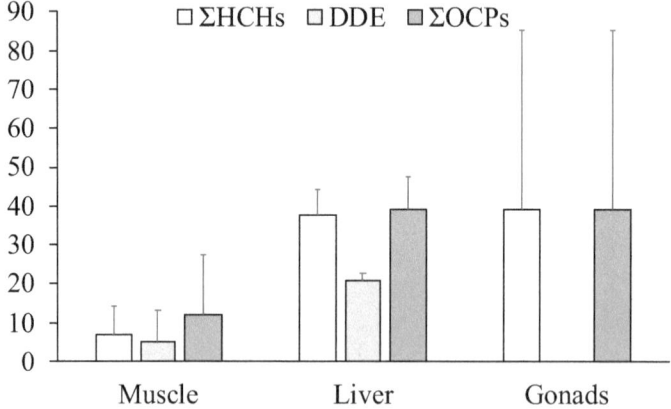

Fig. 7.16 Average concentrations of HCH, DDT, and OCPs in the organs of Chinook salmon from the Kamchatka River, ng/g l.w.

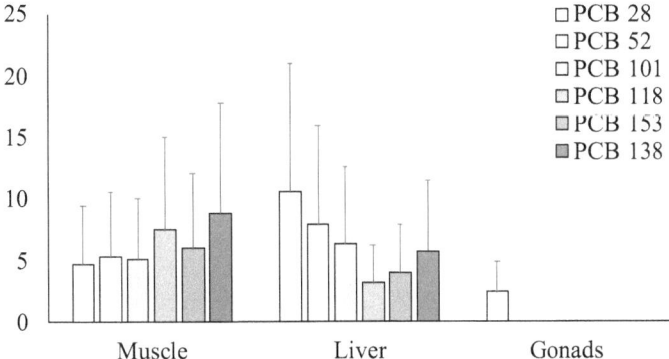

Fig. 7.17 Concentrations of the most detectable PCB congeners in the organs of Chinook salmon from the Kamchatka River, ng/g l.w.

ng/g l.w. Among PCBs, the least detectable congener was PCB 155, found only in muscles at 1.0 ± 0.1 ng/g l.w. PCB 143, 180, and 207 were below the detection limits.

In the Chinook muscles, the total PCB concentrations ranged within 15.2–52.8 (with an average of 34.8 ± 18.9) ng/g l.w.; in the liver, from 15.5 to 38.2 (29.0 ± 11.9) ng/g l.w.; in the gonads, from 2.6 to 14.5 (8.6 ± 8.4) ng/g l.w. Both *light* (PCB 52 and 28) and *heavy* congeners (PCB 101, 118, 153, 138) were found (Fig. 7.17).

The OCP concentrations in all the Chinook salmon samples from the Bering Sea (in 2010) were significantly ($p \leq 0.05$) higher than in fish from the Kamchatka River (in 2018). Thus, the concentrations of pesticides in the fish organs show a clear tendency to decrease over time. There is also a trend of decreasing HCH and DDT concentrations, which indicates the gradual removal of xenobiotics from the ecosystems of the Far Eastern seas of Russia.

The difference in the accumulation of POPs between the fish from the Bering Sea and the Kamchatka River, as in the case of pink salmon, can be explained by the effect of the Great Pacific Garbage Patch (GPGP). When migrating across this area, fish absorb microplastic particles containing POPs, which contributes to the buildup of xenobiotics in their bodies. The Chinook caught in the Kamchatka River in 2018 probably fed in the northern, cleaner areas and, therefore, the concentrations in them were insignificant. Assessment of the pattern of variations in PCB concentrations was impossible, because the analysis of these compounds was carried out only in 2018.

The proportions of certain compounds in fish organs also changed. α-HCH was the dominant isomer in the fish from the Bering Sea, while the β-isomer prevailed in the Chinook from the Kamchatka River. The latter indicates the degradation of the initial compound (lindane) to the most stable form and its long-term circulation in the ecosystem.

Among the DDT metabolites, only *p,p'*-DDE was detected in the Chinook salmon from the Bering Sea, while almost all the metabolites were found sporadically in the fish (mainly in muscles) from the Kamchatka River. This may reflect both the effect of global atmospheric transport and local pollution of the Kamchatka River water, since it is known that Chinook can stay at the bottom of a spawning water body for over a month prior to spawning [12].

High positive correlations ($p \leq 0.01$) between certain compounds were found in the organs of the Chinook salmon from the Bering Sea. The correlations between the HCH isomers and p,p'-DDE are most likely associated with a single source or area of entry of these toxicants, as in the cases of chum and pink salmon.

High positive correlations ($p \leq 0.01$) between α-, β-, and γ-HCH were found in the Chinook from the Kamchatka River (Table 7.14), which also indicates their simultaneous entry into the fish body. o,p'-DDD showed very high negative correlations ($p \leq 0.01$) with PCB 28 and 52. This may be due to the transformation of DDD into these congeners or a possible displacement of the latter during bioaccumulation of xenobiotics. In addition to negative correlations, o,p'-DDD strongly correlated (direct relationship) ($p \leq 0.01$) with PCB 118, 153, and 138. This most likely indicates a single source of entry and also probably the effect of leachates from municipal solid waste landfills during the pre-spawning staying of fish in the water body. PCB 118 showed a high (inverse) correlation with γ-HCH ($p \leq 0.01$) and (direct) with PCB 153 ($p \leq 0.05$). This suggests different sources of OCPs and PCBs.

Sockeye salmon (*Oncorhynchus nerka*). Samples were taken from the western Bering Sea (Kamchatka Peninsula coast) (in 2011), Lake Azabache (Kamchatka Peninsula) (in 2017), and the Kamchatka River (eastern Kamchatka Peninsula coast)

Table 7.14 Correlations between the organic pollutants in the organs of Chinook salmon from the Kamchatka River

	HCH			DDD	PCB				
	α-	β-	γ-	o,p'-	28	52	118	153	138
α-[1]	1.000	*0.829**	**1.000****	− 1.000	− 0.257	− 0.300	− 0.700	0.000	0.000
β-[1]		1.000	**1.000****	− 1.000	0.086	0.200	− 0.029	0.200	− 0.800
γ-[1]			1.000	0	− 0.200	− 0.500	− **1.000****	− 0.500	− 0.500
DDD[2]				1.000	− **1.000****	− **1.000****	**1.000****	**1.000****	**1.000****
28[3]					1.000	0.000	− 0.400	− 0.800	− 0.800
52[3]						1.000	− 0.300	− 0.800	− 0.800
118[3]							1.000	*0.900**	− 0.100
153[3]								1.000	0.000
138[3]									1.000

[**]Correlation is significant at a level of 0.01 (two-way) (bold emphasized values). [*]Correlation is significant at a level of 0.05 (two-way) (italic emphasized values)
[1]HCH; [2]o,p'-DDD; [3]PCB

(2018). In 2010, OCP concentrations were measured in muscles and liver; in 2017 and 2018, OCP and PCB concentrations were measured in muscles, liver, and gonads. POPs were found in all the samples analyzed (Tables 7.15 and 7.16).

In the sockeye salmon from the **Western Bering Sea**, OCP was detected in all the samples within a wide range, from 41.0 to 7103.7 (with an average value of 3052.5 ± 2474.2) ng/g l.w. HCH isomers, with their total concentrations ranging from 41.0 to 6581.9 (2864.4 ± 2336.2) ng/g l.w., constituted the major part of OCPs. Of DDT and its metabolites, only p,p'-DDE was found in a wide range, from 64.3 to 927.8 (332.8 ± 295.0) ng/g l.w.

In muscles, the OCP levels ranged from 165.7 to 3020.1 (with an average concentration of 1640.5 ± 1221.1) ng/g l.w. (Fig. 7.18). The levels of HCH and DDE were 1567.3 ± 1186.8 and 117.1 ± 62.7 ng/g l.w., respectively. In the liver, the OCP range was from 41.0 to 7103.7 (3805.7 ± 2669.6) ng/g l.w. DDE and HCH were found at concentrations of 467.6 ± 304.8 and 3556.2 ± 2529.5 ng/g l.w., respectively.

In the sockeye salmon from **Lake Azabache**, the OCP level in all the organs ranged within 1.4–208.6 (with a mean value of 51.1 ± 77.5) ng/g l.w. Of all HCH isomers, only β-HCH was detected within a range from 1.4 to 150.4 (38.6 ± 55.6) ng/g l.w. DDT and its metabolites were detected in single cases: o,p'-DDT, 6,4; o,p'-DDD, 35.9 and 55.0; p,p'-DDE, 3.1 ng/g l.w. Their total level amounted to 33.5 ± 26.0 ng/g l.w.

In muscles, the OCP concentrations ranged from 6.6 to 208.6 (with an average of 75.5 ± 115.3) ng/g l.w. (Fig. 7.19). The levels of β-HCH were 54.0 ± 83.5 ng/g l.w. Of DDT and its metabolites, o,p'-DDT (at 6.4 ng/g l.w.) and o,p'-DDD (at 35.9 and 55.0 ng/g l.w.) were found. In the liver, the OCP concentrations ranged from 19.4 to 135.8 (58.9 ± 66.6) ng/g l.w. β-HCH amounted to 46.9 ± 45.9 ng/g l.w. Of DDT and its metabolites, only p,p'-DDE was detected at a concentration of 3.1 ng/g l.w. In the gonads, OCPs were represented only by β-HCH at 2.9 ± 2.2 ng/g l.w.

The PCB levels in the organs of sockeye salmon from Lake Azabache ranged from 17.8 to 383.2 (with an average concentration of 134.2 ± 135.6) ng/g l.w. The least detectable congener was PCB 52. Congeners 28, 155, 143, 180, and 207 were below the detection limits.

In muscles of the sockeye salmon from Lake Azabache, the total PCB concentrations ranged within 34.3–110.0 (with an average of 63.9 ± 40.4) ng/g l.w.; in the liver, from 17.8 to 256.0 (121.5 ± 122.0) ng/g l.w. The most detectable congeners were PCB 101, 118, 153, and 138 (Fig. 7.20). In the gonads, PCB congeners were found only in one sample at the following concentrations: PCB 101, 37.7; PCB 118, 174.3; PCB 153, 79.3; and PCB 138, 91.9 ng/g l.w.

In the sockeye salmon from the **Kamchatka River**, the OCP levels ranged from 18.4 to 117.4 (with an average value of 40.7 ± 38.1) ng/g l.w. The total concentrations of HCH isomers were within a range of 2.4–51.4 (21.0 ± 18.8) ng/g l.w. The dominant isomer was β-HCH. DDT and its metabolites, with their concentrations ranging from 1.5 to 66.1 (23.7 ± 26.1) ng/g l.w., constituted the major part of OCPs. The most frequently detected metabolites were o,p'-DDE and p,p'-DDE at concentrations of 3.3 ± 1.8 and 3.4 ± 2.3 ng/g l.w., respectively.

Table 7.15 Average OCP concentrations in the organs of sockeye salmon, ng/g l.w.

	α-HCH	β-HCH	γ-HCH	o,p'-DDT	p,p'-DDT	o,p'-DDD	p,p'-DDD	o,p'-DDE	p,p'-DDE
Western Bering Sea									
Muscles	1274.3 ± 987.0	<DL[1]	380.1 ± 273.0	–[2]	<DL	–	<DL	–	117.1 ± 62.7
Liver	2428.7 ± 1705.7	<DL	1409.5 ± 768.8	–	<DL	–	<DL	–	467.6 ± 304.8
Lake Azabache									
Muscles	<DL	54.0 ± 83.5	<DL	6.4[3]	<DL	55.0	<DL	<DL	3.1
Liver	<DL	46.9 ± 45.9	<DL	<DL	<DL	35.9	<DL	<DL	<DL
Gonads	<DL	2.9 ± 2.2	<DL	<DL	<DL	<DL	<DL	<DL	<DL
Kamchatka River									
Muscles	3.8 ± 4.6	25.8 ± 17.3	4.5 ± 2.5	6.2	23.2	9.2	18.2	1.7 ± 1.7	3.7 ± 3.7
Liver	1.2 ± 0.6	17.5	1.3 ± 0.3	<DL	0.6	6.8	15.3	4.8 ± 0.1	2.2
Gonads	5.4 ± 4.8	21.7	1.4 ± 1.5	6.0	<DL	2.6	<DL	3.5	4.2

[1]Below the detection limit (DL); [2]not measured; [3]found in only a single specimen

Table 7.16 Average concentrations of PCB congeners in the organs of sockeye salmon, ng/g l.w.

	PCB 28	PCB 52	PCB 155	PCB 101	PCB 118	PCB 153	PCB 138
Lake Azabache							
Muscles	< DL[1]	< DL	< DL	16.3 ± 12.8	20.2 ± 17.0	14.1 ± 6.5	13.3 ± 5.4
Liver	< DL	76 ± 98.3	< DL	12.8 ± 5.5	25.3 ± 18.5	23.8 ± 3.8	25.3 ± 9.7
Gonads	< DL	< DL	< DL	37.7	174.3	79.3	91.9
Kamchatka River							
Muscles	13.6 ± 13.0	39.8 ± 31.5	< DL	19.3 ± 12.2	17.8 ± 7.7	16.4 ± 6.7	17.2 ± 4.4
Liver	3.5 ± 0.1	22.5 ± 22.8	1.1[2]	7.8 ± 4.5	7.1 ± 3.6	5.7 ± 2.9	5.8 ± 3.1
Gonads	5.2 ± 5.3	16.5 ± 11.0	< DL	5.4 ± 2.6	6.2 ± 5.1	5.3 ± 2.4	7.6

[1]Below the detection limit (DL); [2]found in only a single specimen

Fig. 7.18 Concentrations of HCH, OCPs (**a**), and DDE (**b**) in the organs of sockeye salmon from the Bering Sea, ng/g l.w.

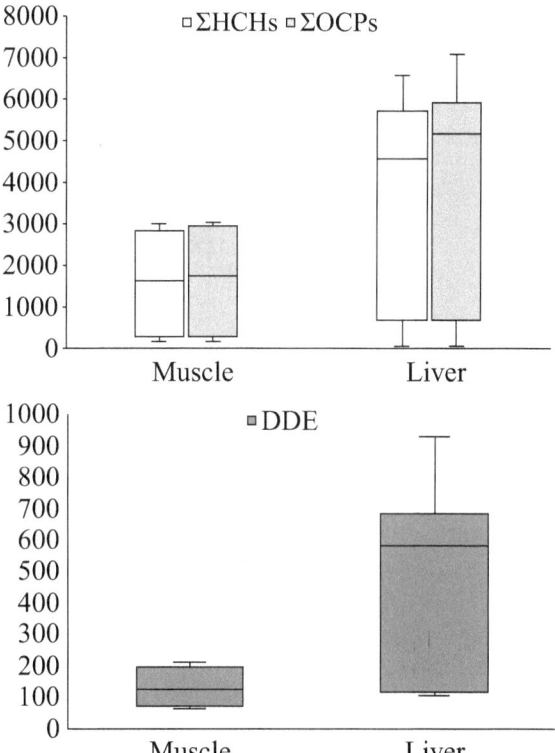

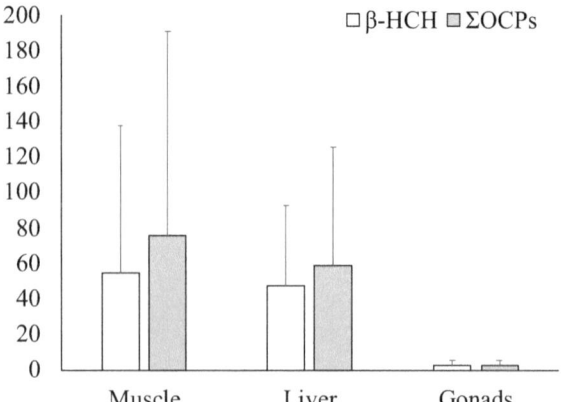

Fig. 7.19 Concentrations of β-HCH and ΣOCP in the organs of sockeye salmon from Lake Azabache, ng/g l.w.

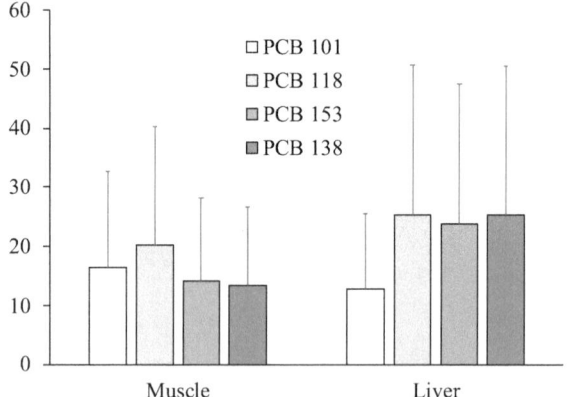

Fig. 7.20 Average concentrations of the most detectable congeners in the organs of sockeye salmon from Lake Azabache, ng/g l.w.

In muscles, the OCP concentrations ranged from 18.4 to 117.4 (with an average value of 67.9 ± 70.0) ng/g l.w. (Fig. 7.21). The HCH isomers and DDT metabolites had approximately equivalent levels, 34.1 ± 24.4 and 33.8 ± 45.6 ng/g l.w., respectively. In the liver, the range of OCP concentrations was from 24.6 to 32.3, with an average value of 28.4 ± 5.5 ng/g l.w. The HCH isomers and DDT metabolites amounted to 11.2 ± 12.0 and 17.2 ± 17.5 ng/g l.w., respectively. In the gonads, the OCP concentration ranged within 18.7–33 (25.8 ± 10.1) ng/g l.w. The HCH level was 17.7 ± 21.6 ng/g l.w. Of the DDT metabolites, o,p'-DDT, o,p'-DDD, o,p'-DDE, and p,p'-DDE were detected at 6.0, 2.6, 3.5, and 4.2 ng/g l.w., respectively.

The total PCB concentrations in the organs of sockeye salmon from the Kamchatka River ranged from 26.3 to 133.4 (with an average value of 73.0 ± 44.5) ng/g l.w. PCB 155 was detected in only a single sample at a concentration of 1.1 ng/g l.w. PCB 143, 180, and 207 were below the detection limits.

In muscles, the PCB levels ranged within 114.4–133.4 (with a total concentration of 123.9 ± 13.4) ng/g l.w. In the liver, PCBs ranged from 26.3 to 79.3 (52.8 ± 37.5)

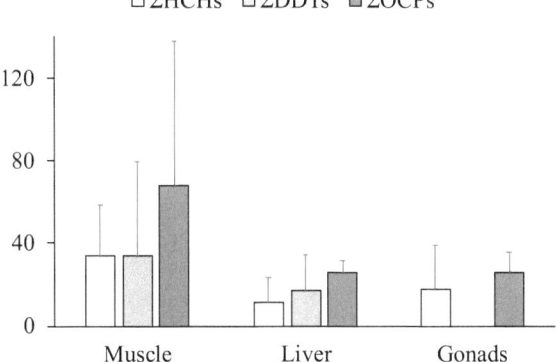

Fig. 7.21 Average concentrations of HCH, DDT and OCPs in the organs of sockeye salmon from the Kamchatka River, ng/g l.w.

ng/g l.w. PCB 155 was detected in only a single liver sample at 1.1 ng/g l.w. In the gonads, the range of PCB concentrations was from 27.3 to 57.3 (42.3 ± 21.2) ng/g l.w. PCB 28, 52, 101, 118, and 153 were found in all the samples (Fig. 7.22). PCB 138 was detected in all the muscle and liver samples.

The levels of xenobiotics in the sockeye salmon organs significantly differed ($p \leq 0.05$) only for the Bering Sea: the OCP concentrations in the liver were higher than in muscles.

A comparison of concentrations in the organs of sockeye salmon between various regions of the Far Eastern seas showed that ΣDDT and α- and γ-isomers of HCH

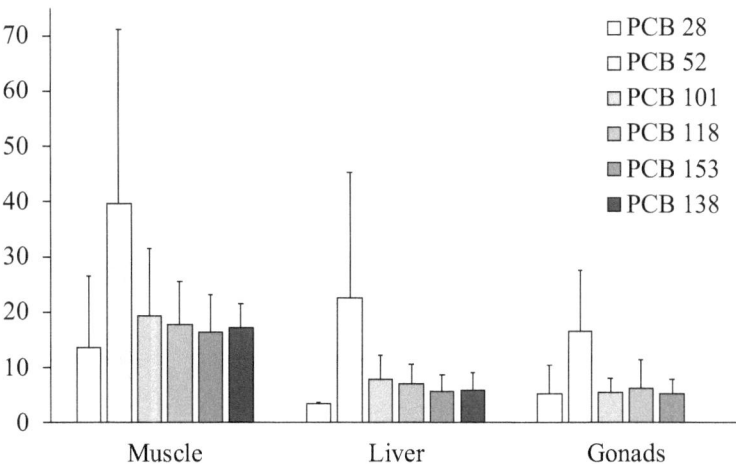

Fig. 7.22 Average concentrations of PCB congeners in the organs of sockeye salmon from the Kamchatka River, ng/g l.w.

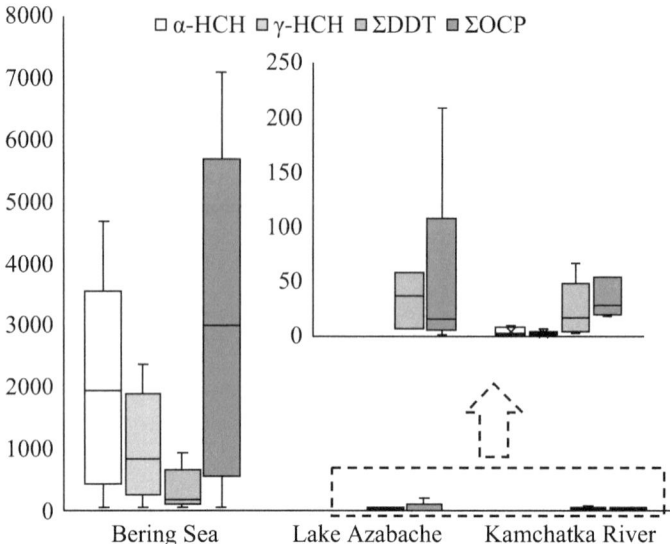

Fig. 7.23 Concentrations of HCH isomers, DDT, and OCPs in sockeye salmon from various regions of the Far Eastern seas, ng/g l.w.

were significantly ($p \leq 0.05$) higher in the sockeye salmon from the Bering Sea than in those from Lake Azabache and the Kamchatka River (Fig. 7.23).

Of the PCB congeners, only PCB 153 showed significant ($p \leq 0.05$) differences in concentrations across the study area: its concentrations were higher in the fish from Lake Azabache than in those from the Kamchatka River (Fig. 7.24).

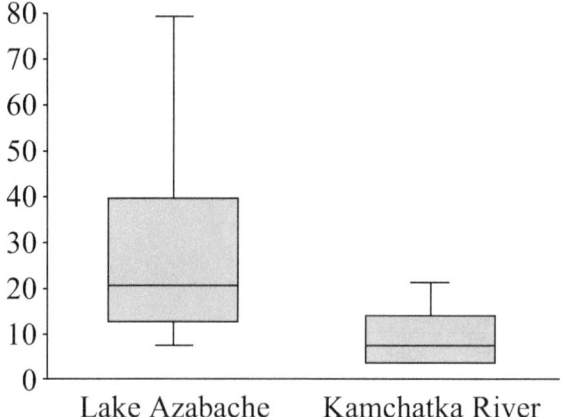

Fig. 7.24 Concentrations (medians) of PCB 153 in sockeye salmon from Lake Azabache and the Kamchatka River, ng/g l.w.

A consideration of variations in POP concentrations revealed a clear trend of decreasing HCH and DDT levels from 2011 to 2018, with the decrease in HCH being even sharper than in DDT. The OCP concentrations also showed a tendency to decrease from 2017 to 2018 (with the HCH levels also reducing more sharply than DDT), but no statistically significant difference between 2017 and 2018 was observed. The total PCB concentrations also decreased from 2017 to 2018 with no statistically significant difference.

Correlations between certain compounds in the organs of the sockeye salmon from the Bering Sea showed a strong relationship between α- and γ-HCH and between γ-HCH and p,p'-DDE, which indicates a single source of the compounds entering the ecosystem and the fish body.

In the sockeye salmon from Lake Azabache, significant correlations were found only between heavy PCB congeners: 101 and 118 ($p \leq 0.01$); 138 and 101 ($p \leq 0.05$); 138 and 118 ($p \leq 0.05$); 138 and 153 ($p \leq 0.01$) (Table 7.17). This indicates a single source of compounds entering the ecosystem and sockeye salmon's body.

Significant correlations were also found in the sockeye salmon from the Kamchatka River (Table 7.18). The positive correlations in the OCP–OCP, OCP–PCB, and PCB–PCB pairs mainly indicate a single source of xenobiotics. Leachates from municipal solid waste landfill sites can be assumed to be the major source of these compounds.

Thus, the pressure of POPs on the Far Eastern seas of Russia has significantly reduced to date. In all the species under study, the POP concentrations show a tendency to decrease, which is most likely a consequence of the measures taken at the state and federal levels to restrict the use of these compounds. The correlations recorded suggest that the xenobiotics actively degrade both under the effect of abiotic environmental factors and due to the protective functions of aquatic organisms. Nevertheless, it is premature to draw any conclusions about the complete removal of POPs from the ecosystems of the world's oceans, since the latter still remain the final receiving water bodies for all pollutants.

Table 7.17 Correlations between organic pollutants in sockeye salmon from Lake Azabache

	β-HCH	o,p'-DDD	PCB 52	PCB 101	PCB 118	PCB 153	PCB 138
β-HCH	1.000	1.000	1.000	0.0714	0.142	− 0.086	− 0.029
o,p'-DDD		1.000	0	1.000	1.000	− 1.000	1.000
PCB 52			1.000	1.000	1.000	0	0
PCB 101				1.000	**0.929****	0.771	*0.886**
PCB 118					1.000	0.771	*0.829**
PCB 153						1.000	**0.943****
PCB 138							1.000

**Correlation is significant at a level of 0.01 (two-way) (bold emphasized values). *Correlation is significant at a level of 0.05 (two-way) (italic emphasized values)

Table 7.18 Correlations between organic pollutants in sockeye salmon from the Kamchatka River

	α-HCH	γ-HCH	o,p′-DDD	o,p′-DDE	p,p′-DDE	PCB 28	PCB 52	PCB 101	PCB 118	PCB 153	PCB 138
α-HCH	1.000	0.429	0.500	0.000	1.000**	−0.086	−0.429	−0.143	−0.200	−0.145	0.300
γ-HCH		1.000	1.000**	−0.500	0.200	0.829*	−0.029	0.314	0.429	0.522	0.700
o,p′-DDD			1.000	−0.500	0.500	1.000**	−0.500	1.000**	0.500	1.000**	1.000**
o,p′-DDE				1.000	0.200	−0.600	−0.600	−0.500	−0.900*	−0.800	−0.800
p,p′-DDE					1.000	−0.400	−1.000**	0.400	0.400	0.200	0.200
PCB 28						1.000	0.086	−0.086	0.086	0.203	0.600
PCB 52							1.000	0.543	0.543	0.638	0.400
PCB 101								1.000	0.829*	0.928**	0.900*
PCB 118									1.000	0.928**	0.900*
PCB 153										1.000	1.000**
PCB 138											1.000

**Correlation is significant at a level of 0.01 (two-way) (bold emphasized values). *Correlation is significant at a level of 0.05 (two-way) (italic emphasized values)

7.1.2 Interspecific Differences in the POP Accumulation by Pacific Salmon from the Sea of Okhotsk and Bering Sea

The Pacific salmon genus, *Oncorhynchus*, comprises six species (pink, chum, masu, sockeye, coho, and Chinook salmon) widely distributed in the northern Pacific Ocean. All these salmons breed in fresh water once and die after spawning. Their marine life-history period with feeding in ocean waters varies between species: pink feed for 1.5 yr; chum, 3 yr; masu, 2.5 yr; sockeye, 4.5 yr; and Chinook, 5 yr [12]. The difference in the levels of POP accumulation between the Pacific salmon species can primarily be explained by the duration of feeding, and, accordingly, by the different patterns of accumulation of xenobiotics during the life history (e.g., 99% of all pollutants in Chinook salmon are accumulated in the marine period) [13]. Another important factor in xenobiotic accumulation is fatness of fish, since POPs are lipophilic xenobiotics.

Recently, there have been increasingly fewer studies that measure POPs in salmon. In the latest publications, a key research object among wild salmons is Chinook, the largest in size and longest-lived species with the highest commercial value in the North American and European countries [2, 13–15]; among the farmed species, it is the Atlantic salmon. The pink and chum salmon are studied either as a food chain component or as the target species of monitoring programs (e.g., the Fish Monitoring Program in Alaska) [16, 17]. The masu salmon, being common only in Asian waters, is studied only by scientists of Japan. The list of samples studied is also extensive, from certain organs to whole body, which makes it difficult to compare results. For this reason, the results of our study (Table 7.19) are presented in terms of average POP concentrations for each organ analyzed.

A comparison of the levels of POP accumulation in all of the salmon organs has revealed the following pattern of variations in OCP concentrations in the samples of 2010–2013 between species:

$$Pink < Chum < Chinook < Sockeye.$$

This may be related to differences in the lipid content of fish organs and the geochemical features of the environment in the study areas. However, this pattern of pollutants' distribution for the species of fish from the Lake Azabache and the Kamchatka River was breached and looked different:

$$Sockeye < Pink; Chinook \sim Chum < Sockeye, \text{respectively}$$

This fact is explained by different ages of fish and lipid contents of their tissues.

The distribution of POPs in lipids of fish organs in almost all samples was in the following order:

$$Muscles < Liver < Eggs < Male\ gonads$$

Table 7.19 Concentrations of POPs (ng/g l.w.) found in Pacific salmon from the Far Eastern seas of Russia

Species	Year	Study area	Organ	ΣHCH	ΣDDT	ΣPCB
Pink salmon	2012	Sea of Okhotsk (off the Kuril Islands)	Muscles	132 ± 49	9 ± 6	–[1]
			Liver	259 ± 144	30 ± 21	–
			Eggs	279 ± 139	10 ± 6	–
			Gonads	600 ± 60	44 ± 5	–
	2017	Lake Azabache (Kamchatka Peninsula)	Muscles	18 ± 12	45 ± 65	33 ± 17
			Liver	69 ± 70	< DL[2]	37 ± 45
			Gonads	52 ± 34	65[3]	596 ± 203
		Poronay River (eastern Sakhalin coast)	Muscles	23 ± 28	6 ± 7	19 ± 8
			Liver	45 ± 40	5 ± 4	18 ± 21
			Eggs	39 ± 53	3 ± 3	25 ± 11
			Gonads	84 ± 37	3 ± 2	6 ± 4
Chum salmon	2013	Sea of Okhotsk (off the Kuril Islands)	Muscles	111 ± 41	14 ± 7	–
			Liver	166 ± 80	21 ± 9	–
			Eggs	1472 ± 587	< DL	–
			Gonads	2628 ± 1342	333 ± 40	–
	2018	Kamchatka River	Muscles	2 ± 2	4 ± 5	41 ± 5
			Liver	6 ± 7	3 ± 3	27 ± 9
			Eggs	4	0.3	68
			Gonads	3 ± 3	< DL	29 ± 27
Masu salmon	2017	Bakhura River (eastern Sakhalin coast)	Muscles	12 ± 12	8 ± 9	24 ± 12
			Liver	75 ± 46	8	25 ± 45
			Eggs	14 ± 2	3 ± 1	31 ± 1
			Gonads	305 ± 179	< DL	30
Chinook salmon	2010	Bering Sea (western part)	Muscles	873 ± 687	367 ± 256	–
			Liver	794 ± 669	510 ± 762	–
	2018	Kamchatka River	Muscles	7 ± 1	5 ± 4	7 ± 1
			Liver	38 ± 12	21 ± 29	29 ± 12
			Gonads	39 ± 45	< DL	9 ± 8
Sockeye salmon	2011	Bering Sea (western part)	Muscles	1567 ± 1187	117 ± 63	–
			Liver	3556 ± 2529	468 ± 305	–
	2017	Lake Azabache (Kamchatka Peninsula)	Muscles	54 ± 84	32 ± 37	64 ± 40
			Liver	47 ± 46	36	122 ± 122
			Gonads	3 ± 2	< DL	383
	2018	Kamchatka River	Muscles	34 ± 24	34 ± 46	124 ± 13
			Liver	11 ± 12	17 ± 17	53 ± 37
			Gonads	18 ± 22	16	42 ± 21

[1]Not studied; [2]below the detection limit (DL); [3]found in only a single specimen

As a comparison of the total OCP amount in the muscles and liver of all four fish species caught in 2010–2013 has shown, the average concentrations in pink and chum salmon did not differ markedly, but were significantly ($p \leq 0.05$) lower than those in Chinook salmon; in the latter, the average values were lower than in sockeye salmon (Fig. 7.25). Pink and chum salmon were collected off the southern Kuril Islands in early summer, during their spawning migrations, when the organs already begin utilizing macronutrients to form sex products. Before spawning, salmon's organism is mobilized entirely and spends up to 80% of its reserve lipids, while the relative concentration of the toxicant per 1 g lipids increases and can reach values hazardous to fish's health [18]. The sockeye and Chinook salmon samples were collected in the western Bering Sea in the autumn, in October and November, when fish are feeding in open sea waters, where they can spend several years. During this period, the concentration of toxicants in their organs consistently increases not only in the metabolically active liver, but also in muscles.

In the muscles and liver of the salmon caught in 2017–2018, the OCP levels did not differ statistically (Fig. 7.26). The highest OCP concentrations in the liver were recorded from masu salmon, which may be explained by the largest lipid content (up to 20%) of this organ compared to all the fish studied. In muscles, the highest concentrations of xenobiotics were recorded from sockeye salmon, which, in turn, may indicate an increase in the specific concentration of toxicants with the redistribution of lipids during the formation of sex cells. This is evidenced by the high *fatness* of the organ, consisting of up to 1.1% lipids. The total PCB concentrations in muscles showed significant ($p \leq 0.05$) differences between all the salmon species under study, with the highest concentrations also found in sockeye salmon (Fig. 7.27). The levels of the toxicant in the liver did not differ statistically.

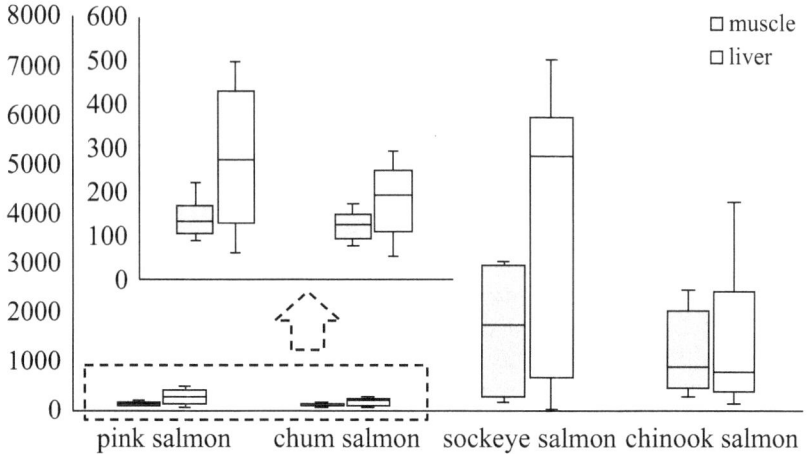

Fig. 7.25 Total OCP (HCH + DDT) concentrations (median) in the organs of Pacific salmon caught in 2010–2013, ng/g l.w.

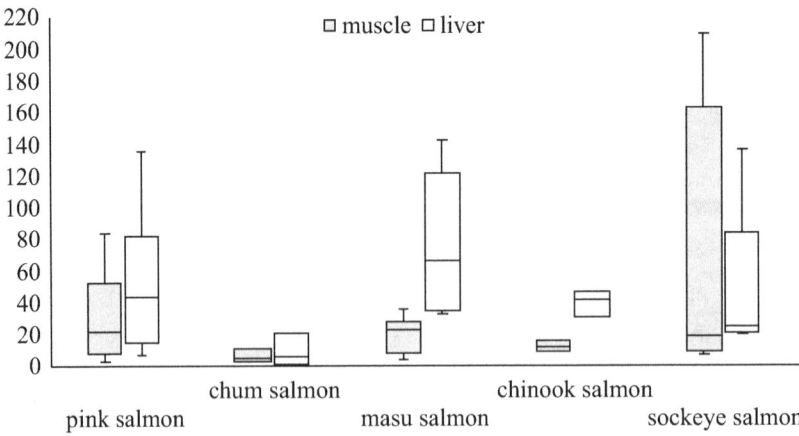

Fig. 7.26 Total OCP (HCH + DDT) concentrations (median) in the organs of Pacific salmon caught in 2017–2018, ng/g l.w.

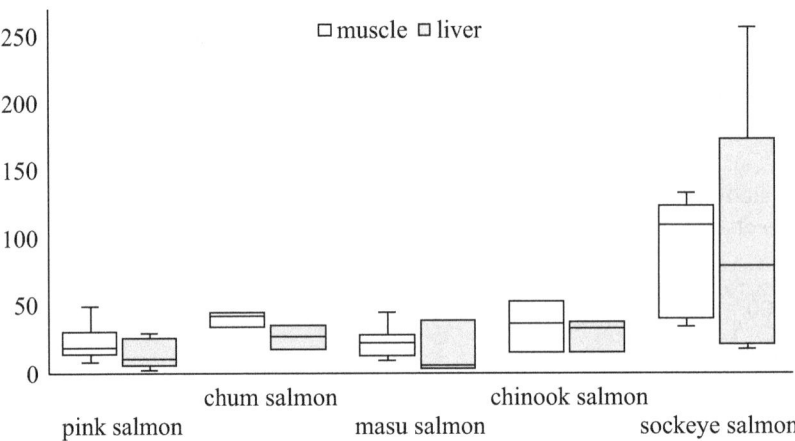

Fig. 7.27 Total PCB concentrations (median) in the organs of Pacific salmon caught in 2017–2018, ng/g l.w.

In sex products, the relationship "more lipids–more POPs" was not observed due to the lower fat content of the male gonads compared to eggs. In the study of [19], the difference in POP accumulation between male and female gonads was explained as follows: when eggs ripen, females mobilize greater amounts of lipids, and eggs constitute up to 20% of total body weight by the spawning time. Thus, POPs are *diluted*, distributed over most of the organ, and, therefore, their concentrations become lower than in the male gonads. Nevertheless, such a distribution of toxicants over organs is not constant. In the study of [20], the PCB levels in muscles were twice as high as those in gonads, whereas other authors [21] found no differences in pollutants' concentrations between the male and female gonads. However, as noted in the latter study, the fish were not ready to spawn during sampling.

Data on POP concentrations in the organs of Pacific and Atlantic salmon from various world's regions are provided in Table 7.20. A comparison of pollutant contents in the organs of salmon from the Far Eastern seas with data for other parts of the world's oceans has shown that HCH in Far Eastern fish prevails over DDT, while the ratio is inverse in almost all other regions. DDT was almost never used on the northern coast of the Russian Far East, but lindane (γ-HCH) or a technical mixture of HCHs (dominated by α-HCH) that are more volatile than DDT were actively applied, and are now commonly detected in northern and temperate latitudes. In Russia, there are several burial sites for pesticides and toxic chemicals located in Sakhalin Island and the Kamchatka Peninsula [11, 22], which can also contribute to the accumulation of pollutants in waters and marine organisms including fish. Atmospheric transport from land is also a source of pesticides entering the marine ecosystems. Technical HCH has long been and is still used in China. Residual amounts of this compound are constantly found in both terrestrial and marine organisms. Thus, the HCH concentrations in bottom sediments from Lake Baikal are threefold higher than the DDT level [23].

A comparison of PCB levels in salmon from the Far Eastern seas and other regions of the world's oceans has clearly shown lower concentrations of these pollutants in almost all samples from the Russian Far East. The only exception is the pink salmon from Lake Azabache, where the total PCB concentration reached 596 ng/g l.w. Note that most of the fish compared were caught from the Atlantic Ocean. When studying the accumulation of pesticides in salmon migrating to the American continent, researchers indicate the high level of urbanization of coastal areas, and, as a consequence, pollution of spawning watercourses. Thus, the egg development and the subsequent growth of fry also occur in polluted water, which significantly affects the viability of juveniles [17] and causes the POP concentration in their organs to increase. As was shown in the study of [29], PCB concentrations in Chinook fry increase significantly with age: from 1300 to 4100 ng/g l.w.

Currently, the Atlantic salmon farmed in net-pens represents a substantial portion of the world's salmon market. The OCP content in the organs of farmed fish is strictly controlled; nevertheless, the habitat pollution leads to the accumulation of toxicants in aquaculture facilities [32]. A comparison of e POP levels in Pacific salmon with those in Atlantic salmon has shown that the PCB concentrations in muscles of the latter are higher than in the species considered in the present study. Note that the Atlantic salmon compared were grown in net-pens, which means they had minimum contact with open ocean waters. The higher PCB levels in tissues of farmed Atlantic salmon can be associated with both a high lipid content in their tissues (up to 20%) [30] and a high anthropogenic pressure on the coastal zone where salmon farms are installed. The DDT concentrations in almost all Pacific salmon are lower than those in Atlantic salmon, with the only exception being the Chinook caught in 2010. The HCH concentrations in Pacific salmon are, on average, higher than those in Atlantic salmon.

Table 7.20 Total POP concentrations (ng/g l.w.) in muscles of Pacific and Atlantic salmon from various regions

Species	Organ	Sampling year	Sampling area	ΣHCH	ΣDDT	ΣPCB	References
Pink salmon	Muscles	1999–2000	Chukchi Sea	22	29	42	[24]
		2013	Alaska	42	38	–	[25]
Chum salmon	Body	2006–2010	Tatoosh Island (USA)	–[1]	100	385	[17]
Masu salmon	Muscles	2003	Otsuchi Bay (Japan)	20	80	–	[26]
	Liver			21	105	–	
	Muscles	2008–2010	Nagara River (Japan)	–	–	600	[27]
Chinook salmon	Muscles	1975–2013	Lake Ontario (USA)	–	–	7701	[15]
	Body[3]	1999–2000	Kitimat Fish Hatchery (Canada)	–	–	397	[2]
	Muscles	2000	Johnstone Strait (Canada)	23	15	91	[13]
	Body		Strait of Georgia (USA)	125	504	1383	
	Muscles	2001	Credit River (Canada)	< DL	170	371	[28]
	Body	2005–2009	Columbia River (USA)	< DL[2]	1800	4100	[29]
	Body	2006–2010	Protection Island (USA)	–	700	944	[17]
	Muscles	2007	Shuswap River (Canada)	10	67	75	[30]
	Eggs			8	46	63	
	Body[3]	2008	Rapid River Hatchery (USA)	135	850	780	[14]
Sockeye salmon	Muscles	2007	Adams River, spawning ground (Canada)	2	830	437	[30]
	Eggs			4	86	134	
	Muscles	2012	Tolmachev Reservoir (Kamchatka Peninsula)	2	23	44	[31]
	Liver			1	11	27	
	Eggs			2	15	29	
	Gonads			3	33	56	

(continued)

Table 7.20 (continued)

Species	Organ	Sampling year	Sampling area	ΣHCH	ΣDDT	ΣPCB	References
Atlantic salmon[3]	Muscles	1999	Scotland	19	250	460	[32]
	Muscles	1999	Norway	10	47	279	
	Muscles	2001	Ireland	39	49	284	
	Muscles	2003–2004	Canada	14	212	115	[33]
Chinook salmon[4]	Muscles	2002	Kenai River (USA)	2	10	11	[34]
	Eggs			4	8	9	
	Body	2013–2014	Great Lakes (Michigan and Huron) (USA)	–	–	380	[35]
	Muscles	2007	Northern Patagonia (Chile)	–	–	28	[36]
	Eggs[3]	2004–2014	Salmon River Fish Hatchery (USA)	–	–	1.6	[37]

[1] Not studied; [2] below the detection limit (DL); [3] farmed; [4] ng/g wet weight

Thus, the generally higher HCH levels in Pacific salmon from the Far Eastern seas of Russia than in salmon from other parts of the world's oceans reflect the wide use of lindane and technical HCHs on the coast of the Russian Far East in the twentieth century. The DDT concentrations are, on the contrary, lower compared to the world's level and most often indicate the global transport of xenobiotics with water and air masses. The PCB concentrations in Pacific salmon from the Far East have not shown a general tendency to increase or decrease compared to the values from other regions of the world. However, as reported in the literature, the level of these pollutants is higher in enclosed water bodies with low water exchange (Great Lakes, USA) [15], in countries that actively applied PCB-containing materials in the past (Japan) [27], in fish farming areas, and at fish hatcheries [2, 18].

7.2 Flounders (Genus *Hippoglossoides*)

Bering flounder (of the genus *Hippoglossoides* Gottsche, 1835) were caught in the eastern (off the Kamchatka coast) and southern (off the coasts of the Kuril Islands) parts of the Sea of Okhotsk, in Nevelskoy Bay off the southwestern Sakhalin coast (Tatar Strait), and in Rifovaya Cove of Peter the Great Bay (Sea of Japan) in the summer seasons of 2016–2018.

Sea of Okhotsk. The OCP concentrations in the fish from the eastern Sea of Okhotsk were within a range of 13.6–433.7 (with an average concentration of 99.8 ± 125.4) ng/g l.w. The total levels of HCH and DDT ranged from 13.6 to 158.0 (50.2 ± 52) and from 0.55 to 276.0 (62.0 ± 89.2) ng/g l.w., respectively. The range of OCP concentrations in the fish from the southern Sea of Okhotsk was 11.4–141 (53.6 ± 40.5) ng/g l.w. The total levels of HCH and DDT ranged from 3.3 to 103.0 (35.5 ± 36.7) and from 1.2 to 44.8 (20.1 ± 16.8) ng/g l.w., respectively.

In the samples from the eastern Sea of Okhotsk, β-HCH, whose concentrations ranged within 13.6–157.8 (with an average of 49.38 ± 50.7) ng/g l.w., was the most detectable of the HCH isomers (Fig. 7.28). α-HCH was found only in one sample at a concentration of 7.9 ng/g l.w. The γ-HCH level was below the detection limits in all samples. Among DDT and its metabolites, o,p'-DDT, p,p'-DDT, and p,p'-DDE were below the detection limits in all the samples analyzed. The concentrations of o,p'-DDD and p,p'-DDD ranged within 6.24–45.21 (23.12 ± 19.1) and 12.8–276.0 (72.3 ± 114.2) ng/g l.w., respectively. o,p'-DDE was detected in two samples at concentrations of 0.55 and 41.68 ng/g l.w.

In the samples from the southern Sea of Okhotsk, HCH was represented by α- and β-isomers whose concentrations ranged within 2.0–11.8 (with an average of 5.4 ± 3.7) and 1.1–96.2 (40.4 ± 37) ng/g l.w., respectively (Fig. 7.28). The levels of γ-HCH were below the detection limits in all the samples analyzed. Among DDT and its metabolites, the concentrations of o,p'-DDT were below the detection limits in all fish. p,p'-DDT, p,p'-DDD, and o,p'-DDE were detected in single cases at concentrations of 7.1, 22.8, and 33.0 ng/g l.w., respectively. The concentrations of o,p'-DDD ranged from 2.0 to 44.8 (with an average concentration of 16.7 ± 17.6) ng/g l.w. p,p'-DDE was found in a concentration range of 1.22–6.8 (4.3 ± 2.3) ng/g l.w.

The total of PCB concentrations in the flounder from the eastern Sea of Okhotsk ranged from 23.5 to 279.1 (with an average of 124.5 ± 90.8) ng/g l.w. PCBs were represented mainly by congeners 101 and 153. PCB 28, 52, 155, 118, 138, and 180

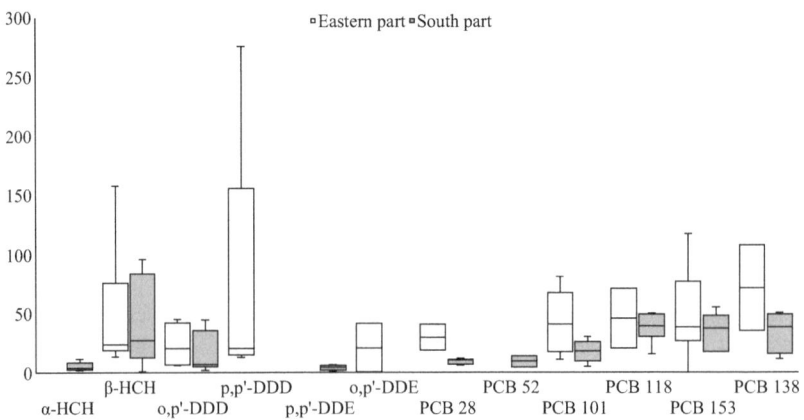

Fig. 7.28 Concentrations of POPs in flounder from the Sea of Okhotsk, ng/g l.w.

were detected sporadically (in one or two samples) at the following concentrations: PCB 28 at 40.5 and 19.1; PCB 52 at 32.7; PCB 155 at 71.2; PCB 118 at 20.6 and 70.6; PCB 138 at 107.8 and 35.3; and PCB 180 at 66.3 ng/g l.w. The concentrations of PCB 207 were below the detection limits in all the samples analyzed. The concentrations of PCB 101 and 153 ranged from 11.0 to 81.0 (42.3 ± 27.1) and from 49.4 to 117.2 (49.4 ± 36.8) ng/g l.w., respectively.

The total of PCB concentrations in the samples from the southern Sea of Okhotsk ranged from 24.7 to 150.0 (with an average of 98.6 ± 42.5) ng/g l.w. PCBs were represented by congeners 28, 101, 118, 153, and 138 at concentrations within ranges of 6.2–11.9, 5.2–30.3, 15.4–50.0, 17.5–55.0, and 34.4–50.7 ng/g l.w., respectively. The average concentrations were 9.4 ± 2.3, 18.0 ± 8.8, 38.0 ± 12.6, 35.0 ± 15.2, and 44.0 ± 7.2 ng/g l.w., respectively (Fig. 7.28). PCB 52, 155, and 180 were detected sporadically at the following concentrations: PCB 52 at 14.2 and 4.6; PCB 155 at 16.9; and PCB 180 at 17.3 and 11.7 ng/g l.w., respectively. The concentrations of PCB 143 and PCB 207 were below the detection limits in all the samples.

Tatar Strait. The OCP concentrations in the samples from the Tatar Strait ranged from 37.4 to 555.1 (with an average of 223.9 ± 180) ng/g l.w. The levels of $\sum$HCH ranged from 37.4 to 555.1 (221.3 ± 181.8) ng/g l.w. DDT and its metabolites were found in three samples and were represented by *p,p'*-DDD and *p,p'*-DDE at the following concentrations: 15 (*p,p'*-DDD), 5.8, and 12.9 ng/g l.w. (*p,p'*-DDE). HCH in the samples from the Tatar Strait was represented only by the β-isomer within a concentration range of 37.4–555.1 ng/g l.w. and at an average level of 221.3 ± 181.8 ng/g l.w. (Fig. 7.29).

The range of the total PCB concentrations was 193.4–1383.9 (with an average value of 454.9 ± 317.1) ng/g l.w. In the flounder from the Tatar Strait, PCB 28, 52, 155, 101, 118, 143, 153, 138, and 180 were found. The concentrations of PCB 207 were below the detection limit in all the samples. The concentration ranges of the

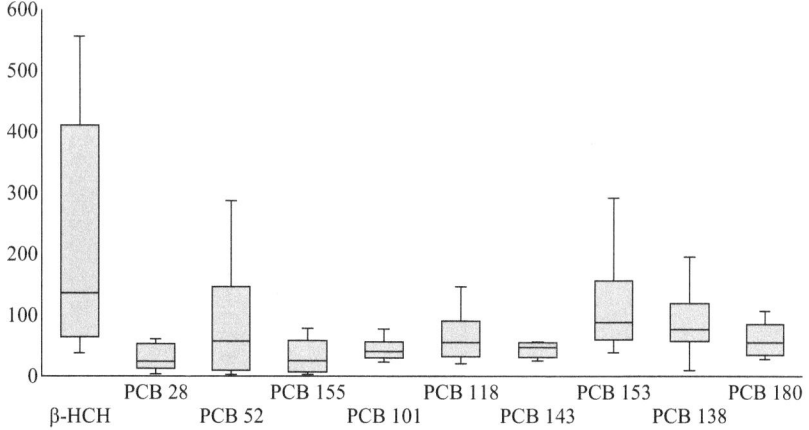

Fig. 7.29 Concentrations of POPs in flounder from the Tatar Strait, ng/g l.w.

congeners were as follows (in ng/g l.w): for PCB 28, from 4.1 to 61.0; for PCB 52, from 3.1 to 286.6; for PCB 155, from 2.7 to 78.0; for PCB 101, from 23.0 to 107.9; for PCB 118, from 19.5 to 325.5; for PCB 143, from 25.4 to 56.3; for PCB 153, from 38.0 to 291.2; for PCB 138, from 8.7 to 423.4; and for PCB 180, from 28.4 to 105.6. The average concentrations were 28.8 ± 20.8, 80.3 ± 82.6, 30.0 ± 28.0, 45.4 ± 22.7, 84.5 ± 79.6, 43.8 ± 12.7, 118.3 ± 72.2, 121.2 ± 109.4, and 58.7 ± 29.4 ng/g l.w., respectively (Fig. 7.29). PCB 101 was detected in all the samples.

Peter the Great Bay, Sea of Japan. The range of OCP concentrations in fish from Rifovaya Cove, Peter the Great Bay, was 37.8–192.8 (with an average value of 102.1 ± 50.0) ng/g l.w. HCH isomers and DDT and its metabolites were found in all the samples. The levels of $\sum$HCH and $\sum$DDT varied from 29.4 to 134.2 and from 8.5 to 87.5 ng/g l.w., respectively. The average concentrations were 62.2 ± 36.3 and 40 ± 29.4 ng/g l.w., respectively.

All the HCH isomers were found in the flounder from the Sea of Japan. The concentrations of α-, β- and γ-HCH ranged within 0.4–4.7, 26.7–126.9, and 0.9–6.1 ng/g l.w., respectively. The average concentration of α-HCH was 2.0 ± 1.4; β-HCH, 59.2 ± 34.6; and γ-HCH, 2.2 ± 2.2 ng/g l.w. (Fig. 7.30). β-HCH was found in all the samples analyzed.

Of DDT and its metabolites, o,p'-DDT was not found, and p,p'-DDT was detected only in one sample (at 5.6 ng/g l.w.). The concentration ranges of o,p'-DDD, p,p'-DDD, o,p'-DDE, and p,p'-DDE were 1.3–37.5, 5.6–52.4, 1.0–33.7, and 3.6–46.6 ng/g l.w., respectively. The average concentrations were 11.6 ± 11.9, 18.7 ± 17.5, 7.4 ± 10.1, and 15.7 ± 16.4 ng/g l.w., respectively (Fig. 7.30).

The PCB concentrations ranged from 421.3 to 3715.6 ng/g l.w. The average concentration was 1615.7 ± 1176.8 ng/g l.w. PCBs were represented by congeners 28, 52, 155, 101, 118, 143, 153, 138, and 180. The PCB 207 level was below the detection limits in all the samples. The concentrations of the congeners ranged as

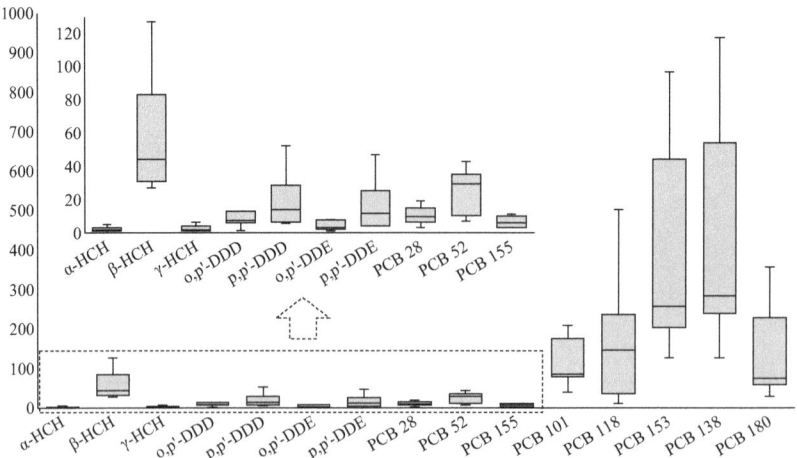

Fig. 7.30 Concentrations of POPs in flounder from the Sea of Japan, ng/g l.w.

follows (in ng/g l.w): PCB 28, from 2.7 to 405.4; PCB 52, from 7.3 to 287.3; PCB 155, from 3.1 to 10.8; PCB 101, from 40.2 to 206.6; PCB 118, from 53.0 to 581.4; PCB 143, from 11.4 to 45.5; PCB 153, from 125.7 to 848.9; PCB 138, from 126.3 to 936.2; and PCB 180, from 27.7 to 1834.7. The average concentrations of the above congeners were 54.0 ± 131.9, 53.8 ± 88.4, 6.4 ± 3.4, 117.1 ± 57.6, 241.4 ± 182.7, 25.5 ± 14.5, 387.4 ± 265.1, 428.7 ± 279.2, and 318.4 ± 579.2 ng/g l.w., respectively (Fig. 7.30).

Righteye flounders (family Pleuronectidae) are among the most common members of the benthic fish fauna, occupying the entire shelf zone and the continental slope. The general biological feature of flounders is their sedentary lifestyle: they lie on the seafloor or swim, remaining most of their time within a narrow range and making only seasonal migrations to deep-sea areas and, thus, can be considered bioindicators of local pollution. By their feeding strategies, flounders are divided into three large groups: predators, species with a mixed diet (bigmouth flounders), and typical benthivores (smallmouth flounders) [12]. The Bering flounder (the genus *Hippoglossoides*) that we studied belong to the flounder group with a mixed type of diet: their food includes both typically benthic (shrimp, bivalve mollusks, etc.) and planktonic animals. Also, flounder often prey on juvenile herring, smelt, and other small fish. Their diet spectrum strongly depends on the habitat. Off the southeastern Sakhalin and in the Sea of Japan, the diet is dominated exclusively by mollusks; in the southeastern Bering Sea, Bering flounder in the lower parts of the shelf and on the slope consume mainly echinoderms and shrimp; in shallow waters, planktonic organisms [12, 38]. Thus, differences in the POP accumulation between flounder from different regions can be associated with both anthropogenic pressure on their habitat and the potential of consumed organisms to bioaccumulate these compounds.

The highest DDT concentrations and moderate HCH concentrations were observed in the fish from the eastern Sea of Okhotsk, which may be associated with the storage facilities for pesticides and toxic chemicals in the Kamchatka Peninsula, where aldrin, dieldrin, hexachlorobenzene, and organochlorine pesticides are buried [22, 39]. DDD was the most common of DDT metabolite and HCH was represented by the most persistent β-isomer, which indicates the long-term circulation of both toxicants in the ecosystem and the degradation of the initial compounds to more stable forms. The damaged tightness of buried tanks and the evaporation of toxicants with subsequent atmospheric transport may be the major sources of pollution in this area, since agriculture on the western side of the Kamchatka Peninsula is poorly developed due to the harsh climatic conditions. PCBs in the Sea of Okhotsk may come from both increased shipping activity and plasticizer evaporation. Currently, there are no solid waste processing plants in Kamchatka Krai and, therefore, garbage is buried in landfills where leaks are probable [39].

The southern Sea of Okhotsk is the cleanest region as regards contamination by the POPs under study and is also characterized by low levels of DDT, HCH, and PCBs in fish. This region is remote from all land-based sources of pollution. Furthermore, the active hydrodynamics and water exchange with the Pacific Ocean through the Kuril straits, observed in the region, can redistribute POPs over the water column. The concentrations of xenobiotics recorded from muscles of flounder from the southern

Sea of Okhotsk can be considered background and, thus, used to assess pollution in other marine areas [6].

The values of OCP level in muscles of the flounder from the Tatar Strait proved to be the most unexpected and noteworthy data. DDT was almost absent from the samples, which allows considering the area as not exposed to serious contamination by this pesticide. However, the same samples showed the highest HCH levels (Fig. 7.31). HCH was represented only by the β-isomer, which indicates a long-term circulation of the toxin in the ecosystem.

Both in Sakhalin Island and in mainland Russia, large agricultural farms are located far from the strait. However, according to the resolution of the Sakhalin Oblast administration (of September 22, 2008), there are several landfills for out-of-use or banned pesticides on the island where chemicals were stored (by the time of the resolution) in a manner that violated the storage rules, which could lead to serious contamination of the environment [11, 40]. Most likely, these landfills became the source of pollution of the strait that resulted in the entry of HCH into the ecosystem of the Tatar Strait. This is also evidenced by the detected β-HCH which is considered the most stable isomer. Another source of HCH may be the sea currents transporting the Sea of Japan water through the Tatar Strait into the Sea of Okhotsk.

The PCB concentrations in the flounder from the Tatar Strait were multifold higher than in the fish from the Sea of Okhotsk, but markedly lower than in those from the Sea of Japan. The high PCB concentrations may be associated with the high shipping activity and the possible impact of municipal solid waste landfills located

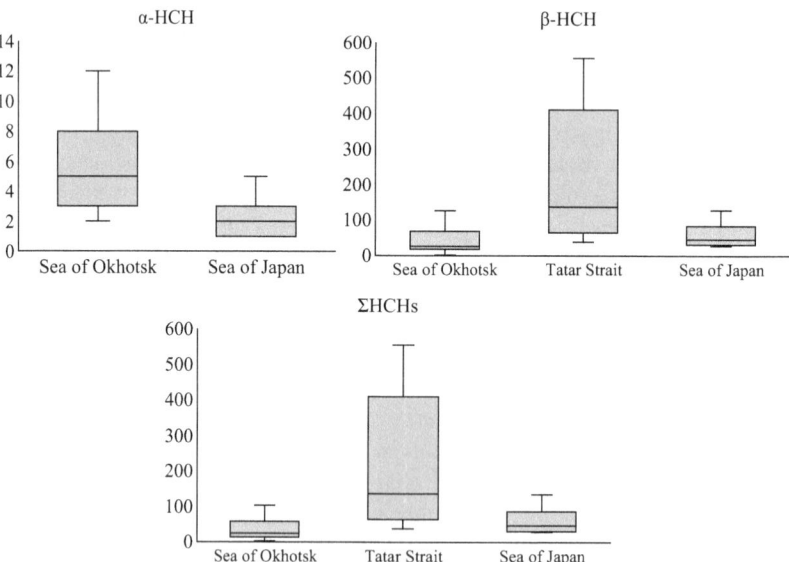

Fig. 7.31 Concentrations of HCH in flounder from the study areas, ng/g l.w.

on the western Sakhalin coast (there are a total of 54 authorized and 37 unauthorized landfill sites in Sakhalin Oblast) [40].

The Tatar Strait is an important fishing area with intensive fishery and transport activities. The warm current from the Sea of Japan, the cold current from the Sea of Okhotsk, and the narrow passage between the mainland and the island are the factors that have a pronounced effect on the accumulation of toxicants in this area. Thus, vessels are probably the major source of PCBs in the waters of the Tatar Strait.

Among the DDT metabolites, DDD and DDE were detected in the flounder from the Sea of Japan; among the HCH isomers, α-, β-, and γ-HCH were found. Primorsky Krai is an agriculturally developed region, where various, including organochlorine, pesticides were widely used in the middle of the twentieth century. These persistent compounds could have remained in soils and are now being leached by river water and surface runoff into the Sea of Japan. Furthermore, the pesticide burial sites created in the region can be another source of toxicants entering the environment. At last, according to the Stockholm Convention, the developing Asian countries still use DDT to protect their populations from malaria vectors and HCH as a medicinal agent against lice and itch mites [41]. The Sea of Japan washes the coasts of China, North and South Korea, and is connected with the East China Sea on the south through the Korea Strait. Pesticides can enter the marine ecosystems via atmospheric transport, river water, and sea currents from the East China Sea carrying their residues from agricultural land, as well as from industrial and surface runoff. China may be the most substantial source of DDT and HCH. Thus, local researchers [42] have shown that DDT from China poses the greatest hazard to biota. In flounder muscles, DDT was represented by its metabolites, DDD and DDE, while DDT was detected only in one case, which indicates a long period since contamination and the decay of the original compound.

The PCB concentrations in muscles of the fish from the Sea of Japan are the most relevant data. The recorded levels of these pollutants in muscles of the Sea of Japan flounder were by an order of magnitude higher than in the fish from the Sea of Okhotsk and the Tatar Strait (Fig. 7.32). On the Sea of Japan coast (south of Primorsky Krai), there are a huge number of resorts and unauthorized, locally called *wild* beaches. Each year thousands of people both from the Far East and from other regions of Russia come to the coast for recreation. Many of them spend some time at such wild beaches and leave tons of garbage and human organic waste that are not timely disposed. Furthermore, an oil refinery and several coal ports that can also have a serious impact on the aquatic environment and its wildlife are operated on the coast. As known, the Sea of Japan is an active transport corridor and a commercial fishing area, which may explain its contamination with biphenyls.

A comparison of the average OCP concentrations that we obtained with the data for other regions of the world's oceans have shown that the DDT levels in the samples from all our regions are significantly lower than in fish from the Atlantic Ocean, Baltic Sea, and Yellow Sea [43–45], but by an order of magnitude higher than in samples from the Bering Sea fish fauna [46] (Table 7.21). The average levels of α- and γ-HCH in fish are comparable in all data considered. However, no β-HCH concentrations were provided in the studies compared [43, 46]. In the study of [44], the level of the

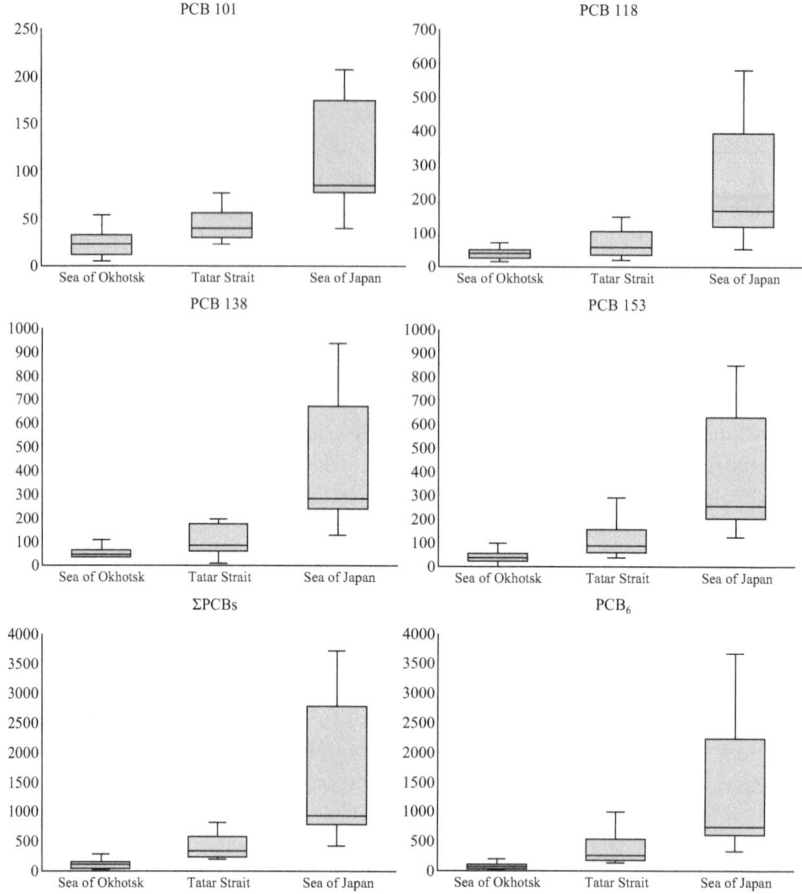

Fig. 7.32 Concentrations of heavy PCB congeners, ΣPCB_6 (indicator PCBs), and ΣPCB in flounder from the study areas, ng/g l.w.

β-isomer was below the detection limit (<0.002 ng/g wet weight); in the study of [45], no levels of the toxicant were indicated. However, in the regions of the Russian Far Eastern seas that we studied, β-HCH was the dominant isomer in flounder and showed a higher concentration than total HCHs in muscles of flounder from the Yellow Sea [43].

The PCB levels in muscles of the fish from the Sea of Okhotsk were not higher than the concentrations in flounder from the Atlantic Ocean and the Baltic Sea [44, 45], but significantly surpassed those in animals from the Yellow and Bering Seas [43, 46]. The total PCB level in the fish from the Tatar Strait was intermediate between those recorded from the Gdansk Bay and the Vistula River estuary (Baltic Sea). In the fish from the Sea of Japan, the maximum PCB levels were significantly higher than in samples from the Baltic, Bering, Yellow Seas, and the Atlantic Ocean. Such a high

Table 7.21 Average concentrations of OCPs and PCBs in muscles of flounders from various regions of the world's oceans, ng/g l.w.

Species	Study area	ΣDDT	HCH			ΣPCB	References
			α-	β-	γ-		
European flounder (*Platichthys flesus*)	Baltic Sea	579	1.3	$-$[1]	0.8	258.5	[45]
	Baltic Sea	732.2	1.4	–	0.73	373.2	[44]
	North Atlantic	140.9	2.3	–	1.8	518.2	
Starry flounder (*Platichthys stellatus*)	Barents Sea	4.5	0			15.1	[46]
Ridged-eye flounder (*Pleuronichthys cornutus*)	Yellow Sea	122	13[2]			7.5	[43]
Bering flounder (*Hippoglossoides robustus*)	Sea of Okhotsk	39.8	5.8	45.4	–	111.6	Present study
	Tatar Strait	–	–	221.3	–	347.5	
	Sea of Japan	40.0	2.0	59.2	2.2	1615.7	

[1]Not found or not measured; [2]total of all HCH isomers

difference in average concentrations between fish from different habitats indicates a serious anthropogenic pressure on the ecosystem of Rifovaya Cove and the Sea of Japan region in general.

During statistical processing, the concentrations of the pollutants found in fish were compared both within the areas and between each other using the two-way Kruskal–Wallis test (Fig. 7.33).

All *heavy* PCB congeners showed significant differences in concentrations in fish from all the study areas at $p \leq 0.05$, which indicates different sources of pollution and different levels of anthropogenic pressure in each of them. The concentrations of α- and β-HCH in fish significantly differed in each of the areas. However, no significant differences were observed in the levels of DDT and its metabolites in fish for the study areas.

A correlation analysis (Tables 7.22, 7.23 and 7.24) between all pollutants in flounders has shown that a number of PCB congeners have strong negative relationships with OCPs, which indicates a probable transformation of some compounds into others. As is known, e.g., DDT and its metabolites can transform into PCBs in the upper atmosphere layers (under UV light) and in seawater (under the effects of UV light and microorganisms) [47]. The positive correlations found between HCH and DDT and various PCB congeners with each other indicate their common sources and joint entry into the environment. The high positive correlation observed in some cases between HCH and PCB, as well as DDT and PCB, may also indicate common sources including, e.g., atmospheric transport [43, 48, 49].

Some researchers have reported a strong relationship between p,p'-DDE and PCB 153 in tissues of crustaceans, flounder, and cod, suggesting that these pollutants

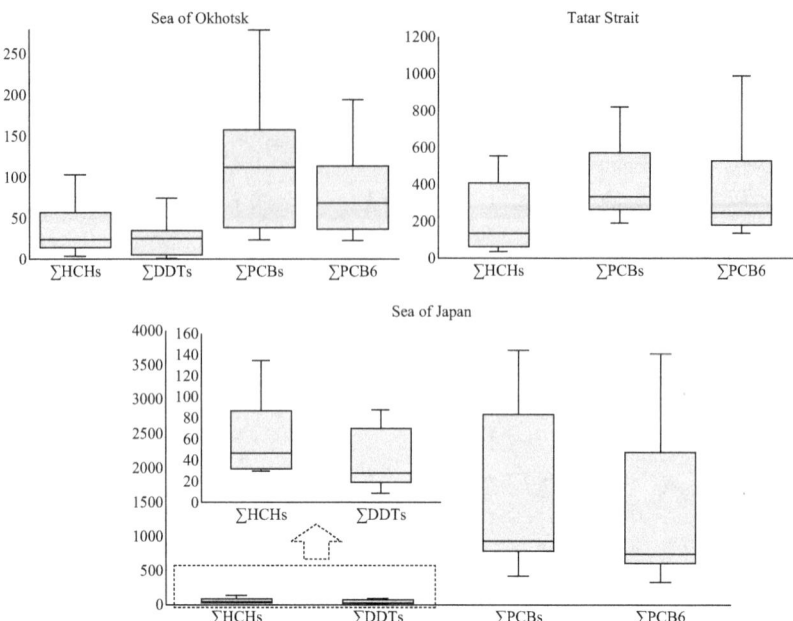

Fig. 7.33 Total concentrations of POPs in flounder from different study areas, ng/g l.w.

have a single source and indicate the contamination background in the observation area [48]. Other authors believe that a positive correlation may indicate not only a single source of these compounds, but also similar pathways of their bioaccumulation and biomagnification [43, 49]. The correlation that we have observed is most likely evidence of a single source of pollution. Although DDT and PCBs are used in various industries and enter ecosystems from different sources, their positive relationship may indicate their substantial and combined flux from municipal solid waste landfills.

Table 7.22 Spearman's coefficient of correlation between certain POPs in flounder from the Sea of Okhotsk

	HCH		DDD		DDE		PCB							
	α-	β-	o,p'-	p,p'-	o,p'-	p,p'-	28	52	155	101	118	153	138	180
α-HCH	1.000													
β-HCH	0.700	1.000												
o,p'-DDD	0.500	−0.059	1.000											
p,p'-DDD	−1.000**	0.371	0.500	1.000										
o,p'-DDE	0.500	0.500	0.500		1.000									
p,p'-DDE	0.000	−1.000**	1.000**	1.000**		1.000								
PCB 28	−0.600	0.100	0.400	−0.400		0.200	1.000							
PCB 52	−1.000**	−0.500	1.000**			1.000**	1.000**	1.000						
PCB 155	1.000**	1.000**							1.000					
PCB 101	−0.429	−0.524	−0.314	−1.000**	−1.000**	−0.400	0.357	−0.500	1.000**	1.000				
PCB 118	0.200	−0.214	−0.500	−1.000**	−1.000**	1.000**	–		1.000**	0.452	1.000			
PCB 153	−0.500	−0.200	−0.143	0.600	0.500	1.000**	0.400	−1.000**	−1.000**	*0.618**	−0.300	1.000		
PCB 138	0.800	−0.400	0.500			0.500	−0.500	−1.000**		0.371	1.000**	1.000**	1.000	
PCB 180	1.000**	−1.000**	1.000**				0.500			−1.000**	1.000**	1.000**	0.500	1.000

*Correlation is significant at $p \le 0.05$ (italic emphasized values); **correlation is significant at $p \le 0.01$ (bold emphasized values)

Table 7.23 Spearman's coefficient of correlation between certain POPs in flounder from the Tatar Strait

	β-HCH	p,p'-DDE	PCB								
			28	52	155	101	118	143	153	138	180
β-HCH	1.000										
p,p'-DDE	**1.000****	1.000									
PCB 28	*0.829**		1.000								
PCB 52	*0.700**	− **1.000****	0.371	1.000							
PCB 155	0.600		0.200	0.600	1.000						
PCB 101	0.445	− **1.000****	0.679	0.368	0.286	1.000					
PCB 118	0.042	**1.000****	0.257	− 0.176	− 0.200	0.424	1.000				
PCB 143	− 0.500		**1.000****	0.600	− **1.000****	0.400	0.800	1.000			
PCB 153	**0.755****	**1.000****	0.657	*0.627**	− 0.214	*0.582**	**0.720****	0.600	1.000		
PCB 138	0.261	**1.000****	0.657	0.285	− 0.214	0.427	**0.918****	0.800	**0.790****	1.000	
PCB 180	0.000	− 0.500	− 0.500	*0.900**	**1.000****	0.700	0.500	**1.000****	0.500	0.500	1.000

*Correlation is significant at $p \leq 0.05$ (italic emphasized values); **correlation is significant at $p \leq 0.01$ (bold emphasized values)

Table 7.24 Spearman's coefficient of correlation between certain POPs in flounder from the Sea of Japan

	HCH			DDD		DDE		PCB								
	α-	β-	γ-	o,p'-	p,p'-	o,p'-	p,p'-	28	52	155	101	118	143	153	138	180
α-HCH	1.000															
β-HCH	0.143	1.000														
γ-HCH	0.200	0.500	1.000													
o,p'-DDD	0.200	0.357	0.300	1.000												
p,p'-DDD	0.600	− 0.314	0.200	0.700	1.000											
o,p'-DDE	0.476	0.300	− 0.100	0.143	0.543	1.000										
p,p'-DDE	*0.900**	0.029	− 0.200	0.300	0.200	0.543	1.000									
PCB 28	0.238	0.100	0.100	0.286	0.314	*0.783**	0.543	1.000								
PCB 52	0.190	− 0.217	0.400	0.286	0.429	0.350	*0.886**	*0.667**	1.000							
PCB 155	− 0.600	0.300	− 0.500	− 0.400	**1.000***	0.100	− 0.600	0.500	− 0.400	1.000						
PCB 101	0.381	0.200	− 0.200	0.357	0.200	0.533	0.543	**0.850***	0.467	0.500	1.000					
PCB 118	0.548	0.333	−0.200	0.179	− 0.371	0.083	0.714	0.300	0.050	0.300	*0.717**	1.000				
PCB 143	−0.600	0.400	*1.000***	0.600	**1.000***	0.400	**1.000***	0.800	0.200	**1.000***	0.800	0.000	1.000			
PCB 153	0.643	− 0.050	0.000	0.286	0.429	0.500	*0.829**	*0.767**	0.600	− 0.100	**0.900***	0.633	0.000	1.000		

(continued)

Table 7.24 (continued)

	HCH			DDD		DDE		PCB								
	α-	β-	γ-	o,p'-	p,p'-	o,p'-	p,p'-	28	52	155	101	118	143	153	138	180
PCB 138	0.643	−0.050	0.000	0.286	0.429	0.500	0.829*	0.767*	0.600	−0.100	**0.900****	0.633	0.000	**1.000****	1.000	
PCB 180	0.524	−0.133	0.000	0.286	0.486	0.633	0.829*	**0.867****	0.733*	−0.100	**0.833****	0.383	0.000	**0.950****	**0.950****	1.000

*Correlation is significant at $p \leq 0.05$ (italic emphasized values); ** correlation is significant at $p \leq 0.01$ (bold emphasized values)

7.3 Conclusion

Accumulation of highly toxic persistent compounds can affect the health of adult individuals, their reproductive success, and the survival rate of offspring. Further monitoring the POP bioaccumulation in salmon directly at spawning grounds will provide new information about this poorly studied factor that can affect the total stock and catch of such a valuable group of commercial fishes. Xenobiotics should not have background concentrations in the environment. However, the accumulation of POPs in the fish organs is indicative of the global background of toxicants that has formed both on the planet in general and in the world's oceans in particular. Therefore, the levels of these compounds in flounder from the southern Sea of Okhotsk (waters off the Kuril Islands), which is characterized by the lack of direct sources of pollution and by active hydrodynamics, can be considered as background. The toxic compounds found in the organs and tissues of flounder are a reflection of local pollution, because these animals lead an almost sedentary life.

Acknowledgements The work was supported by the Russian Science Foundation (no. 23-74-10032)

References

1. Kelly BC, Ikonomou MG, Higgs DA et al (2008) Mercury and other trace elements in farmed and wild salmon from British Columbia, Canada. Environ Toxicol Chem 27:1361. https://doi.org/10.1897/07-527.1
2. Kelly BC, Fernandez MP, Ikonomou MG, Knapp W (2008) Persistent organic pollutants in aquafeed and Pacific salmon smolts from fish hatcheries in British Columbia, Canada. Aquaculture 285:224–233. https://doi.org/10.1016/j.aquaculture.2008.08.035
3. Malde MK, Bügel S, Kristensen M et al (2010) Calcium from salmon and cod bone is well absorbed in young healthy men: a double-blinded randomised crossover design. Nutr Metab 7:61. https://doi.org/10.1186/1743-7075-7-61
4. Khristoforova NK, Tsygankov VYu, Boyarova MD, Lukyanova ON (2015) Concentrations of trace elements in Pacific and Atlantic salmon. Oceanology 55:679–685. https://doi.org/10.1134/S0001437015050057
5. Tsygankov VYu, Boyarova MD, Lukyanova ON, Khristoforova NK (2017) Bioindicators of Organochlorine Pesticides in the Sea of Okhotsk and the Western Bering Sea. Arch Environ Contam Toxicol 73:176–184. https://doi.org/10.1007/s00244-017-0380-2
6. Lukyanova ON, Tsygankov VYu, Boyarova MD (2018) Organochlorine pesticides and polychlorinated biphenyls in the Bering flounder (Hippoglossoides robustus) from the Sea of Okhotsk. Mar Pollut Bull 137:152–156. https://doi.org/10.1016/j.marpolbul.2018.10.017
7. Tsygankov VYu (2019) Organochlorine pesticides in marine ecosystems of the Far Eastern Seas of Russia (2000–2017). Water Res 161:43–53. https://doi.org/10.1016/j.watres.2019.05.103
8. Vorozhbit OYu, Danilovskikh TE, Kuzmicheva IA et al (2016) The fishing industry of the Russian Far East: current state, problems and prospects for increasing competitiveness. Vladivostok State University of Economics and Service, Vladivostok
9. Rios LM, Jones PR, Moore C, Narayan UV (2010) Quantitation of persistent organic pollutants adsorbed on plastic debris from the Northern Pacific Gyre's "eastern garbage patch." J Environ Monit 12:2226. https://doi.org/10.1039/c0em00239a

10. Donets MM, Tsygankov VYu, Boyarova MD et al (2020) Organochlorine compounds in floun-
    ders of genus Hippoglossoides Gottsche, 1835 from the Far Eastern seas of Russia. Mar Biol
    J 5:29–42. https://doi.org/10.21072/mbj.2020.05.1.04
11. About the long-term regional… (2008) About the long-term regional target program "Production
    and consumption waste of the Sakhalin region (2009–2015)"
12. Fadeev NS (2005) Handbook of biology and fishing of fish of the North Pacific. TINRO-Center,
    Vladivostok
13. Cullon DL, Yunker MB, Alleyne C et al (2009) Persistent organic pollutants in Chinook salmon
    (Oncorhynchus tshawytscha): implications for resident killer whales of British Columbia and
    adjacent waters. Environ Toxicol Chem 28:148. https://doi.org/10.1897/08-125.1
14. Arkoosh MR, Strickland S, Van Gaest A et al (2011) Trends in organic pollutants and lipids in
    juvenile Snake River spring Chinook salmon with different outmigrating histories through the
    Lower Snake and Middle Columbia Rivers. Sci Total Environ 409:5086–5100. https://doi.org/
    10.1016/j.scitotenv.2011.08.031
15. Visha A, Gandhi N, Bhavsar SP, Arhonditsis GB (2018) A Bayesian assessment of polychlori-
    nated biphenyl contamination of fish communities in the Laurentian Great Lakes. Chemosphere
    210:1193–1206. https://doi.org/10.1016/j.chemosphere.2018.07.070
16. Apeti DA, Hartwell SI, Myers SM et al (2013) Assessment of contaminant body burdens and
    histopathology of fish and shellfish species frequently used for subsistence food by Alaskan
    Native communities
17. Good TP, Pearson SF, Hodum P et al (2014) Persistent organic pollutants in forage fish prey
    of rhinoceros auklets breeding in Puget Sound and the northern California Current. Mar Pollut
    Bull 86:367–378. https://doi.org/10.1016/j.marpolbul.2014.06.042
18. Brett R (1995) Energetics. In: Groot C, Margolis L, Clarke WC (eds) Physiological ecology
    of Pacific salmon. University of British Columbia Press, Vancouver, pp 1–65
19. Hendry AP, Berg OK (1999) Secondary sexual characters, energy use, senescence, and the cost
    of reproduction in sockeye salmon. Can J Zool 77:1663–1675. https://doi.org/10.1139/z99-158
20. Debruyn AMH, Ikonomou MG, Gobas FAPC (2004) Magnification and toxicity of PCBs,
    PCDDs, and PCDFs in upriver-migrating Pacific salmon. Environ Sci Technol 38:6217–6224.
    https://doi.org/10.1021/es049607w
21. Jackson LJ, Carpenter SR, Manchester-Neesvig J, Stow CA (2001) PCB congeners in Lake
    Michigan coho (Oncorhynchus kisutch) and Chinook (Oncorhynchus tshawytscha) salmon.
    Environ Sci Technol 35:856–862. https://doi.org/10.1021/es001558+
22. Nedra (2018) Information report on the object "Monitoring of the Kozelsky landfill for disposal
    of pesticides and pesticides." Nedra, Yelizovo
23. Tsydenova OV, Batoev VB, Weissflog L, Wenzel K-D (2003) Pollution of Lake Baikal Basin:
    organochlorine pesticides. Chem Sustain Dev 11:349–352
24. Hoekstra PF, O'Hara TM, Fisk AT et al (2003) Trophic transfer of persistent organochlorine
    contaminants (OCs) within an Arctic marine food web from the southern Beaufort-Chukchi
    Seas. Environ Pollut 124:509–522. https://doi.org/10.1016/S0269-7491(02)00482-7
25. Gerlach B (2013) Fish monitoring program. Alaska Department of Environmental Conservation
26. Oka M, Arai T, Shibata Y, Miyazaki N (2009) Concentrations of persistent organic pollutants
    in Masu Salmon, Oncorhynchus masou. Bull Environ Contam Toxicol 83:393–397. https://doi.
    org/10.1007/s00128-009-9785-6
27. Matsumoto R, Tu NPC, Haruta S et al (2014) Polychlorinated biphenyl (PCB) concentrations
    and congener composition in masu salmon from Japan: a study of all 209 PCB congeners by
    high-resolution gas chromatography/high-resolution mass spectrometry (HRGC/HRMS). Mar
    Pollut Bull 85:549–557. https://doi.org/10.1016/j.marpolbul.2014.04.021
28. O'Toole S, Metcalfe C, Craine I, Gross M (2006) Release of persistent organic contaminants
    from carcasses of Lake Ontario Chinook salmon (Oncorhynchus tshawytscha). Environ Pollut
    140:102–113. https://doi.org/10.1016/j.envpol.2005.06.019
29. Johnson L, Anulacion B, Arkoosh M et al (2013) Persistent organic pollutants in juvenile
    Chinook salmon in the Columbia River Basin: implications for stock recovery. Trans Am Fish
    Soc 142:21–40. https://doi.org/10.1080/00028487.2012.720627

30. Kelly BC, Ikonomou MG, MacPherson N et al (2011) Tissue residue concentrations of organohalogens and trace elements in adult Pacific salmon returning to the Fraser River, British Columbia, Canada. Environ Toxicol Chem 30:367–376. https://doi.org/10.1002/etc.410

31. Mamontova EA, Lepskaya EV, Tarasova EN et al (2018) Organochlorine pesticides and polychlorinated biphenyls in tissues of landlocked Kokanee salmon from Tolmachevskoye Reservoir, Kamchatka. Inland Water Biol 11:219–226. https://doi.org/10.1134/S1995082918020128

32. Jacobs MN, Covaci A, Schepens P (2002) Investigation of selected persistent organic pollutants in farmed Atlantic Salmon (*Salmo salar*), Salmon Aquaculture Feed, and fish oil components of the feed. Environ Sci Technol 36:2797–2805. https://doi.org/10.1021/es011287i

33. Shaw SD, Brenner D, Berger ML et al (2006) PCBs, PCDD/Fs, and organochlorine pesticides in farmed Atlantic salmon from Maine, Eastern Canada, and Norway, and wild salmon from Alaska. Environ Sci Technol 40:5347–5354. https://doi.org/10.1021/es061006c

34. Rice S, Moles A (2006) Assessing the potential for remote delivery of persistent organic pollutants to the Kenai River in Alaska. Alsk Fish Res Bull 12:153–157

35. Gerig BS, Chaloner DT, Janetski DJ et al (2018) Environmental context and contaminant biotransport by Pacific salmon interact to mediate the bioaccumulation of contaminants by stream-resident fish. J Appl Ecol 55:1846–1859. https://doi.org/10.1111/1365-2664.13123

36. Montory M, Habit E, Fernandez P et al (2010) PCBs and PBDEs in wild Chinook salmon (Oncorhynchus tshawytscha) in the Northern Patagonia, Chile. Chemosphere 78:1193–1199. https://doi.org/10.1016/j.chemosphere.2009.12.072

37. Garner AJ (1987) Pagano JJ (2019) Trends of polychlorinated dioxins, polychlorinated furans, and dioxin-like polychlorinated biphenyls in Chinook and Coho salmonid eggs from a Great Lakes tributary. Environ Pollut Barking Essex 247:1039–1045. https://doi.org/10.1016/j.envpol.2019.01.117

38. Napazakov VV, Chuchukalo VI (2002) Feeding and food relations of flounders in the western part of the Bering Sea in the summer-autumn period. Izv TINRO 130:595–617

39. Report on the state... (2019) Report on the state of the environment in the Kamchatka Territory in 2018. Ministry of Natural Resources and Ecology of the Kamchatka Territory, Petropavlovsk-Kamchatsky

40. Zubtsova IL (2008) Disposal of waste on the territory of the Sakhalin region. Vladivostok, pp 109–118

41. Tripathi V, Edrisi SA, Chaurasia R et al (2019) Restoring HCHs polluted land as one of the priority activities during the UN-International Decade on Ecosystem Restoration (2021–2030): a call for global action. Sci Total Environ 689:1304–1315. https://doi.org/10.1016/j.scitotenv.2019.06.444

42. Grung M, Lin Y, Zhang H et al (2015) Pesticide levels and environmental risk in aquatic environments in China—a review. Environ Int 81:87–97. https://doi.org/10.1016/j.envint.2015.04.013

43. Byun G-H, Moon H-B, Choi J-H et al (2013) Biomagnification of persistent chlorinated and brominated contaminants in food web components of the Yellow Sea. Mar Pollut Bull 73:210–219. https://doi.org/10.1016/j.marpolbul.2013.05.017

44. Waszak I, Dabrowska H, Komar-Szymczak K (2014) Comparison of common persistent organic pollutants (POPs) in flounder (Platichthys flesus) from the Vistula (Poland) and Douro (Portugal) River estuaries. Mar Pollut Bull 81:225–233. https://doi.org/10.1016/j.marpolbul.2014.01.044

45. Kopko O, Dabrowska H (2018) Variability of biological indices, biomarkers, and organochlorine contaminants in flounder (Platichthys flesus) in the Gulf of Gdańsk, southern Baltic Sea. Chemosphere 194:701–713. https://doi.org/10.1016/j.chemosphere.2017.12.039

46. Hartwell SI, Apeti AD, Pait AS et al (2018) Benthic habitat contaminant status and sediment toxicity in Bristol Bay, Alaska. Reg Stud Mar Sci 24:343–354. https://doi.org/10.1016/j.rsma.2018.09.009

47. Crosby DC (1983) Atmospheric reactions of pesticides. Pestic Chem Hum Welf Environ 3:327–332

48. Voorspoels S, Covaci A, Maervoet J et al (2004) Levels and profiles of PCBs and OCPs in marine benthic species from the Belgian North Sea and the Western Scheldt Estuary. Mar Pollut Bull 49:393–404. https://doi.org/10.1016/j.marpolbul.2004.02.024
49. Moon H-B, Kim H-S, Choi M et al (2009) Human health risk of polychlorinated biphenyls and organochlorine pesticides resulting from seafood consumption in South Korea, 2005–2007. Food Chem Toxicol 47:1819–1825. https://doi.org/10.1016/j.fct.2009.04.028

Chapter 8
Biotransport of Persistent Organic Pollutants and Heavy Metals by Pacific Salmon in the Northwestern Pacific Ocean

Abstract Marine animals making long-range migrations biotransport some amounts of persistent organic pollutants (POPs) and heavy metals (HMs) that can remain in ecosystems and be involved in food chains. The most common transport vectors are salmon whose biomass in the subarctic region of the Pacific Ocean has reached 1.5 million tons over the past decade. Salmon that die after spawning in rivers and lakes leave sometimes substantial quantities of pollutants at their spawning grounds. In this study, we calculate the total amount of POPs and HMs brought by salmon to the Russian coast of the northwestern Pacific Ocean and consider the role of these fishes in the sea-to-land transport of toxicants.

Keywords POPs · Heavy metals · Biotransport · Pacific salmon · Northwestern Pacific Ocean

8.1 Introduction

Biogeochemical cycles of chemical elements in the biosphere are driven by major environmental factors. A unique example of directional transport of nutrients in the ocean is the spawning migration of anadromous fish, including Pacific salmon, that feed in marine waters and arrive to breed and spend early development in fresh water of rivers and lakes. Millions of fish that die after spawning leave substantial amounts of mineral components and organic matter as part of their carcasses and skeletons at spawning grounds. Such a "marine pump" for the transport of nutrients is considered as an evolutionary mechanism that provides success of egg development and survivability of juveniles in fresh water [1, 2].

During feeding in the ocean and, especially, prior to spawning migration, salmon accumulate lipids reserves both to meet energy costs and for gonad maturation. Lipophilic persistent organic pollutants (POPs) that come from the marine environment are accumulated in animal's body along with lipids [3, 4]. POPs can build up to high levels in the organs and exert various negative biological effects by disrupting the major metabolic processes, which eventually reduces the breeding success [5]. The presence of toxicants at salmon feeding grounds remote from areas of high industrial activities is evidence of the increase in the global background of pesticides that are carried by winds over great distances from areas of use (in the tropical and subtropical zones) to temperate latitudes. Migrating fish can act as vectors that provide biotransport of organic pollutants from subtropical and southern boreal ecosystems to boreal and subarctic ones.

Wild salmon gain more than 70% of their biomass in the ocean and when migrating to their natal habitats [6]. Intake of trace elements by wild fish depends on concentrations of these elements in the waters where migration routes and feeding grounds are located; for fish raised in aquaculture facilities, it depends on the feed formula consumed by them and the ecological condition of the habitat. Aquatic systems are, in fact, collectors of all types of pollutants, including those caused by human activity, on both regional and global scale. Extracted from the bowels of the Earth and enriched in technological cycles, many elements acquire toxic properties and form technogenic biogeochemical provinces [7].

Pacific salmon spawners arriving at their spawning grounds bring huge amounts of matter and energy, thereby providing the existence of salmon river ecosystems. They constitute a major part in the diet of many terrestrial mammals (from bears to chipmunks) and birds. Along with nutrients, salmon also bring large quantities of substances accumulated during the marine life-history period. The amount of trace elements and POPs transported from sea to land varies significantly between years, depending on the abundance and biomass of running spawners.

In this chapter, we calculate amounts of POPs (2008–2018) and heavy metals (2018–2021) transported to the Russian part of the Northwest Pacific.

We determined POPs by the methods described in Chap. 2 and heavy metals (HM) by the classical atomic absorption spectroscopy method [8].

8.2   Biotransport of POPs in 2008–2012

There are two major vectors of pollutants' migration in the ocean: atmospheric transport and biotransport, which differ in several important parameters. For example, in the case of biotransport, toxicants brought by fish can be directly involved in food chains and contribute to biomagnification, because salmon eggs, carcasses, and skeletons are used as food by predators. During atmospheric transport, pollutants are subject to various physical and chemical transformations, which changes the efficiency of their transfer along food chains, as well as the qualitative composition of xenobiotics. However, biotransport contributes to distribution of chemically unstable substances or various compounds that are not capable of atmospheric transport (nonvolatile substances, e.g., chlorinated fatty acids that rapidly build up in fatty tissues of fish) [4].

Pacific salmon, due to the specifics of their life strategy, continuously carry out and will always do the POP transport to the spawning grounds on the mainland.

This phenomenon is of both fundamental (since it changes the major biogeochemical cycles of the sea-to-land transport of matter) and applied interest: the entry of pesticides to spawning grounds can affect the breeding success of salmon, which will ultimately influence their commercial stocks [9].

We calculated amounts of pesticides transported in 2008–2010 using the data on OCP content of whole body of chum and pink salmon as the most common species. The proportion of this fishes in the total run during spawning migrations is greater than 80%. The OCP level in them was converted from ng/g lipid weight (l.w.) to ng/g wet weight (w.w.). To calculate the biotransport of 2011 and 2012, sockeye and Chinook salmon were also taken into account in addition to the former two species.

The number of salmon migrating to the Russian coast varies from year to year, but the structure of runs remains unchanged: pink salmon accounts for 60–65%; chum salmon, 20–25%; sockeye salmon, 10–12%; and coho and Chinook salmon together make up only a small proportion. Chum and pink salmon constitute the major part of runs to the eastern Kamchatka coast, eastern Sakhalin, the mainland coast of the Sea of Okhotsk, and in the Amur River basin [10, 11]. The number of fish that escaped to spawn in some regions of the Far East in 2008–2012 is shown in Tables 8.1, 8.2 and 8.3.

Fish that have escaped to their spawning grounds eventually die there and serve as food for many organisms, thus, linking the marine and terrestrial food chains, or, in other words, transporting organic matter from the ocean to the land. Data on occupancy of salmon spawning grounds in the North Pacific are provided in the free-access reports of the North Pacific Anadromous Fish Commission (NPAFC).

The OCP content as an average of the HCH + DDE sum was 68.85 ng/g w.w. in pink salmon and 182.5 ng/g w.w. in chum salmon. The average body weight of one pink salmon spawner is 1.3 kg and one chum salmon spawner 3.5 kg. Calculations show that one pink salmon contains up to 90 μg pesticides and one chum salmon

Table 8.1 Occupancy of chum and pink salmon spawning grounds (thousand fish) in the Russian zone of the Far Eastern seas in 2008–2010

Region	Chum salmon			Pink salmon		
	2008	2009	2010	2008	2009	2010
Eastern Kamchatka	399.56	2953.67	404.95	2914.9	94 497	8757
Western Kamchatka	722	715.8	391.6	38 948.8	118.6	46 441
Sea of Okhotsk (Magadan Oblast)	3860	8827	5931	3225.4	28 042.5	1926
Kuril Islands	105	72.6	261.2	1431.9	1216.4	2009
Eastern Sakhalin	478.5	513.5	481.5	8782.7	18 478.7	11 708
Amur River basin	2747	20 079	10 827	1132.9	2927.0	9444
Western Bering Sea	3999.9	1942.9	2191.5	70.2	170.0	60

Table 8.2 Occupancy of salmon spawning grounds (thousand fish) on the Russian coast of the Far Eastern seas in 2011

Region	Pink salmon	Chum salmon	Sockeye salmon	Chinook salmon
Western Bering sea	10 000	2310	500	
Eastern Kamchatka	53 549	393.1	308.8	7.7
Kuril Islands	1387.8	265.1		
Western Kamchatka	1212.7	479.9	2342.7	16.1
Sea of Okhotsk (Magadan Oblast)	18 744.5	4790	22	
Eastern Sakhalin	12 926.1	156.0		
Amur River basin	7982.7	6137.1		
Primorsky Krai	213.7	60.2		
Southwestern Sakhalin	176.6	2.6		

Table 8.3 Occupancy of salmon spawning grounds (thousand fish) on the Russian coast of the Far Eastern seas in 2012

Region	Pink salmon	Chum salmon	Sockeye salmon	Chinook salmon
Eastern Kamchatka	8077.5	223.1	634.7	7.9
Kuril Islands	1481.0			
Western Kamchatka	19 636.7	235.1	2133.3	26.7
Sea of Okhotsk (Magadan Oblast)	19 755.5	4643.2	47.4	
Eastern Sakhalin	12 141.2			
Amur River basin	9347.4	6545.2		
Primorsky Krai	2046.2	61.9		
Southwestern Sakhalin	1394.2			

does up to 640 μg. We could not carry out whole-body analysis for sockeye and Chinook salmon. Therefore, we estimated the average fat content of fish: 1.6% for sockeye salmon (according to our data) and 10% for Chinook (according to published reference data). Based on these contents, we calculated the total amounts of pesticides in muscles of these fish species: 124.1 ng/g w.w. for Chinook salmon and 20.33 ng/g w.w. for sockeye salmon. Fish body consists of more than 75% muscle tissue and, therefore, the total level of pesticides in muscles would be approximately equivalent to that in whole body. Since a body weight of one sockeye salmon is 4 kg, then the total amount of pesticide in whole fish is estimated at 81.3 μg. It is worth noting that the sockeye salmon samples were collected in July, during spawning migrations, when fish weighed about 1 kg, and the pesticide content was minimum. For one Chinook salmon weighing 7 kg, the total OCP amount will be 870 μg.

Calculations showed that the total amount of OCPs transported, e.g., in 2009 by pink and chum salmon only to eastern Kamchatka amounted to 10.4 kg; to the Amur

River basin, more than 13 kg; to the Sea of Okhotsk mainland coast, 8.1 kg (Tables 8.4 and 8.5). In 2011, pink salmon transported the maximum amount of OCPs to eastern Kamchatka, 4.8 kg. In 2012, chum salmon transported a maximum of 4.2 kg to the Amur River basin (Table 8.6).

In 2008, the total amount of pesticides brought by salmon to various regions of the Pacific coast of Russia ranged from 0.52 to 4 kg; in 2009, from 0.47 to 13 kg; in 2010, from 0.35 to 7.75 kg. For those three years, the Amur River basin received the largest amount of pesticides, approximately 23 kg (Fig. 8.1). The high salmon runs in recent years provided an annual sea-to-land transport of pesticides ranging from 13 to 30 kg. Thus, the total level of pesticides at each spawning ground increases.

In 2011, the total amount of pesticides was 19.1 kg; in 2012, 14.4 kg. The Amur River basin, eastern Kamchatka, and the mainland Sea of Okhotsk coast received the largest amounts (Fig. 8.2). Thus, the total level of pesticides at each spawning ground gradually decreases each year, but depends on the overall size of fish run.

Salmon catches in Russian waters remained consistently high in recent years: 542 000 t in 2009, 325 000 t in 2010, 504 000 t in 2011, 438 000 t in 2012, and about

Table 8.4 Amounts of organochlorine pesticides (kg) in chum and pink salmon migrating to the Pacific coast of Russia (2008–2010)

Region	Chum salmon			Pink salmon		
	2008	2009	2010	2008	2009	2010
Eastern Kamchatka	0.26	1.89	0.26	0.26	8.51	0.79
Western Kamchatka	0.46	0.46	0.25	3.50	0.01	4.10
Sea of Okhotsk (Magadan Oblast)	2.47	5.65	3.80	0.29	2.52	0.17
Kuril Islands	0.07	0.05	0.17	0.13	0.11	0.18
Eastern Sakhalin	0.31	0.33	0.31	0.79	1.66	1.05
Amur River basin	1.76	12.81	6.90	0.10	0.26	0.85
Western Bering sea	2.56	1.24	1.40	0.01	0.02	0.01
Total	7.9	22.4	13.1	5.1	13.1	7.1

Table 8.5 Total amount of organochlorine pesticides (kg) transported by chum and pink salmon to the Pacific coast of Russia in 2008–2010

Region	2008	2009	2010
Eastern Kamchatka	0.52	10.39	1.05
Western Kamchatka	3.96	0.47	4.35
Sea of Okhotsk (Magadan Oblast)	2.76	8.17	3.97
Kuril Islands	0.20	0.16	0.35
Eastern Sakhalin	1.10	1.99	1.36
Amur River basin	1.86	13.07	7.75
Western Bering sea	2.57	1.26	1.41
Total	13.0	35.5	20.2

Table 8.6 Total amount of pesticides (kg) transported by salmon to the Russian coast of the Far Eastern seas in 2011–2012

Region	Pink salmon	Chum salmon	Sockeye salmon	Chinook salmon	ΣOCP
2011					
Western Bering sea	0.9	1.5	0.04		2.44
Eastern Kamchatka	4.8	0.3	0.03	0.007	5.137
Kuril Islands	0.1	0.2			0.3
Western Kamchatka	0.1	0.3	0.2	0.01	0.61
Sea of Okhotsk (Magadan Oblast)	1.7	3.1	0.002		4.802
Eastern Sakhalin	1.2	0.1			1.3
Amur River basin	0.7	3.9			4.6
Primorsky Krai	0.02	0.04			0.06
Southwestern Sakhalin	0.02	0.002			0.022
Total	9.54	9.442	0.272	0.017	19.271
2012					
Eastern Kamchatka	0.73	0.14	0.05	0.01	0.93
Kuril Islands	0.13				0.13
Western Kamchatka	1.77	0.15	0.17	0.02	2.11
Sea of Okhotsk (Magadan Oblast)	1.78	2.97	0.004		4.75
Eastern Sakhalin	1.09				1.09
Amur River basin	0.84	4.19			5.03
Primorsky Krai	0.18	0.04			0.22
Southwestern Sakhalin	0.13				0.13
Total	6.65	7.49	0.23	0.03	14.40

400 000 t in 2013. The latter value is equivalent to 40–60 kg of pesticides, which are transferred up the food chain and eventually spread over land.

Currently, salmon are the most common fish of the upper epipelagic layer in the North Pacific. The food resources of this water layer, which is relatively poorly occupied by other species, are sufficient to support large numbers of fish [11]. A study of migrating fish has shown that the carbon and nitrogen isotope composition of organs of salmon from spawning grounds is similar to that of fish from feeding grounds in the seas or ocean waters [12], which means that salmon transport nutrients from the ocean to rivers and lakes.

In general, the distribution of salmon in marine waters matches the seasonal distribution of water masses within a range of water temperatures optimum for salmon, from 2–4 to 15–20 °C. However, their migration processes depend on a variety of

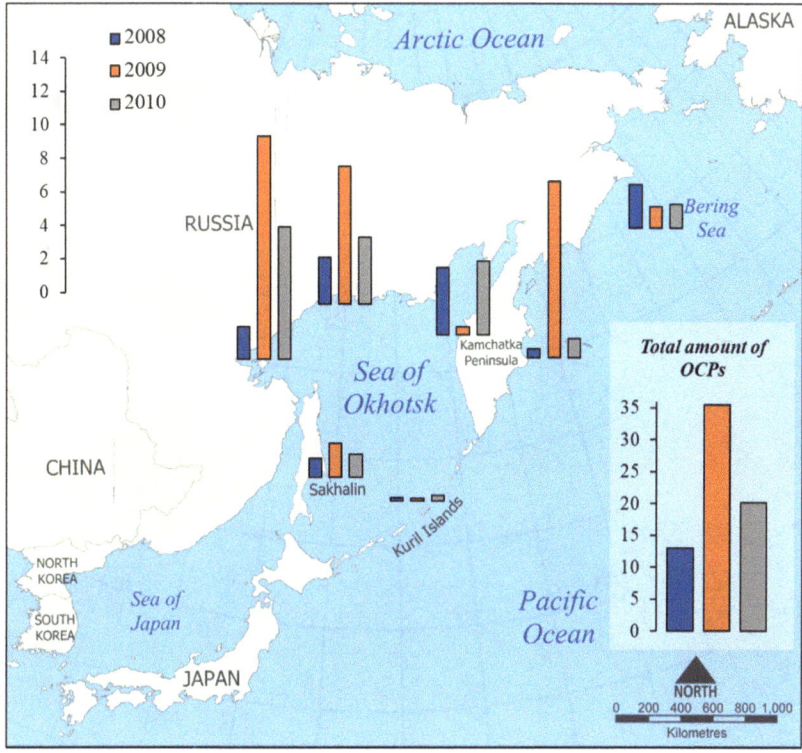

Fig. 8.1 Distribution of pesticides (kg) transported by salmon over the major areas of spawning grounds on the Russian coast of the Far Eastern seas in 2008–2010

factors, e.g., the season, climatic features of the year, biological and physiological condition of fish, their age, availability of food items, etc. Accumulation of pollutants can also make a certain contribution to the migration behavior and spawning success.

It is the layer where atmospheric precipitation and pollutants adsorbed on mineral and organic particles can be concentrated. In the southern part, the salmon distribution range is bounded by latitudes 38°–40°N. This zone is adjacent to the Great Pacific garbage patch located roughly from 135° to 155°W and within 35°–42°N. Floating plastic debris and other solid waste brought by the waters of the North Pacific Gyre are concentrated in this area. The photodegradation of plastic causes organochlorine, polycyclic aromatic compounds, and estrogen-like substances to be released into the water, where these are adsorbed on suspended particles in the epipelagic zone and can be ingested and accumulated by salmon [13–15].

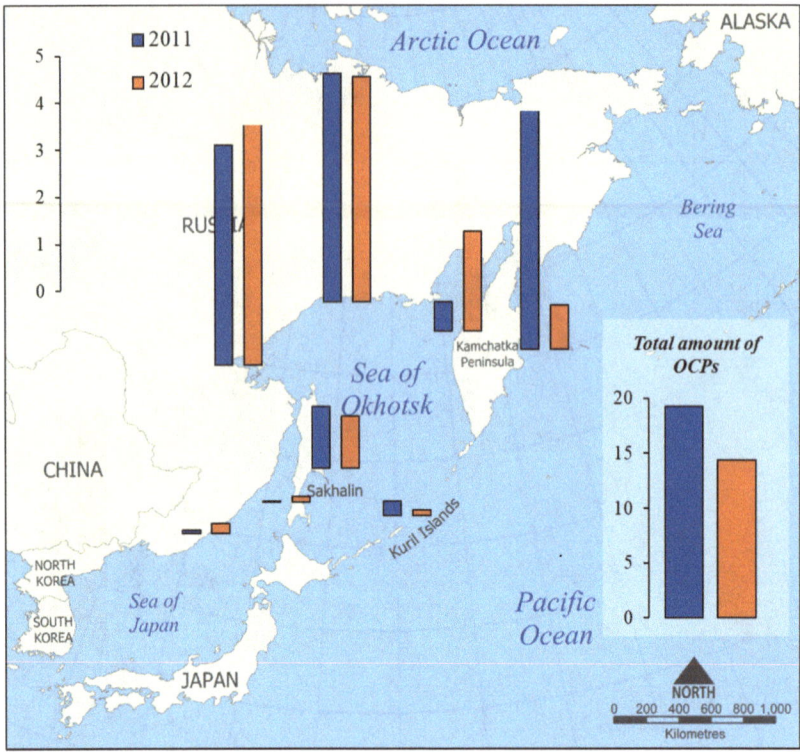

Fig. 8.2 Distribution of pesticides (kg) transported by salmon over the major areas of spawning grounds on the Russian coast of the Far Eastern seas in 2011–2012

8.3 Biotransport of POPs in 2018

In 2018, the biomass of salmon that arrived to spawn in the Far Eastern fishery basin amounted to 677 200 t. The distribution of salmon over the regions is shown in Table 8.7. A feature of that year was the abundant salmon runs to western Kamchatka (Sea of Okhotsk) and eastern Kamchatka coasts (northwestern Pacific Ocean). High runs were also recorded from Sakhalin and the continental coast of the Sea of Okhotsk (Magadan Oblast). According to publications in local mass media, another feature of 2018 was the absence of salmon at spawning grounds in the Amur River: with a normal density of 50 chum salmon per 100 m, the actual value ranged from 0.07 to 0.12 per 100 m.

During spawning migrations, salmon are partially caught by fisheries and partially reach their spawning grounds in rivers and lakes. These successful individuals carry out the transport of nutrients to low-food spawning grounds and, thus, link oceanic and terrestrial ecosystems. Salmon carcasses left after spawning serve as food for bears and other animals of the coastal zone and, after being broken down by decomposers, eventually enter biogeochemical cycles along the coast.

Table 8.7 Occupancy of salmon spawning grounds (thousand fish) on the Russian coast of the Far Eastern seas in 2018

Region	Sockeye salmon	Chum salmon	Pink salmon
Western Bering sea	450	2700	1000
Eastern Kamchatka	686.1	696	41 396
Kuril Islands	–	307.5	1955.5
Western Kamchatka	2415	1184	112 362
Sea of Okhotsk (Magadan Oblast)	–	1430	8071
Eastern Sakhalin	–	218.3	10 287.9
Primorsky Krai	–	80	550
Southwestern Sakhalin	–	294.5	10.6

– No data

Salmon are known to accumulate pollutants, including POPs such as OCPs and PCBs, along with nutrients in their organs. Intake of these substances from the marine environment occurs during salmon's migrations in open ocean waters where POPs are precipitated to the surface after being transported by winds from the regions of use in subtropical and tropical latitudes [3, 4, 16]. POPs can build up in the fish's organs and exert various negative biological effects, disrupting major metabolic processes and reducing the breeding success [5].

In 2018, pesticides and PCBs were detected only in salmon collected near the estuary of the Kamchatka River that empties on the eastern coast of the Kamchatka Peninsula into the northwestern Pacific Ocean. According to our data, the average concentration of OCPs in sockeye salmon muscles was 10.1 ng/g w.w.; in chum salmon muscles, 2.5 ng/g; in pink salmon muscles, 4.0 ng/g. One sockeye salmon weighs an average of 2.2 kg; one chum salmon, 2.8 kg; and one pink salmon, 1.2 kg (according to the NPAFC report 2018). Consequently, the OCP amount in one sockeye salmon was estimated at 22.2 µg; one chum salmon, 7 µg; one pink salmon, 4.8 µg.

Thus, the total amount of OCPs transported by sockeye salmon to the Russian coast in 2018 (Table 8.8) was 0.08 kg; by chum salmon, 0.05 kg; by pink salmon 0.084 kg. These three fish species together brought almost 1 kg of OCPs. Eastern and western Kamchatka received the largest amounts.

Calculated based on the total OCP level in fish for the years under consideration, the transport in 2008 amounted to 13 kg; in 2009, 35.5 kg; in 2010, 20 kg; in 2011, 19 kg; and in 2012, 14 kg [17]. In 2018, the transport reached only 1 kg, and the total OCP level in fish muscles was significantly lower than in previous years. This gives reason to conclude that the reduction in the OCP transport by salmon indicates the general tendency of the *pesticide* background to decrease across the globe and in the Pacific Ocean in particular due to the restrictions imposed on the production and use of pesticides in most countries in accordance with the resolutions of the Stockholm Convention [18].

Table 8.8 Total amount of pesticides (kg) transported by salmon to the Russian coast of the Far Eastern seas in 2018

Region	Sockeye salmon	Chum salmon	Pink salmon	Total
Western Bering sea	0.010	0.019	0.005	0.034
Eastern Kamchatka	0.015	0.005	0.199	0.219
Kuril Islands	–	0.002	0.009	0.012
Western Kamchatka	0.054	0.008	0.539	0.601
Sea of Okhotsk (Magadan Oblast)	–	0.010	0.039	0.049
Eastern Sakhalin	–	0.002	0.049	0.051
Primorsky Krai	–	0.001	0.003	0.003
Southwestern Sakhalin	–	0.002	0.000	0.002
Total	0.08	0.05	0.84	0.97

– No data

Unlike OCPs, polychlorinated biphenyls continue to be applied in some countries and accumulate in the biota. The continuous influx of PCBs into the marine environment in various regions is associated with marine shipping activities. According to our data, the average amount of PCBs in salmon body is higher than that of OCPs (Table 8.9).

For sockeye salmon, the average concentration amounted to 18.4 µg/kg w.w.; for chum salmon, 9.1 µg/kg; for pink salmon, 86 µg/kg. The distribution of these compounds between species differed from that for OCPs, with the highest concentration recorded from sockeye salmon. In general, sockeye salmon brings a total 0.07 kg of PCBs to the Russian coast each year; chum salmon, 0.06 kg; pink salmon, 15.1 kg.

Table 8.9 Total amount of PCBs (kg) transported by salmon to the Russian coast of the Far Eastern seas in 2018

Region	Sockeye salmon	Chum salmon	Pink salmon	Total
Western Bering sea	0.008	0.025	0.086	0.119
Eastern Kamchatka	0.013	0.006	3.560	3.579
Kuril Islands	–	0.003	0.168	0.171
Western Kamchatka	0.044	0.011	9.663	9.718
Sea of Okhotsk (Magadan Oblast)	–	0.013	0.694	0.707
Eastern Sakhalin	–	0.002	0.885	0.887
Primorsky Krai	–	0.001	0.047	0.048
Southwestern Sakhalin	–	0.003	0.001	0.004
Total	0.07	0.06	15.10	15.23

– No data

Eastern and western Kamchatka received the largest amounts. The total amount of PCBs annually transported by salmon to the Russian coast is estimated at 15.23 kg, which is 15-fold higher than the total OCP amount. We neither measured the PCB level in the organs of salmon from the Far Eastern basin nor, accordingly, considered the biotransport of these compounds before and, therefore, it is impossible to assess long-term variations in these parameters. It is also worth mentioning that in the landlocked form of sockeye salmon (kokanee) from the Tolmachevo Reservoir, Kamchatka Peninsula, the PCB concentration in muscles ranged from 23 to 43 ng/g l.w., which also confirms the accumulation of PCBs by salmon in this region [19].

Our results show (Fig. 8.3) that presence of OCPs and PCBs in the marine environment inevitably leads to their accumulation in the biota. Coastal water pollution has decreased significantly in recent years (as evidenced by the measurements of pollutants in salmon in 2010 and 2018) due to the ban on the use of POPs. However, this process is much less pronounced in open ocean waters [20]. The oceans still remain the terminal reservoirs that receive persistent toxicants and, therefore, the bioaccumulation of POPs up the food chains in them continues. Salmon are considered a convenient object for monitoring the POP cycles in the biosphere. Organic pollutants appear to be firmly integrated in the directional transport of nutrients by salmon that links the oceanic and terrestrial ecosystems. The annual influx of POPs to spawning grounds and the steady increase in their concentrations in local areas potentially pose an environmental risk to certain populations whose spawning success may be reduced due to the toxicity of the habitat.

8.4 Biotransport of Heavy Metals in 2018–2021

To determine levels of heavy metals in organs and tissues of anadromous pink and chum salmon, we collected and processed samples from several rivers of the Sakhalin–Kuril region such as the Firsovka (Gulf of Patience, Sea of Okhotsk) and the Reidovaya (Prostor Bay, Sea of Okhotsk), from Aniva Bay, Sea of Okhotsk, and also from the southwestern Sakhalin coast, Sea of Japan.

We assessed the involvement of the major commercial species of Pacific salmon in the biotransport of various trace elements from oceanic ecosystems to coastal regions in relationship with the biological and ecological features of fish.

Unlike pink salmon that spend only one winter at sea before returning to the natal rivers, from where they previously migrated to the sea being smolts, chum salmon return to the rivers of the Sakhalin–Kuril basin at age 2+, 3+, 4+, and 5+ yr. According to official data by the Sakhalin branch of VNIRO, chum salmon from the Reidovaya River return to spawn mostly (54.5% of all fish in spawning run) at age 3+ yr. According to our data, 80% of examined chum salmon spawners from Primorsky Krai are represented by individuals aged 5+, i.e. the feeding grounds of Sea of Japan chum salmon are, apparently, less rich in food (according to the Primorsky branch of Glavrybvod, 53% of mature individuals return at age 4+ and 44% at age 3+ yr). We estimated the age of chum salmon from the Firsovka River at 3+ yr. As is known,

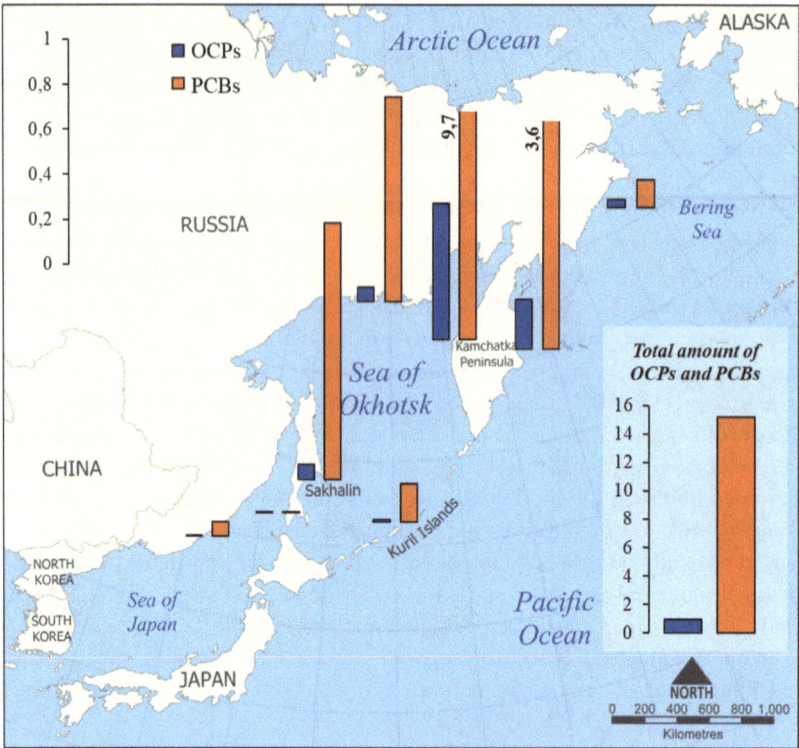

Fig. 8.3 Distribution of OCPs and PCBs (kg) transported by salmon over the major areas of spawning grounds on the Russian coast of the Far Eastern seas in 2018

individuals of older ages or larger, having much greater weight than individuals of younger ages within the same species show higher concentrations of trace elements. It is likely that the greater age (5+ yr) and, consequently, the longer feeding season, and also the greater body weight of Sea of Japan chum salmon than those in Sakhalin and Iturup chum salmon explain the higher concentrations of trace elements in them.

The concentrations of metals in the organs and tissues of pink and chum salmon in 2018 and 2019 varied greatly. Pre-spawning fish from river estuaries on the eastern coast of Sakhalin were distinguished by a significantly higher lead content in tissues and particularly in the liver. Concentrations of this toxic element in the eastern Sakhalin and Kuril chum salmon ranged from 0.875 µg/g w.w. (in the liver of fish from the Reidovaya River, Iturup Island) to 1.176 µg/g w.w. (in the liver of fish from the Firsovka River, Gulf of Patience, Sea of Okhotsk). Pink salmon from the Firsovka River also had a high level of Pb in the liver (0.963 µg/g w.w.) despite their smaller body size and shorter time spent at sea while feeding. In some cases, the levels of lead and cadmium approached the safety standards adopted in the Russian Federation (1.0 mg/g w.w. for Pb and 0.2 mg/g w.w. for Cd) [21].

High concentrations of lead, one of the toxic elements whose level in foods is strictly regulated by the State's regulatory documents (Sanitary Rules and Norms, SanPiN), are always observed in the organs and tissues of fish that migrate across the biogeochemical province of the Kuril–Kamchatka region once or more times during their life history. In this region, the high activity of volcanoes in Kamchatka, Kuril and Japan islands, as well as the more southerly located island arcs in the West Pacific, exerts a powerful geochemical effect on marine ecosystems. The undersea and above-surface volcanism and the Kuril–Kamchatka Trench are major suppliers of chemical elements in the environment that bring nutrients and other elements to the surface via upwelling events, thus, forming impact geochemical zones in the northwestern Pacific Ocean [22–25]. In earlier publications, we estimated and described the levels of Pb and Cd in pink and chum salmon that migrate in the waters of the geochemical province of the Kuril–Kamchatka region at least twice in their life history [26–28].

Pink and chum salmon from Aniva Bay and bays of the Sea of Japan (the mainland coast of southern Primorsky Krai and the southwestern Sakhalin coast) are character-ized by increased concentrations of metals that are indicators of anthropogenic (Zn, Cu), technogenic (Ni), and terrigenous (Fe) inputs to the environment. The highest concentrations of Zn (60.45 µg/g w.w.) and Cu (84.03 µg/g w.w.) were found in the liver of chum salmon from the Lovetskaya River (southwestern Sakhalin coast); the highest Ni concentrations (1.81 µg/g w.w.), in eggs of pink salmon from the Sea of Japan (southwestern Sakhalin coast). The Fe level (300.08 µg/g w.w.) was also the highest in the liver of chum salmon from the Lovetskaya River (Table 8.10).

Number of salmon that escaped to spawn in certain regions of the Far East in 2018, 2019, and 2021 are provided in Tables 8.11 and 8.12.

Pink and chum salmon running from the sea to watercourses for spawning each year bring organic matter containing heavy metals from the ocean to the land as part of biotransport. After spawning, fish die at their spawning grounds serving as food for many organisms. Some portion of trace elements contained in their carcasses will be passed down to the offspring, some will remain in the soil, and some will be involved in the food chain and transferred to the upper levels. Data on occupancy of salmon spawning grounds in the Russian part of the North Pacific in 2019 and 2021 are available in free-access reports of the North Pacific Anadromous Fish Commission (NPAFC).

Studies of the water areas where heavy metals may enter pink salmon body contribute to our knowledge of the migration routes and feeding grounds of this species. For this purpose, we identified the geochemical features of the feeding grounds, dietary preferences, and duration of the marine life-history period of several Pacific salmon species in relationship with the trace elements accumulated.

As known, aquatic systems are a kind of depot for all types of pollutants, both on a regional scale and a global one. Extracted from the bowels of the Earth and enriched in technological cycles, many elements in the environment form technogenic biogeochemical provinces.

Kuril and eastern Sakhalin chum salmon that overwinter in the ocean and feed in the food-rich zone of the Kuril Islands build up high levels of lead and cadmium in their organs and tissues. This is explained, on the one hand, by nutrients supplied

Table 8.10 Average concentrations of trace elements in the organs and tissues of pink and chum salmon in 2018 and 2019, µg/g w.w.

Organs and tissues	Zn	Cu	Ni	Cd	Pb	Fe	Mn
Pink salmon, Gulf of patience, Sakhalin Island, Firsovka River, 2018							
Muscles	1.925	0.576	0.369	0.056	0.754	5.38	
Liver	3.276	0.588	0.344	0.694	0.963	57.68	
Male gonads	1.874	0.397	0.288	0.054	0.64	7.33	
Eggs	2.133	0.476	0.209	0.044	0.512	13.47	
Pink salmon, southwestern Sakhalin coast, Nevelsk, 2019							
Muscles	6.23	0.56	0.99	0.005	0.113	3.73	0.25
Liver	35.8	44.73	1.00	0.104	0.191	46.83	0.78
Eggs	28.22	12.84	1.81	0.02	0.184	41.43	0.96
Male gonads	23.32	0.83	0.93	0.018	0.195	22.97	0.36
Pink salmon, Sakhalin Island, Aniva Bay, 2019							
Muscles	6.45	0.21	0.83	0.008	0.146	4.26	0.2
Liver	37.29	44.2	0.77	0.571	0.193	84.66	1.18
Eggs	24.55	5.39	1.2	0.006	0.246	22.07	0.81
Male gonads	16.57	0.29	0.71	0.016	0.237	9.79	0.43
Chum salmon, Iturup Island, Reidovaya River, 2018							
Muscles	1.741	0.489	0.256	0.055	0.448	11.43	
Liver	3.353	0.553	0.309	0.723	0.875	48.09	
Male gonads	2.184	0.377	0.229	0.093	0.642	10.42	
Eggs	2.603	0.464	0.198	0.075	0.573	16.18	
Chum salmon, Gulf of patience, Sakhalin Island, Firsovka River, 2018							

(continued)

Table 8.10 (continued)

Muscles	1.656	0.524	0.288	0.057	0.397	7.59
Liver	3.48	0.541	0.228	0.659	1.176	60.24
Male gonads	1.742	0.389	0.195	0.12	0.481	13.04
Eggs	1.923	0.327	0.258	0.036	0.453	15.57
Chum salmon, southwestern Sakhalin coast, Lovetskaya River, 2018						
Muscles	4.22	0.39	0.55	0.006	0.105	5.01	0.3
Liver	60.45	84.03	0.71	0.855	0.155	300.07	1.66
Male gonads	7.82	0.7	1.02	0.031	0.304	14.7	0.28
Eggs	18.99	5.49	0.57	0.006	0.144	16.15	0.62
Chum salmon, southern Primorsky Krai, Poima River, 2019						
Muscles	5.206	0.76	0.476			10.54
Liver	33.463	5.49	0.511			121.9
Male gonads	7.169	0.66	0.975			20.02
Eggs	20.672	5.681	0.7498			23.38

Table 8.11 Number of pink salmon and chum salmon that arrived at spawning grounds in the regions of the Russian Far East in 2018, thousand fish (NPAFC data)

Region	Pink salmon	Chum salmon
Western Bering sea	1000	2700
Eastern Kamchatka	41 396	696
Kuril Islands	1955.5	307.5
Western Kamchatka	112 362	1184
Sea of Okhotsk (Magadan Oblast)	8071	1430
Eastern Sakhalin	10 287.9	218.3
Primorsky Krai	550	80
Southwestern Sakhalin	10.6	294.5
Total	175 633.0	6910.3

Table 8.12 Number of pink and chum salmon that arrived at spawning grounds in the regions of the Russian Far East in 2019 and 2021, thousand fish (NPAFC data)

Region	Pink salmon		Chum salmon	
	2019	2021	2019	2021
Bering sea (western part)	3052.4	9000.0	930.6	1450.000
Eastern Kamchatka coast	202 275.0	82 737.6	5413.4	1595.710
Kuril Islands	14 139.8	1555.3	9157.7	245.120
Western Kamchatka coast	41 384.8	69 886.5	6155.1	563.050
Sakhalin	5111.4	10 970.0	6070.8	184.960
Amur River basin	4.2	323.0	3245.1	3723.000
Primorsky Krai	27.4	528.0	74.8	141.400
Southwestern Sakhalin	0.3	22.1	417.0	15.6
Total	274 272.7	176 705.3	36 660.6	12 024.800

by volcanism and upwelling that rise to the surface and cause rapid proliferation of plankton and, on the other hand, by the geochemical impact potential of the region associated, in particular, with the high sorption of these elements on food particles suspended in the water [29].

The Sea of Japan is an enclosed body of water that has poor water exchange with the ocean and is subject to terrigenous sediment input and anthropogenic and technogenic pressures. Chum salmon accumulate iron, zinc, copper, and nickel in their organs during overwintering and feeding at sea. All these elements are tracers of the above-mentioned impacts that depend on the high shipping activity, household wastewater discharges, and the *tight* position of the water body between the continent and Sakhalin Island [27].

We investigated and described the heavy metal content of the organs and tissues of pink salmon, which actively move in the surface layer of the ocean and feed in marine waters for only one year, by analyzing fish from the Gulf of Patience and Prostor Bay, Sea of Okhotsk [26]. In all cases, we recorded increased levels of lead, cadmium, and mercury from muscles, gonads, and liver of pink salmon that had passed through the Kuril straits once or twice and stayed for some time in food-rich waters off the Kuril Islands. By comparing the levels of trace elements in Sea of Okhotsk fish with values for pink salmon from Sea of Japan waters, we found significant differences ($p \leq 0.05$).

The Zn level was higher in Sea of Japan pink salmon, which was likely a result of the anthropogenic pressure on the almost enclosed sea. The Cd and Pb concentrations in pink salmon from waters off the Kuril Islands and especially from the Sakhalin–Kuril waters was 40–60-fold higher than the values for fish from the Sea of Japan [30].

The feeding ground in the Sea of Japan is quite extensive: it stretches from the Korean coast to the northern boundary of the warm Tsushima Current. From this area, pink salmon migrate to the rivers of Primorsky Krai, western Sakhalin, the Amur River basin, eastern Sakhalin, the southern Kuril Islands, and rivers of Aniva Bay [10, 11]. Fish running to spawn in the rivers of southwestern Sakhalin and Aniva Bay in June and July belong to the summer-run group from the Sea of Japan and, therefore, the trace element composition of their organs and tissues will differ significantly from the mineral composition of the late-run, autumn group of Sakhalin–Kuril pink salmon that passed across the impact Kuril region at least twice in their life history. The Sea of Japan summer-run pink salmon group differs from the Sakhalin–Kuril autumn-run group in the level of trace elements in the organs and tissues. The high concentrations of Zn and Cu in pink salmon from the Sea of Japan and Pb in chum salmon from the Kuril waters are caused by natural geochemical conditions of the environment. The ecological situation in the Sea of Japan, with its waters enriched in zinc, copper, and nickel, results from human activity, whereas the trace element composition of waters along the Kuril–Kamchatka ridge is formed by effects of natural factors: modern volcanism, post-volcanism, and upwelling events [27].

The western Sakhalin coast, unlike the southeastern coast and waters off the Kuril Islands, is characterized by pronounced anthropogenic pressure from high shipping activity and due to the almost enclosed system of the Sea of Japan. The situation here is similar to that observed along the mainland coast of the sea. Since the heavy metal content of the organs and tissues of Sea of Japan pink salmon has increased multifold compared to the values recorded 20 years before, regular monitoring of the environment becomes increasingly important. Pacific salmon, whose feeding grounds and migration routes are confined to certain North Pacific seas, can also be used as bioindicators for this (Table 8.10).

The total amount of essential (Zn, Cu, and Fe) elements transported by pink and chum salmon to the Pacific coast of Russia in 2018 alone reached approximately 6887 kg. Of these, the eastern Kamchatka coast received about 1426 kg; the western Kamchatka coast, about 3722 kg; and the mainland coast of the Sea of Okhotsk, more than 551 kg. In 2019, the two salmon species transported about 18 400 kg of zinc, copper, and iron to the eastern Kamchatka coast, about 3700 kg to the western Kamchatka coast, and 3245 kg to the Kuril Islands. The total amount of essential metals transported by salmon to the Russian coast of the Northwest Pacific in 2019 was approximately 31.5 t. In 2021, in addition to the eastern and western coasts of Kamchatka (7.4 and 6.0 t, respectively), salmon brought the largest amount of metals to Sakhalin Island, more than 1.8 t.

The largest amount of nickel (which is evidence of human's impact on the habitat) biotransported by pink and chum salmon was recorded from the eastern Kamchatka coast in 2019, more than 280 kg (Tables 8.13, 8.14 and 8.15).

As the data on the biotransport of toxic elements in total catch show, pink and chum salmon together transported the largest total amounts of Cd and Pb in the territorial waters of the Sea of Okhotsk, to the western and eastern Kamchatka coasts, Sakhalin Island, and the southern Kuril Islands. Such a distribution can be explained by the impact geochemical conditions characteristic of the Sea of Okhotsk. These

Table 8.13 Amounts of heavy metals (kg) transported by pink and chum salmon to the Russian coast of the Far Eastern seas in 2018

Region	Zn	Cu	Ni	Cd	Pb	Fe	Mn
Pink salmon							
Western Bering sea	3.0	0.7	0.4	0.28	0.93	27.3	–*
Eastern Kamchatka	123.9	27.4	16.3	11.4	38.6	1128.2	–
Kuril Islands	5.9	1.3	0.8	0.54	1.8	53.3	–
Western Kamchatka	336.3	74.4	44.2	31.0	104.8	3062.4	–
Sea of Okhotsk (Magadan Oblast)	24.2	5.3	3.2	2.2	7.5	220.0	–
Eastern Sakhalin	30.8	6.8	4.0	2.8	9.6	280.4	–
Primorsky Krai	1.6	0.36	0.22	0.15	0.51	15.0	–
Southwestern Sakhalin	0.032	0.007	0.004	0.003	0.01	0.29	–
Total	525.60	116.3	69.1	48.4	163.8	4786.8	–
Chum salmon							
Western Bering sea	86.8	74.2	3.8	2.1	4.5	408.3	6.8
Eastern Kamchatka	22.4	19.1	1.0	0.55	1.2	105.3	1.7
Kuril Islands	9.9	8.5	0.43	0.24	0.52	46.5	0.77
Western Kamchatka	38.0	32.6	1.7	0.94	2.0	179.1	3.0
Sea of Okhotsk (Magadan Oblast)	45.9	39.3	2.0	1.1	2.4	216.3	3.6
Eastern Sakhalin	7.0	6.0	0.31	0.17	0.37	33.0	0.55
Primorsky Krai	2.6	2.2	0.11	0.06	0.13	12.1	0.20
Southwestern Sakhalin	9.5	8.1	0.41	0.23	0.49	44.5	0.74
Total	222.0	190.0	9.7	5.5	11.6	1045.0	17.3

–* No data

Table 8.14 Amounts of heavy metals (kg) transported by pink and chum salmon to the Russian coast of the Far Eastern seas in 2019

Region	Zn	Cu	Ni	Cd	Pb	Fe	Mn
Pink salmon							
Bering sea (western part)	88.5	54.1	4.1	0.371	0.746	116.9	2.5
Eastern Kamchatka coast	5864.9	3584.4	270.8	24.6	49.5	7748.7	163.4
Kuril Islands	410.0	250.6	18.9	1.72	3.5	541.7	11.4
Western Kamchatka coast	1199.9	733.4	55.4	5.03	10.1	1585.4	33.4
Sakhalin	148.2	90.6	6.84	0.62	1.3	195.8	4.1
Amur river basin	0.12	0.07	0.01	0.001	0.001	0.16	0.003
Primorsky Krai	0.80	0.49	0.04	0.003	0.007	1.1	0.02
Southwestern Sakhalin	0.007	0.004	0.0003	0.00003	0.0001	0.01	0.0002
Total	7712.5	4713.6	356.2	32.3	65.1	10 189.7	214.8
Chum salmon							
Bering sea (western part)	54.2	10.3	2.2	–*	–	143.2	–
Eastern Kamchatka coast	315.0	59.6	12.8	–	–	832.9	–
Kuril Islands	532.9	100.9	21.7	–	–	1409.0	–
Western Kamchatka coast	358.2	67.8	14.6	–	–	947.0	–
Sakhalin	353.3	66.9	14.4	–	–	934.1	–
Amur river basin	188.9	35.8	7.7	–	–	499.3	–
Primorsky Krai	4.4	0.82	0.18	–	–	11.5	–
Southwestern Sakhalin	24.3	4.6	1.0	–	–	64.2	–
Total	1831.1	346.6	74.7	–	–	4841.1	–

–* No data

Table 8.15 Amounts of heavy metals (kg) transported by pink and chum salmon to the Russian coast of the Far Eastern seas in 2021

Region	Zn	Cu	Ni	Cd	Pb	Fe	Mn
Pink salmon							
Bering sea (western part)	261.0	159.5	12.1	1.1	2.2	344.8	7.3
Eastern Kamchatka coast	2399.0	1466.2	110.8	10.1	20.2	3169.5	66.8
Kuril Islands	45.1	27.6	2.1	0.19	0.38	59.6	1.3
Western Kamchatka coast	2026.4	1238.4	93.6	8.5	17.1	2677.2	56.4
Sakhalin	318.1	194.4	14.7	1.3	2.7	420.2	8.9
Amur river basin	9.4	5.7	0.4	0.04	0.08	12.4	0.26
Primorsky Krai	15.3	9.4	0.71	0.06	0.13	20.2	0.43
Southwestern Sakhalin	0.64	0.39	0.030	0.003	0.01	0.85	0.02
Total	5074.8	3101.5	234.4	21.3	42.8	6704.7	141.4
Chum salmon							
Bering sea (western part)	84.4	16.0	3.4	–*	–	223.1	–
Eastern Kamchatka coast	92.9	17.6	3.8	–	–	245.5	–
Kuril Islands	14.3	2.7	0.58	–	–	37.7	–
Western Kamchatka coast	32.8	6.2	1.3	–	–	86.6	–
Sakhalin	10.8	2.0	0.44	–	–	28.5	–
Amur river basin	216.7	41.0	8.8	–	–	572.8	–
Primorsky Krai	8.2	1.6	0.34	–	–	21.8	–
Southwestern Sakhalin	0.91	0.17	0.037	–	–	2.4	–
Total	460.8	87.2	18.8	–	–	1218.4	–

–* No data

geochemical provinces are created by undersea and above-water volcanism, post-volcanism of the Kuril Islands, and also by upwelling events that raise a large set of chemical elements from the abyss of the Kuril–Kamchatka Trench to the surface layer [26]. The water of the Amur River brings large amounts of pollutants drained from the mainland coast into the Sea of Okhotsk. Furthermore, the shipping activity in the river is high, and oil production is carried out in the shelf zone. All these factors may have negative effects on the quantitative trace element composition of Pacific salmon.

In 2018, we recorded the largest amounts of heavy metals transported to land by salmon: the total weight of Pb reached values higher than 170 kg, and the weight of

Cd higher than 50 kg (Fig. 8.4). This was due to the unexpectedly abundant run of salmon (especially pink salmon) spawners to Russian coasts that year. In 2018, more than 500 000 t of salmon came to the Kamchatka coasts. Fishermen could not fully take up the quotas and anyhow reduce the biomass of spawning runs, and a huge number of spawners entered the rivers [31].

In 2018, the largest amounts of toxic trace elements in the Bering Sea waters were recorded from off the western Kamchatka coast: Pb, about 106 kg; Cd, more than 30 kg. The eastern coast of Kamchatka in 2018 received about 40 kg of lead and about 12 kg of cadmium (Fig. 8.4).

In 2019, salmon transported about 10 kg of lead and about 5 kg of cadmium to the western Kamchatka coast; the eastern coast received almost five-fold more toxic metals, about 50 and 25 kg, respectively. In 2019, biotransport was responsible for a total of 65 kg of lead and 32 kg of cadmium brought to the Russian coast (Fig. 8.5).

The largest amounts of toxic elements in 2021 were recorded from the Bering Sea coast of Kamchatka: Pb, more than 20 kg; Cd, about 10 kg. On the Sea of Okhotsk coast of Kamchatka, the amounts were slightly lower: about 17 kg of Pb and more than 7 kg of Cd (Fig. 8.6). Such a distribution can be explained by the

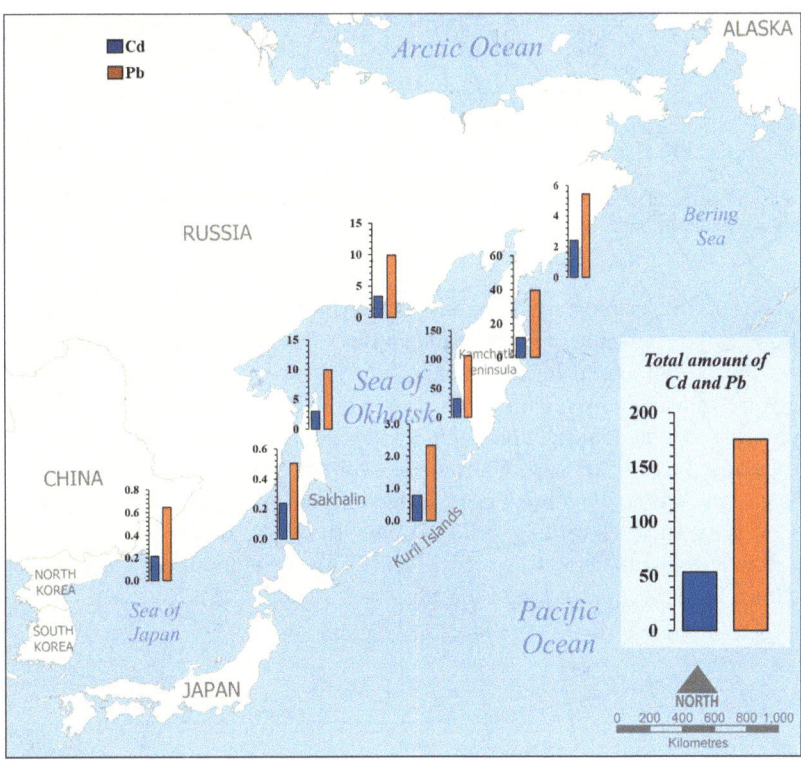

Fig. 8.4 Amounts of toxic heavy metals transported by salmon to the Russian coast of the Far Eastern seas in 2018, kg

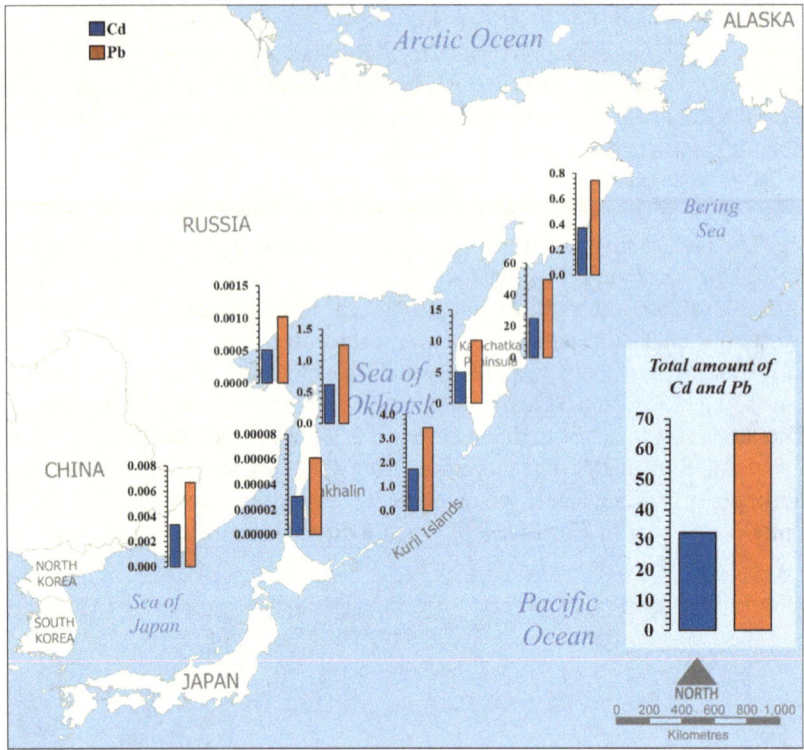

Fig. 8.5 Amounts of toxic heavy metals transported by salmon to the Russian coast of the Far Eastern seas in 2019, kg

anthropogenic pressure on some zones of the Bering Sea, in particular Ugolnaya Bay, the Kamchatka Peninsula shelf (Kamchatka Bay). The trace element composition of waters can be influenced by undersea volcanism of, e.g., Piip, one of the largest submarine volcanoes. We cannot but mention here the coal mines near the town of Anadyr, where industrial and household wastewater is the major source of pollution. In some years, this water carried discharged oil and petroleum products, sulfurous and hydrogen sulfide-containing gases, mineralized produced water and wastewater from oil fields and well drilling, drilling fluids, etc. to the Anadyr Estuary.

In 2021, a total of almost 43 kg of lead and 21 kg of cadmium were biotransported by salmon to the Russian coast of the Northwest Pacific (Fig. 8.6).

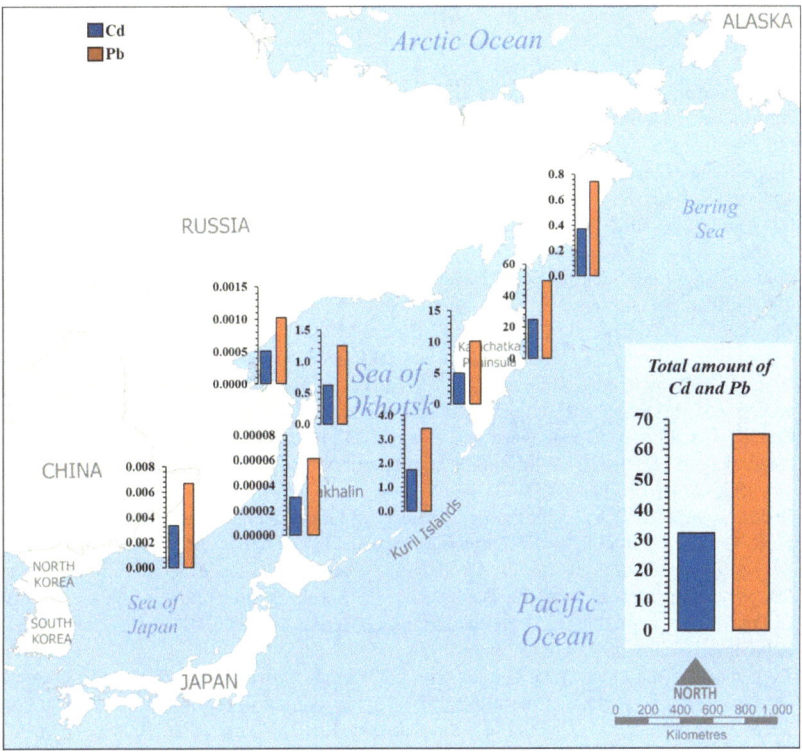

Fig. 8.6 Amounts of toxic heavy metals transported by salmon to the Russian coast of the Far Eastern seas in 2021, kg

8.5 Conclusion

Thus, Pacific salmon of the northwestern Pacific Ocean, as a biotransport vector, bring substantial amounts of POPs and heavy metals. This is a fundamental natural process that requires in-depth study and analysis. Based on the long-term trends that we have identified, one can observe a significant reduction in POPs transported from oceanic ecosystems to continental ones (probably as a consequence of the Stockholm Convention). However, heavy metals enter this region not only from anthropogenic sources but also from natural ones such as undersea volcanism and upwelling. Therefore, further study of biotransport by migratory species of marine organisms is highly relevant.

Acknowledgements The work was supported by the Russian Science Foundation (no. 23-74-10032)

References

1. Cederholm CJ, Kunze MD, Murota T, Sibatani A (1999) Pacific Salmon carcasses: essential contributions of nutrients and energy for aquatic and terrestrial ecosystems. Fisheries 24:6–15. https://doi.org/10.1577/1548-8446(1999)024%3c0006:PSC%3e2.0.CO;2
2. Helfield JM, Naiman RJ (2001) Effects of Salmon-derived nitrogen on Riparian forest growth and implications for stream productivity. Ecology 82:2403–2409. https://doi.org/10.1890/0012-9658(2001)082[2403:EOSDNO]2.0.CO;2
3. Ewald G, Larsson P, Linge H et al (1998) Biotransport of organic pollutants to an Inland Alaska Lake by migrating Sockeye Salmon (*Oncorhynchus nerka*). Arctic 51:40–47. https://doi.org/10.14430/arctic1043
4. Krümmel EM, Macdonald RW, Kimpe LE et al (2003) Delivery of pollutants by spawning salmon. Nature 425:255–256. https://doi.org/10.1038/425255a
5. Wong MH, Armour M-A, Naidu R, Man M (2012) Persistent toxic substances: sources, fates and effects. Rev Environ Health 27. https://doi.org/10.1515/reveh-2012-0040
6. Devotta DA, Fraterrigo JM, Walsh PB et al (2021) Watershed Alnus cover alters N: P stoichiometry and intensifies P limitation in subarctic streams. Biogeochemistry 153:155–176. https://doi.org/10.1007/s10533-021-00776-w
7. Ermakov VV (2017) A.P. Vinogradov's concept of biogeochemical provinces and its development. Geochem Int 55:872–886. https://doi.org/10.1134/S0016702917100044
8. Khristoforova NK, Litvinenko AV, Kovalchuk MV et al (2023) Trace elements in Masu Salmon Oncorhynchus Masou from the Bakhura River, Southeastern Sakhalin Island, the Sea of Okhotsk. In: Chaplina T (ed) Processes in GeoMedia—Volume VI. Springer International Publishing, Cham, pp 459–466
9. Tsygankov VYu, Donets MM, Gumovskiy AN, Khristoforova NK (2022) Temporal trends of persistent organic pollutants biotransport by Pacific salmon in the Northwest Pacific (2008–2018). Mar Pollut Bull 185:114256. https://doi.org/10.1016/j.marpolbul.2022.114256
10. Shuntov VP, Temnykh OS (2008) Pacific salmon in marine and ocean ecosystems. TINRO Center, Vladivostok, Russia
11. Shuntov VP, Temnykh OS (2011) Pacific salmon in marine and ocean ecosystems. TINRO-Center, Vladivostok, Russia
12. Veldhoen N, Ikonomou MG, Dubetz C et al (2010) Gene expression profiling and environmental contaminant assessment of migrating Pacific salmon in the Fraser River watershed of British Columbia. Aquat Toxicol 97:212–225. https://doi.org/10.1016/j.aquatox.2009.09.009
13. Moore CJ, Moore SL, Leecaster MK, Weisberg SB (2001) A comparison of plastic and plankton in the north Pacific central gyre. Mar Pollut Bull 42:1297–1300
14. Derraik JGB (2002) The pollution of the marine environment by plastic debris: a review. Mar Pollut Bull 44:842–852
15. Choy C, Drazen J (2013) Plastic for dinner? Observations of frequent debris ingestion by pelagic predatory fishes from the central North Pacific. Mar Ecol Prog Ser 485:155–163. https://doi.org/10.3354/meps10342
16. Lukyanova ON, Tsygankov VYu, Boyarova MD, Khristoforova NK (2014) Pesticide biotransport by Pacific salmon in the northwestern Pacific Ocean. Doklady Biol Sci 456:188–190. https://doi.org/10.1134/S0012496614030089
17. Lukyanova ON, Tsygankov VYu, Boyarova MD, Khristoforova NK (2015) Pacific salmon as a vector in the trasnsfer of persistent organic pollutants in the Ocean. J Chthyol 55:425–429. https://doi.org/10.1134/S0032945215030078
18. UNEP (2020) Stockholm convention on persistent organic pollutants
19. Mamontova EA, Lepskaya EV, Tarasova EN et al (2018) Organochlorine pesticides and polychlorinated biphenyls in tissues of landlocked Kokanee Salmon from Tolmachevskoye Reservoir, Kamchatka. Inland Water Biol 11:219–226. https://doi.org/10.1134/S1995082918020128

20. Tanabe S (2007) Chapter 18 Contamination by persistent toxic substances in the Asia-Pacific Region. In: Li A, Tanabe S, Jiang G et al (eds) Developments in environmental science. Elsevier, pp 773–817
21. SanPin 2.3.2.1078-01 (2001) SanPin 2.3.2.1078-01. Hygienic requirements of safety and nutritional value of food products
22. Propp MV, Propp LN (1988) Hydrochemical indicators and chlorophyll a content in the coastal waters of the Kuril Islands. Biologiya Morya 68–70
23. Sapozhnikov VV (1994) Integrated environmental studies of the ecosystems of the Bering and Okhotsk Seas. Oceanology 34:309–312
24. Malinovskaya TM, Khristoforova NK (1997) Characteristics of the coastal waters of the southern Kuriles on the content of trace elements in indicator organisms. Biologiya Morya 23:239–246
25. Kobzar AD, Khristoforova NK (2015) Monitoring heavy-metal pollution of the coastal waters of Amursky Bay (Sea of Japan) using the brown alga Sargassum miyabei Yendo, 1907. Russ J Mar Biol 41:384–388. https://doi.org/10.1134/S1063074015050065
26. Khristoforova NK, Litvinenko AV, Tsygankov VYu et al (2019) Trace element content in the pink Salmon Oncorhynchus gorbuscha Walbaum, 1792 from the Sakhalin–Kuril Region. Russ J Mar Biol 45:221–227. https://doi.org/10.1134/S1063074019030064
27. Khristoforova NK, Litvinenko AV, Tsygankov VYu, Kovalchuk MV (2021) Comparative characteristics of the trace elemental composition of chum salmon Oncorhynchus keta Walbaum, 1792 from the Sea of Japan and the Sea of Okhotsk. Mar Biol J 6:92–104. https://doi.org/10.21072/mbj.2021.06.4.08
28. Khristoforova NK, Alekseev MYu, Litninenko AV, Tsygankov VYu (2022) Heavy metal content in pink Salmon from the Euro-Arctic and Sakhalin-Kuril Regions, pp 476–488
29. Khristoforova NK, Tsygankov VYu, Boyarova MD, Lukyanova ON (2015) Concentrations of trace elements in Pacific and Atlantic salmon. Oceanology 55:679–685. https://doi.org/10.1134/S0001437015050057
30. Litvinenko A, Khristoforova N, Tsygankov V, Kovalchuk M (2021) Trace element composition of the Southwestern Sakhalin Chum Salmon. IOP Conf Ser Earth Environ Sci 937:022074. https://doi.org/10.1088/1755-1315/937/2/022074
31. Makoedov AN, Makoedov AA (2022) Artificial reproduction and status of stock for pacific salmon. Izvestiya TINRO 202:661–678. https://doi.org/10.26428/1606-9919-2022-202-661-678

Chapter 9
Organochlorine Pesticides and Polychlorinated Biphenyls in Fish from Lake Khanka, the Sea Okhotsk Basin

Abstract Lake Khanka is the largest freshwater body of water in Northeast Asia. Located in the agriculturally most developed regions of China and Primorsky Krai of Russia, this lake is of paramount international importance. To assess accumulation and transformation of organochlorine compounds in Lake Khanka using indicator fish species, we analyzed organs of the Prussian carp (*Carassius gibelio*), Amur carp (*Cyprinus rubrofuscus*), northern snakehead (*Channa argus*), lake skygazer (*Chanodichthys oxycephalus*), Ussuri sharpbelly (*Hemiculter lucidus*), Mongolian redfin (*Chanodichthys mongolicus*), predatory carp (*Chanodichthys erythropterus*), and pike-perch (*Sander lucioperca*). DDT and its metabolites, hexachlorocyclohexane (HCH) isomers, and various polychlorinated biphenyl (PCB) congeners were found in the fish organs under study. The organs of benthos feeders showed significant ($p \leq 0.05$) differences in DDT concentrations, while the organs of predatory fishes had differences in HCH concentrations in relationship with the size and weight characteristics and feeding specialization of fish. When comparing the results obtained with the requirements in regulatory documents, we did not find any cases where pollutants' levels exceeded the maximum permissible concentrations.

Keywords POPs · OCPs · PCBs · Lake Khanka · Sea of Okhotsk basin · Freshwater fish

9.1 Introduction

Of all pollutants that enter the environment as a result of human activity, persistent organic pollutants (POPs) [1] are the most hazardous ones. Although certain POP groups have been banned in many developed and some developing countries since the 1970s, the potential of these compounds to be spread via atmospheric transport and biotransport, as well as the biomagnification potential, contributes to their ubiquitous distribution. These substances are found in all abiotic and biotic environmental components [2–4].

Lake Khanka is the largest freshwater body of water in Northeast Asia. Its location and climate largely determine the high productivity of local ecosystems and their biological and genetic diversity [5]. Since the mid-twentieth century, the Khanka Lowland has become one of the major agricultural regions of the Russian Far East and the Heilongjiang Province of China. However, the anthropogenic pressure on this region substantially reduced as a result of the closure of many agricultural farms during the economic crisis in the 1990s. Since 2000, there has been a rise of economic activity in the areas adjacent to the lake. The main crops grown here are soybeans, corn, and rice. In addition to the agricultural pressure, the waters of the lake are also exposed to the pressure from the local human population that totals approximately 605 800 people (on the Russian and Chinese sides), as shown by the 2017 census [6]. According to the Ramsar Convention on Wetlands of International Importance Especially as Waterfowl Habitat (Ramsar Convention) and the agreement between the Governments of Russia and China, Lake Khanka has been given the status of international nature reserve. This even more increases the importance of monitoring of these superecotoxicants not only to determine probability of POP emissions into the Sea of Okhotsk basin (via the Amur River), but also to control the condition of this specially protected natural area.

Thus, we aimed to study the accumulation and transformation of organochlorine compounds (OCCs) in Lake Khanka (Sea of Okhotsk basin) using the local indicator fish species.

9.2 Physical and Geographical Characteristics of Lake Khanka

Lake Khanka is the largest freshwater lake in Northeast Asia (Fig. 9.1) and is part of the Sea of Okhotsk basin. The adjacent Khanka Lowland differs significantly from other regions of the Far East in terms of topography. Wetlands of Lake Khanka are a unique natural ecosystem, and the reed swamps of the southern and eastern banks of the lake are characterized by the flora and fauna that do not exist anywhere else in the region. The highly productive ecosystems of the Lake Khanka basin are actually a hotspot of species and population-genetic diversity, they are rich in valuable and rare plants and animals species, and are also of great landscape-forming, climatic, and aesthetic significance [7].

In 1976, Lake Khanka was included in the List of the Ramsar Convention (1971) of international importance mainly as habitats of waterfowl. It is located in the southern part of the Sanjiang Plain in the Amur Basin at the Chinese–Russian border and is the subject of economic use by the two countries. The lake is a shallow (on average, 2–3 m) water body of a loess type having an area of 4070 km^2. In summer, the transparency of the water is not greater than 15–18 cm. The altitude of the water surface is 69 m a.s.l. About two dozen rivers empty into the lake (the Muling, Jinyinku, Xiaohei, Ilistaya, Spassovka, etc.) with only one outflow, the Songacha

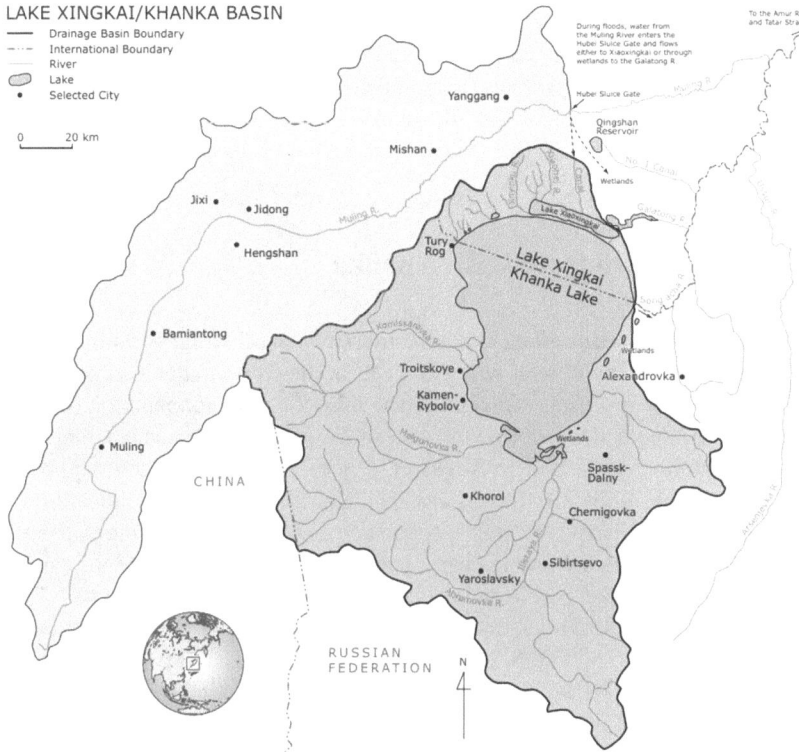

Fig. 9.1 Map of Khanka/Xingkai Lake Basin [8]

River, connecting the lake with the Amur River. There are several large district centers in the lake basin, including the city of Spassk-Dalny, where a number of mining and woodworking companies operate.

The lake is a Russian–Chinese cross-border basin-type body of water. It sensitively responds, at all levels of self-organization of ecosystems and the geosystem in general, to the processes of various geneses occurring both on the Russian and Chinese sides within a single natural system. This explains a number of features and problems in the functioning of the system and its condition that largely determine the environmental situation [8].

The Lake Khanka basin is the most economically important agricultural region, where major rice-growing areas are concentrated, both in Primorsky Krai and in China. Various other industries are also developed, and the human population concentrated here is very high, especially on the Chinese side of the basin. This exerts a pronounced pressure on the local landscapes and on the coastal and aquatic ecosystems of Lake Khanka in general [8].

We studied eight species representing the fish fauna of the lake: the Prussian carp (*Carassius gibelio* Bloch, 1782), Amur carp (*Cyprinus rubrofuscus* Lacepède, 1803), lake skygazer (*Chanodichthys oxycephalus* Bleeker, 1871), Ussuri sharpbelly

(*Hemiculter lucidus* Dybowski, 1872), Mongolian redfin (*Chanodichthys mongolicus* Basilewsky, 1855), predatory carp (*Chanodichthys erythropterus* Basilewsky, 1855), northern snakehead (*Channa argus*), and pike-perch (*Sander lucioperca*). Samples were collected in the vicinity of the village of Novoselskoye during an expedition of the Pacific Geographical Institute (PGI), FEB RAS, to Lake Khanka in 2018.

9.3 OCPs in Fish from Lake Khanka

Prussian carp (*Carassius gibelio*) is a freshwater, warm-loving, non-migratory fish living in ponds and small, well-warmed water bodies. The major part of its diet is made up of benthic organisms (mainly larval chironomids), zooplankton (mainly of the genera *Bosmina* and *Chydorus*, and also various Copepoda species), filamentous algae and diatoms (of the genera *Navicula*, *Cymbella*, *Melosira*, etc.), and parts of plants; it also ingests silt with food. It is capable of interspecific hybridization when eggs are fertilized by closely related species such as Amur carp, common bream, etc. The major competitors are crucian carp, tench, common carp, and Amur carp. The major predators are pike, perch, and eel [9].

We analyzed muscles, liver, and eggs of Prussian carp. In muscles, the concentrations of $\sum$HCH (total of α, β, γ, and δ isomers), $\sum$DDT (DDT + DDD + DDE), and $\sum$PCB (total of PCBs 28, 52, 155, 101, 118, 143, 153, 138, and 180) varied in ranges of 94–3372, 74–1038, and 15–1798 ng/g lipid weight (l.w.), respectively (Tables 9.1 and 9.2). In liver, the levels of these pollutants were found in ranges of 193–1557, 21–183, and 35–129 ng/g l.w., respectively. In eggs, the concentrations of $\sum$HCH ranged rather widely, from 6 to 7593 ng/g l.w.; $\sum$DDT and $\sum$PCB had significantly narrower ranges, 4–66 and 24–239 ng/g l.w., respectively.

A comparison of qualitative compositions of pollutants did not show any statistically significant differences. In all the organs analyzed, we found α-, β-, γ- and δ-HCH, 2,4- and 4,4-DDE (in single cases 2,4- and 4,4-DDD and endrin), and also mainly lower chlorinated PCB congeners (28 and 52). Significant ($p \leq 0.05$) differences in concentrations of pollutants were recorded only for $\sum$DDT and 2,4-DDE in the following order: *muscles > liver > eggs* (Fig. 9.2). The lack of significant differences in levels of toxicants' accumulation may indicate a long-term chronic entry of OCCs from the environment into the fish body.

Amur carp (*Cyprinus rubrofuscus*) is a freshwater fish also found in brackish waters of the Caspian and Aral seas. It is very hardy, usually living in slow-moving water bodies with abundant vegetation, but can also inhabit fast-flowing rivers with rocky or sandy sediments. Adult fish feed on benthic organisms. In the sea, the food items are larval chironomids, mollusks, crustaceans, worms, etc.; in heavily overgrown lakes, oxbow lakes, and ponds, the diet consists largely of aquatic plants (higher and algae). Fry up to 3–4 cm in length feed mainly on planktonic crustaceans. The major competitors are common bream, tench, Prussian carp, and, in some cases, eel and benthivorous whitefish. The major fish predators are catfish, pike, to a lesser extent, common perch, zander, and asp; bird predators are cormorant, heron, sea

Table 9.1 Concentrations of certain OCPs in the organs of fish analyzed, ng/g l.w. (M ± SD)

Species	Organs	HCH isomers				DDT		DDD		DDE		Dieldrin	Endrin
		α-	β-	γ-	δ-	2,4-	4,4-	2,4-	4,4-	2,4-	4,4-		
Prussian carp (*Carassius gibelio*)	Muscles	402 ± 568	149 ± 173	598 ± 1145	35 ± 24	<DL[1]	<DL	<DL	<DL	258 ± 348	301 ± 154	<DL	143[2]
	Liver	613 ± 409	163 ± 27	82 ± 71	76 ± 65	<DL	<DL	21	<DL	91 ± 63	45 ± 34	<DL	<DL
	Eggs	1208 ± 2071	348 ± 643	98 ± 196	392 ± 540	<DL	<DL	<DL	10	25 ± 23	11	<DL	263
Amur carp (*Cyprinus rubrofuscus*)	Muscles	134 ± 111	130 ± 255	19 ± 21	89 ± 97	<DL	<DL	<DL	3.4	29 ± 37	24 ± 35	<DL	<DL
	Liver	296 ± 343	39 ± 26	25 ± 19	19 ± 13	<DL	<DL	<DL	<DL	45	<DL	<DL	<DL
	Eggs	357 ± 377	<DL	7	19 ± 16	<DL	<DL	<DL	<DL	<DL	<DL	<DL	<DL
	Male gonads	48 ± 8	33	9 ± 8	11 ± 13	<DL	<DL	<DL	<DL	16	<DL	<DL	<DL
Ussuri sharpbelly (*Hemiculter lucidus*)	Muscles	<DL	<DL	9	<DL	<DL	<DL	<DL	<DL	43 ± 30	<DL	<DL	<DL
Lake skygazer (*Chanodichthys oxycephalus*)	Muscles	488 ± 656	70 ± 67	33 ± 34	144 ± 127	<DL	<DL	<DL	<DL	88 ± 90	24	<DL	<DL
	Liver	357 ± 403	53 ± 65	51 ± 44	32	<DL	<DL	<DL	<DL	283	<DL	<DL	<DL
Mongolian redfin (*Chanodichthys mongolicus*)	Muscles	225 ± 349	3.3	6.2 ± 3.3	78 ± 107	<DL	<DL	<DL	<DL	37 ± 26	<DL	<DL	<DL
	Liver	211 ± 221	4.0 ± 1.6	8.8 ± 7.7	23	<DL	<DL	<DL	10 ± 8	10	4.1	<DL	<DL
	Eggs	44	15	7.1 ± 7.3	<DL	<DL	<DL	<DL	<DL	13	<DL	<DL	<DL
	Fat	21 ± 23	2.5 ± 1.7	1.4 ± 0.6	1.2	<DL	<DL	1.5	5.8	<DL	4.0	<DL	<DL
Predatory carp (*Chanodichthys erythropterus*)	Muscles	290 ± 381	314 ± 341	402 ± 540	15	123	<DL	<DL	<DL	30 ± 34	140	<DL	<DL
	Liver	280 ± 387	17 ± 9	8.1	53	<DL	<DL	<DL	<DL	<DL	11	<DL	<DL
	Male gonads	92 ± 120	15 ± 9	15 ± 2	4.4	<DL	36	<DL	<DL	52	33	<DL	<DL

(continued)

Table 9.1 (continued)

Species	Organs	HCH isomers				DDT		DDD		DDE		Dieldrin	Endrin
		α-	β-	γ-	δ-	2,4-	4,4-	2,4-	4,4-	2,4-	4,4-		
	Fat	6.6 ± 6.9	10 ± 13	0.6	0.4 ± 0.2	< DL	< DL	2.5	3.6	0.9	42	< DL	< DL
Northern snakehead (*Channa argus*)	Muscles	443 ± 715	81 ± 64	17 ± 2		< DL	< DL	< DL	< DL	29 ± 19	137	< DL	40
	Liver	124 ± 152	24 ± 20	26 ± 14	9.0 ± 5.3	< DL	< DL	< DL	< DL	9.4 ± 11.1	12.3	< DL	< DL
Pike-perch (*Sander lucioperca*)	Muscles	428 ± 632	1823 ± 2234	570 ± 796	2578	< DL	< DL	< DL	489	54 ± 33	569	< DL	212
	Liver	1192 ± 1350	130 ± 110	85 ± 115	138 ± 130	< DL	< DL	6	24.6	147 ± 18	34 ± 42	5.8	8

¹Below the detection limit (DL); ²found in only a single specimen

Table 9.2 Concentrations of PCB congeners in the organs of fish analyzed, ng/g l.w. (M ± SD)

Species	Organs	PCB 28	PCB 52	PCB 101	PCB 118	PCB 153	PCB 138
Prussian carp (*Carassius gibelio*)	Muscles	372 ± 630	251 ± 301	18[1]	< DL[2]	36	< DL
	Liver	66 ± 33	52 ± 33	< DL	< DL	< DL	< DL
	Eggs	63 ± 99	13 ± 14	1.6	3.4	< DL	4.2
Amur carp (*Cyprinus rubrofuscus*)	Muscles	36 ± 41	39 ± 52	< DL	< DL	34	< DL
	Liver	49 ± 47	6.1	< DL	< DL	< DL	< DL
	Eggs	22	5.7 ± 0.4	4.4	16	< DL	14
	Male gonads	16	5.8	< DL	< DL	< DL	< DL
Ussuri sharpbelly (*Hemiculter lucidus*)	Muscles	22 ± 11	7.5 ± 1.3	< DL	6.6 ± 1.1	< DL	13 ± 1
Lake skygazer (*Chanodichthys oxycephalus*)	Muscles	42 ± 58	70 ± 59	< DL	< DL	< DL	< DL
	Liver	20	26	< DL	16	23	26
Mongolian redfin (*Chanodichthys mongolicus*)	Muscles	11 ± 12	58 ± 70	< DL	4.7	4.2	< DL
	Liver	11	3	< DL	< DL	< DL	< DL
	Eggs	21	< DL	< DL	< DL	< DL	< DL
	Fat	4.5 ± 0.3	1.2 ± 0.6	1.4 ± 0.1	2.2 ± 0.9	3.2	3.4 ± 0.5
Predatory carp (*Chanodichthys erythropterus*)	Muscles	6	47 ± 15	< DL	24	91	56
	Liver	10 ± 4	19 ± 11	7.1	14	35	14
	Male gonads	44	14	< DL	< DL	33	19
	Fat	2.3 ± 1.3	1.7 ± 1.1	4.4 ± 2.4	8.9 ± 2.8	13	13 ± 2
Northern snakehead (*Channa argus*)	Muscles	955 ± 1332	119 ± 99	< DL	< DL	< DL	< DL
	Liver	33 ± 25	96 ± 158	1.2	3.8	5.3	4.7
Pike-perch (*Sander lucioperca*)	Muscles	228 ± 446	1547 ± 2341	< DL	< DL	< DL	< DL
	Liver	639 ± 1241	323 ± 443	289.9	17 ± 8	12 ± 9	12 ± 8

[1]Below the detection limit (DL); [2]found in only a single specimen

eagle; predators among reptiles and amphibians are dice snake and frogs. Fry suffer predation also from diving beetles and their larvae, backswimmers, and dragonfly larvae [9].

Amur carp is one of the most popular commercial species. It is caught almost all the year round (except for fishing closure periods). In the winter, spring, and summer seasons, the catch size is insubstantial. The major catch is recorded in late winter, i.e., during the overwintering season [10].

Fig. 9.2 Average total POP concentrations in the organs of Prussian carp

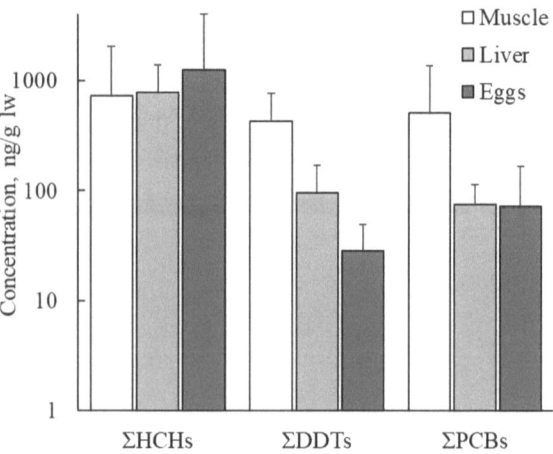

In the organs of Amur carp, HCHs made up the major part of all detected POPs. Their total range was from 1 to 705 ng/g l.w. in muscles, from 21 to 797 ng/g l.w. in liver, from 97 to 661 ng/g l.w. in eggs, and from 59 to 110 ng/g l.w. in male gonads (Tables 9.1 and 9.2; Fig. 9.3). DDT and its metabolites were mainly represented by DDE isomers, with their highest concentrations recorded from muscles within a range of 10–147 ng/g l.w. 2,4-DDT was detected in one liver sample. Lower chlorinated PCBs (28 and 52) dominated the PCB congeners, while detections of higher chlorinated congeners were rare (Table 9.2). In general, the total range of PCB concentrations was from 2 to 130 ng/g l.w. in muscles, from 16 to 129 ng/g l.w. in liver, and from 6 to 61 ng/g l.w. in eggs; in male gonads, the PCB concentration was 21.4 ng/g l.w. We did not find any statistically significant differences in concentrations of pollutants between organs, which may also indicate a long-term chronic entry of pollutants into the fish body.

Ussuri sharpbelly (*Hemiculter lucidus*) is the smallest representative of bleak-like fishes and, therefore, makes a very limited contribution to the commercial catch from the Russian part of Lake Khanka [11]. However, the closely related fishes in

Fig. 9.3 Comparison of average total POP levels in the organs of Amur carp

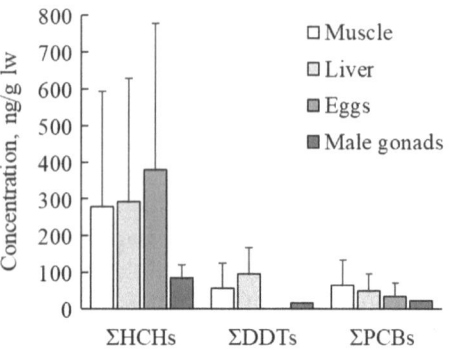

China are important target species of commercial fisheries [9]. Sharpbelly is the most important consumer of zooplankton in the lake. Another its ecosystem role is providing a food supply for valuable commercial predatory fishes such as the predatory carp (*Chanodichthys erythropterus*), yellowcheek (*Elopichthys bambusa*), and pike-perch (*Sander lucioperca*) [11]. Organisms consumed by sharpbelly are almost free of exposure to historical POP pollution and, therefore, accumulate pollutants coming mainly via atmospheric transport. Due to its feeding habits, this fish lives in the upper part of the lake, which significantly reduces contacts with bottom sediments that often become a kind of *natural burial* of POPs [12].

Due to the small body size of sharpbelly (Tables 9.1 and 9.2), we collected only muscles for analysis. Among pollutants, we found only 2,4-DDE within a concentration range from 12 to 81 ng/g l.w., and also PCB congeners whose total level ranged from 14 to 70 ng/g l.w. (Tables 9.1 and 9.2). HCH, represented by the γ-isomer, was detected in a single sample. The DDE levels were significantly ($p \leq 0.05$) higher than those of PCBs. However, the finding of exclusively this form of compound, which is a product of aerobic degradation of DDT, indicates long-ago contamination, while the lower PCB concentration (not exceeding a level of 100 ng/g l.w.) indicates a low rate of entry of pollutants via atmospheric transport.

Lake skygazer (*Chanodichthys oxycephalus*) is a small fish of limited commercial value. At a juvenile age, the major part of its diet is made up of shrimp, mysids, and water fleas *Daphnia*; adult fish feed mainly on sharpbelly [9]. Thus, skygazer do not have close contacts with bottom sediments and accumulate pollutants mainly either through biomagnification or from the water column.

We analyzed muscles and liver from lake skygazer. HCH isomers in muscles, dominated by α-HCH, showed the highest concentrations that varied within a very wide range, from 19 to 2225 ng/g l.w. Levels of $\sum$DDT ranged from 25 to 247 ng/g l.w. with a significant predominance of 2,4-DDE. The total PCB concentrations ranged from 9 to 255 ng/g l.w. with a predominance of PCB 28. In the liver, POPs were actually represented only by HCH isomers, from 29 to 810 ng/g l.w. Only 2,4-DDE and some of PCB congeners were found in one sample (Tables 9.1 and 9.2; Fig. 9.4). We did not find any statistically significant differences in concentrations of pollutants between organs, which may also indicate a long-term chronic entry of these compounds into the fish body.

Mongolian redfin (*Chanodichthys mongolicus*) is a freshwater fish species found both in rivers and lakes. In spring, it enters lakes to spawn. Its body length reaches 66 cm, the weight is 3.7 kg, and the maximum age is 10 yr. The main diet items are crustaceans and small fish. The transition to piscivorous feeding occurs at age 5 or 6 yr. It consumes large amounts of eggs of cyprinid fishes, preferring Amur carp eggs. Mongolian redfin is harvested mainly as by-catch during fishing for other freshwater species. It is caught all year round (except for fishing closure periods). Being a species of low commercial value, it is not in demand on the domestic market [9]. The most closely related species of the same genus are the lake skygazer (*C. oxycephalus*) and predatory carp (*C. erythropterus*).

In Mongolian redfin, we analyzed muscles, liver, eggs, and visceral fat. In these organs, the levels of $\sum$HCH ranged within 7–781, 5–411, 3–75, and 11–40 ng/g

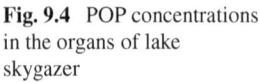

Fig. 9.4 POP concentrations in the organs of lake skygazer

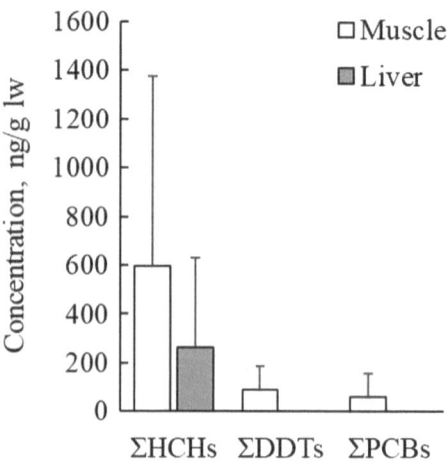

l.w., respectively (Tables 9.1 and 9.2; Fig. 9.5). Among the DDT metabolites, we detected only DDE in muscles, and also DDD and DDE in liver and fat. The $\sum$DDT concentrations ranged as follows: in muscles, from 11 to 67 ng/g l.w.; in liver, from 15 to 18 ng/g l.w.; in fat, from 7 to 12 ng/g l.w. In one egg sample, we found 2,4-DDE. PCBs were mainly represented by PCB 28 and PCB 52. In muscles, the total concentrations of biphenyls ranged from 3 to 107 ng/g l.w.; in fat, from 10 to 18 ng/g l.w. In liver and eggs, the pollutants were detected in single cases (Tables 9.1 and 9.2). Despite the significantly lower level of contamination, fat showed a much more diverse qualitative composition of PCB congeners including higher chlorinated ones, PCBs 118, 153, and 138. We did not find any statistically significant differences in concentrations of pollutants between the organs of Mongolian redfin.

Predatory carp (*Chanodichthys erythropterus*) is a freshwater fish that prefers clean, running water. It is a predator preying on small fish, insect larvae, crustaceans, and also mayflies during their mass migration. The commercial value of this species is low [9].

The total levels of HCHs in muscles, liver, male gonads, and visceral fat ranged from 15 to 1899, from 25 to 614, from 48 to 199, and from 13 to 22 ng/g l.w., respectively (Tables 9.1 and 9.2; Fig. 9.6). Of DDT and its metabolites, the degradation products (DDD and DDE) of the initial compound were most frequently detected. The total concentrations of DDT varied within the following ranges: 6–318 ng/g l.w. in muscles, 33–88 ng/g l.w. in male gonads, and 1–48 ng/g l.w. in fat. *o,p'*-DDT and *p,p'*-DDT were recorded in single cases (123 and 36 ng/g l.w., respectively). Compared to other species, predatory carp showed the greatest qualitative diversity of PCB congeners with a relatively low level of contamination: the total concentrations of pollutants ranged within 37–178 ng/g l.w. in muscles, 18–109 ng/g l.w. in liver, 52–58 ng/g l.w. in male gonads, and 34–54 ng/g l.w. in fat. We did not find any statistically significant differences in concentrations of pollutants between the organs of predatory carp.

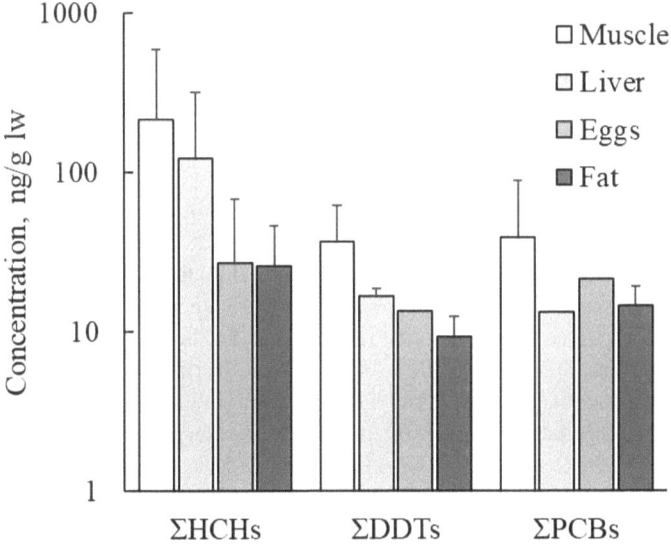

Fig. 9.5 POP concentrations in the organs of Mongolian redfin

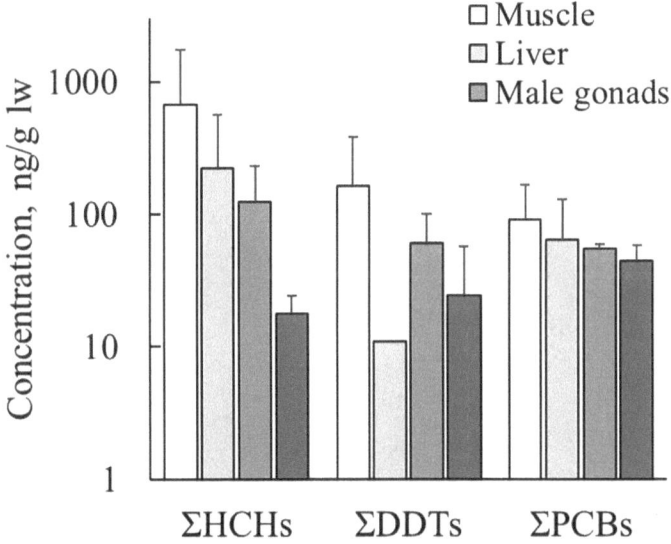

Fig. 9.6 POP concentrations in the organs of predatory carp

Pike-perch (*Sander lucioperca*) is a freshwater and semi-anadromous fish. It is very sensitive to water quality. This species is distributed almost ubiquitously, from West Europe to Southeast Asia. Juveniles feed on zooplankton, and then on benthic organisms and fry of other fish species. Adults mainly consume small fish (up to 10 cm) with the highest local abundance. Due to the specific dietary habits, it is considered a good *reclamator* of freshwater bodies of water [13]. Pike-perch is one of the major valuable commercial fishes of natural water bodies, however, providing low catches. This species is reared in Lake Khanka by Chinese fish farmers, and, therefore, its stock does not decline. There is no available data on allowable pike-perch catches in the lake.

We analyzed muscles and liver from pike-perch. In muscles, HCH isomers dominated other pollutants and were detected in a very wide concentration range, from 16 to 9787 ng/g l.w. The $\sum$DDT levels ranged from 23 to 1219 ng/g l.w. Total PCB concentrations ranged from 9 to 5266 ng/g l.w., with a predominance of PCB 28. In the liver, HCH concentrations ranged from 419 to 4304 ng/g l.w.; DDT, from 17 to 246 ng/g l.w.; and PCBs, from 426 to 8154 ng/g l.w. (Tables 9.1 and 9.2; Fig. 9.7). We did not find any significant differences between pollutants' concentrations.

Northern snakehead (*Channa argus*) is a freshwater lungfish widely distributed in the Asia–Pacific region. It is most abundant in the Songacha River and common in the Ussuri River and Lake Khanka. The diet items of snakehead are small fish such as gudgeons, bitterlings, juvenile bighead carp, etc. It also feeds on benthic animals and ingests silt. The commercial value is low, as it is caught in small quantities. Nevertheless, it is a promising aquaculture species due to its high resistance to adverse environmental factors and high growth rate. It is cultured in South Korea and China [9].

We analyzed muscles and liver from northern snakehead. The levels of $\sum$HCH in muscles ranged from 66 to 1450 ng/g l.w.; $\sum$DDT, from 16 to 180 ng/g l.w.; $\sum$PCB, from 13 to 1946 ng/g l.w. (Tables 9.1 and 9.2; Fig. 9.8). In the liver, the concentration ranges of these pollutants were 13–327, 1–17, and 19–336 ng/g l.w.,

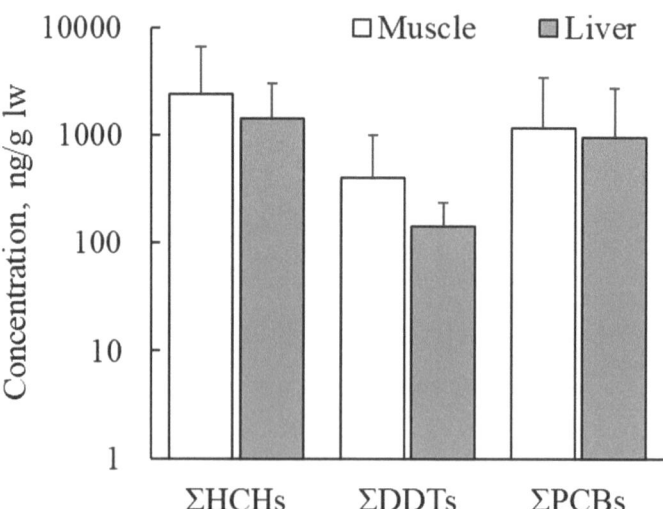

Fig. 9.7 POP concentrations in the organs of pike-perch

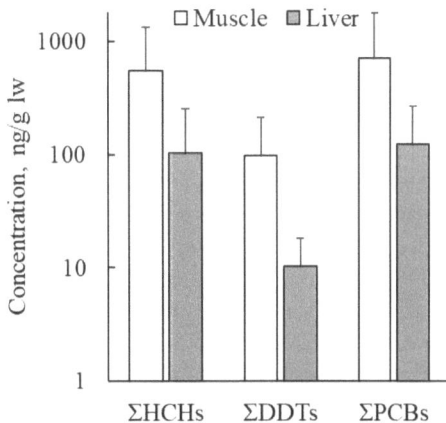

Fig. 9.8 POP concentrations in the organs of northern snakehead

respectively. We did not find any statistically significant differences between organs in accumulation of toxicants.

In almost all cases, the OCC concentrations in the organs of the fish species under study did not differ statistically, which may indicate a chronic low-level contamination of Lake Khanka and the fish analyzed. Two major sources of POPs can be distinguished in the study area in 2018: (1) atmospheric transport, characteristic primarily of PCBs (which is indicated by sharpbelly feeding on zooplankton); (2) bottom sediments, *burying* toxic compounds (as evidenced by the levels of pollutants in such benthos feeders as Amur carp and Prussian carp). In general, the detected OCP isomers and metabolites indicate the lack of entry of maternal compounds (DDT and lindane) previously used as insecticides.

9.4 Interspecific Differences in POP Accumulation in Fishes from Lake Khanka

A survey of concentrations of various OCPs in the waters of Lake Khanka was conducted in 2020 [14]. According to the data collected, the ranges of DDT and γ-HCH concentrations in the lake water were 24.9–49.6 and 3.0–14.9 ng/L, respectively (with the permissible levels of 1 and 10 ng/L, respectively). There is evidence of a long period since the entry of HCHs to the ecosystem, while the *fresh* entry of DDT most likely indicates a continuous use, despite the ratification of the Stockholm Convention. Furthermore, the authors detected a serious excess of the permissible levels for DDT and HCH (315- and 38-fold, respectively) in the Astrakhanka River. The predominance of γ-HCH indicates a leakage of this banned compound from storage facilities or an intentional use of them in agriculture. The authors also noted the free drainage of water from agricultural lands into the network of rivers feeding the lake. As a result, among DDT metabolites, only the original compound was found in some of water bodies within a concentration range from 43.7 to 61.9 ng/L. Nevertheless, there is no available information about OCP levels in the northern part of the lake.

Thus, for a better understanding of the processes that took place in the lake in 2018, we needed interspecific comparison of concentrations, with categorizing the studied fishes by their lifestyle and feeding habits. We divided the fishes into three groups: predators (lake skygazer, predatory carp, Mongolian redfin, pike-perch, and northern snakehead), benthos feeders (Amur carp, Prussian carp, and northern snakehead), and plankton eaters (Ussuri sharpbelly).

Amur carp and Prussian carp are the most similar species in their biological characteristics, capable of interspecific hybridization in natural conditions. They are very similar in feeding habits and diet and are competitors. A substantial portion in their food spectrum is made up of benthic organisms, especially chironomid larvae [15]. Another fish that has close contacts with bottom sediments is the northern snakehead. All three species prefer soft silty sediments that well adsorb pollutants [9]. When feeding, these fishes ingest silt which is, most likely, the major source of pollutants in their bodies.

When comparing the total concentrations of pollutants in the organs of Prussian carp, Amur carp, and northern snakehead, we found significant differences ($p \leq$ 0.05) only for DDT in muscles that can be arranged in the following order: *Prussian carp > northern snakehead > Amur carp* (Fig. 9.9). The rest of the organs showed no statistically significant differences, and the concentrations were approximately at the same level (by median). This may indicate similar sources of HCHs and PCBs entering the lake waters (according to our theory, it is the atmospheric transport). The significant differences in DDT concentrations may be associated with the burial of the largest portion of this pesticide in the bottom sediments of the lake. The higher levels of this compound in muscles of snakehead, compared to those of Amur carp, look quite logical, since the latter species is not so much subject to POP biomagnification (as it is not a predator). However, the higher levels of the pesticide in Prussian carp, compared to snakehead and Amur carp, seems illogical. This is probably explained by the increasing entry of DDT and its metabolites into the lake waters or by effects of any other factors that require further in-depth research.

Since the oral apparatus allows Prussian carp to reach a depth of bottom sediments of no more than 1–2 cm (3–5 cm for Amur carp), it ingests the upper, more polluted layer and accumulates pollutants in its organs [16]. This theory requires more evidence to be verified and a survey of POP concentrations in bottom sediments of the lake.

When comparing the concentrations of pollutants in muscles and liver of predatory fish and Ussuri sharpbelly, we found significant ($p \leq 0.05$) differences for HCH (Fig. 9.10) that can be arranged in the following orders: *pike-perch > predatory carp > lake skygazer > northern snakehead > Mongolian redfin* and *pike-perch > lake skygazer > predatory carp > Mongolian redfin > northern snakehead*, which is consistent with the size and weight characteristics of the fishes and with the predatory feeding behavior. The levels of DDT and PCBs were approximately equal between all fish, which indicates a low degree of contamination of the water body by these

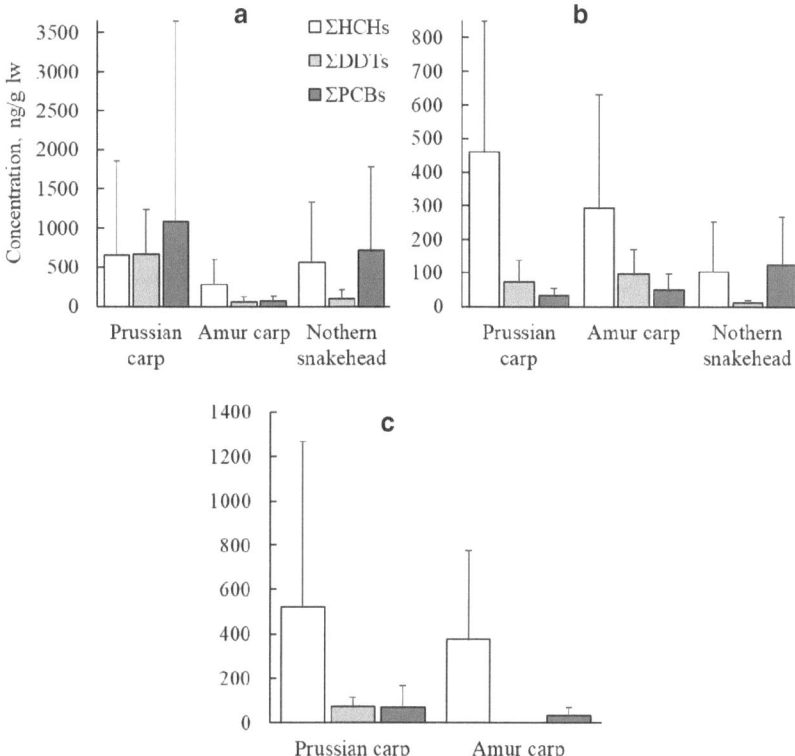

Fig. 9.9 Comparison of POP concentrations in the organs of Prussian carp, Amur carp, and northern snakehead: **a** muscles; **b** liver; **c** eggs

pollutants. Taking into account the fact that most of POPs enter the fish body with food, and the main food supply for predators in Lake Khanka is Ussuri sharpbelly [11] which almost does not contain POPs, the low levels of fish contamination in 2018 become quite explainable.

The results of the interspecific comparison generally indicate a chronic low-level pollution of Lake Khanka. In 2018, two major sources of pollutants were identified: atmospheric transport (for HCHs and PCBs) and bottom sediments (for DDT). However, some uncertainty as regards the difference in concentrations between the organs of benthos feeders, as well as data from other authors [14] showing the increasing anthropogenic pressure on the waters of the lake, necessitates further monitoring of this water body of paramount international importance and also study of abiotic components of the ecosystem to draw a more detailed conclusion about the condition of the lake.

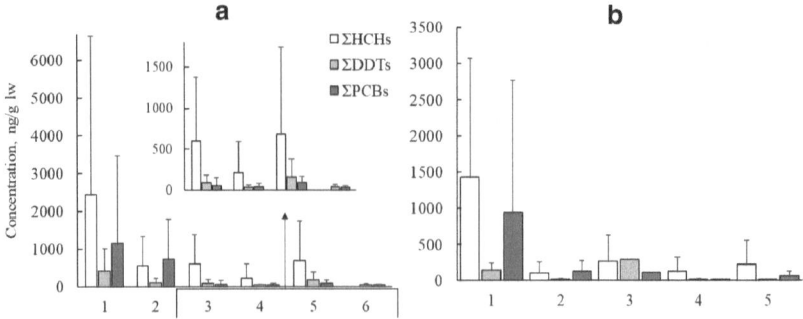

Fig. 9.10 Comparison of POP concentrations in muscles (**a**) and liver (**b**) of predators and Ussuri sharpbelly: 1—pike-perch; 2—northern snakehead; 3—lake skygazer; 4—Mongolian redfin; 5— predatory carp; 6—Ussuri sharpbelly

9.5 Possible Sources of POP Entering the Waters of Lake Khanka

According to experts, the total stockpiles of PCBs within the territory of the Russian Federation are estimated at 28 000–35 000 t [17]. Approximately 180 t of PCBs are present in the Far Eastern Federal District (a region of North and Northeast Asia), which is the lowest value compared to other regions. Nevertheless, these data are based on official statistics. In the Chukotka Autonomous Okrug, there are landfills of abandoned barrels containing, among other substances, PCB impurities that leak and enter the adjacent seas which are fisheries zones of the Russian Federation [18]. Stockpiles of pesticides unsuitable for use are estimated at no less than 24 000 t [19]. A significant part of them are DDT group compounds and HCH isomers.

The USSR and the Russian Federation are not the only countries where OCPs and PCBs were widely applied. From 1965 to 1974, about 10 000 t of PCB mixtures, which consisted of 90% trichlorobiphenyls and 10% pentachlorobiphenyls, were produced in China. After the ban on the production and use of PCBs (1974), most of obsolete PCB-containing equipment was decommissioned and put in dead storage. However, further surveys have shown leaks from part of this equipment, which still poses threat to the environment and public health. Other sources of PCB contamination are illegal dismantling and/or disposal of PCB-containing equipment, undesirable by-products of combustion of chlorine-containing waste without anti-pollution filters, and some other chemical processes involving OCCs [20].

China is an agriculturally developed country where the consumption of pesticides accounted for a significant part of the global total consumption. Among OCPs, DDT was most widely applied in various sectors of economy including agriculture (crop protection), healthcare (control of malaria vectors), and commercial shipping (anti-fouling paints for vessels). According to official statistics, by the time of the ban (1983), approximately 270 000 t of DDT were produced in China [21]. Moreover, according to some authors [21–23], anti-fouling ship hull coatings containing up to

20% technical DDT are still actively applied and make a significant contribution to the pollution of fresh water bodies in the country. Since many water bodies and courses (mainly rivers and large lakes) are located at the Chinese–Russian border, their pollution poses danger and have a negative impact on people's health on both sides. As regards POPs, freshwater bodies of water in the Far East and Asia have been studied only partially to date. The major studies have estimated concentrations of these pollutants in the central provinces of China and some prefectures of Japan [23–25].

POPs can enter ecosystems in two ways: as a result of *historical* pollution or current use of banned substances. The qualitative composition of pollutants significantly varies depending on the time of entry into the ecosystem. For example, in the case of a recent entry of HCH and DDT into the ecosystem, the dominant forms of compounds would be lindane (γ-isomer) and DDT, respectively. However, when studying the biota, it should be taken into account that aquatic organisms are capable of metabolizing certain pollutants to the forms that indicate a long-ago contamination, while the influx of xenobiotics into the ecosystem actually continues. In the case of the fish species that we studied, the ratio of (α- + β-HCH)/γ-HCH and (DDD + DDE)/DDT is mainly greater than unity, which indicates that these pesticides entered long ago.

The territory of Heilongjiang Province is one of the major rice-growing regions in China. Agricultural foods are produced here mostly by small household farms. Due to their greater *self-dependence*, farmers often make decisions about crops to be planted and chemicals to treat them on the hit-or-miss principle. This can facilitate the use of banned pesticides, as older farmers (who previously practiced active use of OCPs) pass down their knowledge and experience to younger ones [16]. Furthermore, recent studies by Chinese authors [26, 27] note a significant overuse of pesticides without, nevertheless, specifying their qualitative composition. Amounts of pesticides applied to rice fields in northeastern China are on average 1.2–2.3-fold greater than necessary [27]. Thus, DDT and HCH can potentially be used at household farms in Heilongjiang Province and contribute, to a certain extent, to the OCP concentrations recorded from the waters of Lake Khanka.

The studied PCBs were dominated by lower chlorinated congeners, 28 and 52. This may indicate the entry of PCB predominantly as a result of the activity of fishing vessels catching local fish or via the atmospheric transport [28, 29]. Higher chlorinated congeners make a negligible contribution to the total POP contamination of Lake Khanka and, most likely, almost do not enter this body of water.

9.6   Comparison of POP Concentrations with Other Regions and Regulatory Documents

Food safety is one of the most critical issues across the world [30]. It is guaranteed by compliance with regulatory documents where permissible levels of toxic substances in foods are established.

Fish and non-fish products of marine fisheries are distributed worldwide and in some of world's regions constitute the major part of the diet of local residents (e.g., in Japan and Alaska). The most essential nutrients from fatty fish and fish products that proved to be important for the health and normal functions of the human body are omega-3 fatty acids, in particular eicosapentaenoic and docosahexaenoic acids. These are found in fish in greater quantities than in other foods [31, 32].

Aquatic ecosystems are an important source of high-value protein and various nutrients and, therefore, are actively exploited by humans. Due to their lipophilicity, POPs build up in tissues of aquatic organisms. These compounds are regulated worldwide by determination of their maximum permissible concentrations (MPC) in tissues of aquatic biological resources. In Russia, there are numerous sanitary and hygienic standards aimed at providing safety of food products. The most important regulatory documents that establish safety requirements for fish and non-fish objects of fisheries are as follows: SanPin 2.3.2.1078-01, TR CU 021/2011 and TR EAEU 040/2016 [33–35].

To assess the compliance of the detected POP concentrations with the standards in the regulatory documents, we recalculated these concentrations from ng/g l.w. into ng/g wet weight (Table 9.3). For this, we used POP values only in muscles.

When comparing the values that we obtained to the standards in the respective regulatory documents, we did not find any single case of MPC excess, which indicates the compliance of the quality of fish products from Lake Khanka with the requirements of the regulatory laws. However, some of the compounds (e.g., dieldrin

Table 9.3 Average total concentrations of HCHs, DDT, and PCBs in fish from Lake Khanka, ng/g wet weight

Species	ΣHCH	ΣDDT	ΣPCB
Prussian carp	2.9 ± 2.9	0.81 ± 0.61	3.0 ± 3.1
Amur carp	1.9 ± 1.8	0.43 ± 0.34	0.55 ± 0.45
Northern snakehead	2.5 ± 3.6	0.44 ± 0.54	2.5 ± 4.0
Lake skygazer	2.4 ± 4.2	0.31 ± 0.22	0.35 ± 0.21
Ussuri sharpbelly	< DL*	0.45 ± 0.27	0.38 ± 0.28
Mongolian redfin	1.2 ± 1.7	0.38 ± 0.20	0.37 ± 0.14
Predatory carp	2.4 ± 3.9	0.82 ± 0.49	0.51 ± 0.21
Pike-perch	1.8 ± 3.5	0.47 ± 0.41	0.18 ± 0.14

*Below the detection limit (DL)

and endrin) are not regulated, which poses potential danger to consumers. Furthermore, POPs are capable of accumulating in the organs of living organisms, including humans, due to their lipophilicity [36]. As a certain critical mass of a toxicant is built up, it may cause poisoning or increase the risk of cancer during life, which necessitates further studies including assessment of such probability.

9.7   Conclusions

We have analyzed the patterns of POP accumulation in various fish species from Lake Khanka. When comparing the total concentrations of pollutants in the organs of benthos feeders (Prussian carp, Amur carp, and northern snakehead), we found significant ($p \leq 0.05$) differences only for DDT concentrations in muscles that can be arranged in the following order: Prussian carp > northern snakehead > Amur carp. When comparing the levels of pollutants in the organs of predatory fish and Ussuri sharpbelly, we recorded significant ($p \leq 0.05$) differences for HCHs in muscles (pike-perch > predatory carp > lake skygazer > northern snakehead > Mongolian redfin) and liver (pike-perch > lake skygazer > predatory carp > Mongolian redfin > northern snakehead), which is consistent with the size and weight characteristics of fish and their feeding specializations. The qualitative composition of pollutants is characteristic of *historical* contamination by OCCs and does not indicate *fresh* entries of toxicants to the ecosystem. When we compared the results obtained with the standards in the respective regulatory documents, we did not find a single case of excess of MPCs, which shows compliance of the quality of fish products from Lake Khanka with the requirements of the regulatory laws.

Acknowledgements   The authors are grateful to E.V. Brunevskaya for her assistance in collecting the material.

References

1. UNEP (2020) Stockholm convention on persistent organic pollutants
2. Covaci A, Gheorghe A, Hulea O, Schepens P (2006) Levels and distribution of organochlorine pesticides, polychlorinated biphenyls and polybrominated diphenyl ethers in sediments and biota from the Danube Delta, Romania. Environ Pollut 140:136–149. https://doi.org/10.1016/j.envpol.2005.06.008
3. Tsygankov VYu (2019) Organochlorine pesticides in marine ecosystems of the Far Eastern Seas of Russia (2000–2017). Water Research 161:43–53. https://doi.org/10.1016/j.watres.2019.05.103
4. Tsygankov VYu, Gumovskaya YP, Gumovskiy AN et al (2020) Bioaccumulation of POPs in human breast milk from south of the Russian Far East and exposure risk to breastfed infants. Environ Sci Pollut Res 27:5951–5957. https://doi.org/10.1007/s11356-019-07394-y
5. Fedun IN (2017) Cadastral data about the State Nature Biosphere Reserve "Khankaisky" for 2013–2016

6. Egidarev EG, Mishina NV, Bazarov KY (2019) Current land use in the Khanka Lake basin. In: Collection of scientific articles. FSBIS PGI FEB RAS, Vladivostok, pp 197–203
7. Boyarova MD (2008) Current levels of organochlorine pesticides in aquatic organisms of peter the Great Bay (Sea of Japan) and Lake Khanka. Dissertation for the degree of candidate of biological sciences, Far Eastern Federal University
8. Xiangcan J, Xia J (2007) Experience and lessons learned brief for Lake Xingkai/Khanka. In: Materials of the international conference. Ministry of Natural Resources of Russia, Moscow, pp 81–108
9. Berg LS (1949) Fishes of fresh waters of the USSR and adjacent countries. Publishing House of the USSR Academy of Sciences, Moscow, Leningrad
10. Ostrovskiy VI (2020) Materials of the total allowable catch of aquatic biological resources in the inland waters of the Khabarovsk Territory, the Amur Region and the Jewish Autonomous Region for 2021 (except for the inland sea waters of the Russian Federation)
11. Kurdyaeva VP, Shapovalov ME, Rachek EI (2002) To the biology and diagnostics of the Ussuri (Hemiculter lucidus (Dybowski, 1872)) and Korean (H. leucisculus (Basilewsky, 1855)) sharp-bellies from the Lake Khanka and reservoir-cooler of Primorskaya SDPP. Izvestiya TINRO 131:208–227
12. Lu X, Chen C, Zhang S et al (2013) Concentration levels and ecological risks of persistent organic pollutants in the surface sediments of Tianjin Coastal Area, China. Sci World J 2013:1–8. https://doi.org/10.1155/2013/417435
13. Zhukov PI (1988) Handbook of freshwater fish ecology. Science and Technology, Minsk
14. Liagusha MS, Cherniaev AP (2020) Chapter 5. Current levels of organochlorine pesticides (OCPs) in the abiotic components of the Northwest Pacific ecosystems. In: Tsygankov VY (ed) Persistent organic pollutants (POPs) in the Far Eastern region: seas, organisms, human: monograph. Publishing House of the Far Eastern Federal University, Vladivostok, pp 101–127
15. Burik VN (2010) Common carp (Cyprinus carpio haemotopterus) in the basins of the Tunguska and Zabelovka rivers. Reg Probl 13:62–66
16. Yang X, Lin E, Ma S et al (2007) Adaptation of agriculture to warming in Northeast China. Clim Change 84:45–58. https://doi.org/10.1007/s10584-007-9265-0
17. Treger YuA (2011) Stable organic contaminants. Problems and ways of solving thems. Tonkie Khimicheskie Tekhnologii [Fine Chemical Technologies] 6:87–97
18. Dudarev AA (2009) Persistent polychlorinated hydrocarbons and heavy metals in Arctic biosphere: the main regularities of exposure and reproductive health of indigenous people. Biosfera 1:186–202
19. Treger YuA, Chagir KA (2010) Partial inventory of obsolete pesticides in the Russian Federation. Encyclopedia of the Chemical Engineer, pp 21–26
20. Xing Y, Lu Y, Dawson RW et al (2005) A spatial temporal assessment of pollution from PCBs in China. Chemosphere 60:731–739. https://doi.org/10.1016/j.chemosphere.2005.05.001
21. Qiu X, Zhu T, Yao B et al (2005) Contribution of Dicofol to the current DDT pollution in China. Environ Sci Technol 39:4385–4390. https://doi.org/10.1021/es050342a
22. Lin T, Hu Z, Zhang G et al (2009) Levels and mass burden of DDTs in sediments from fishing harbors: the importance of DDT-containing antifouling paint to the coastal environment of China. Environ Sci Technol 43:8033–8038. https://doi.org/10.1021/es901827b
23. Liu R, Tan R, Li B et al (2015) Overview of POPs and heavy metals in Liao River Basin. Environ Earth Sci 73:5007–5017. https://doi.org/10.1007/s12665-015-4317-7
24. Guan Y-F, Wang J-Z, Ni H-G, Zeng EY (2009) Organochlorine pesticides and polychlorinated biphenyls in riverine runoff of the Pearl River Delta, China: assessment of mass loading, input source and environmental fate. Environ Pollut 157:618–624. https://doi.org/10.1016/j.envpol.2008.08.011
25. Li C, Huo S, Yu Z et al (2017) National investigation of semi-volatile organic compounds (PAHs, OCPs, and PCBs) in lake sediments of China: occurrence, spatial variation and risk assessment. Sci Total Environ 579:325–336. https://doi.org/10.1016/j.scitotenv.2016.11.097
26. Hu Y, Fan L, Liu Z et al (2019) Rice production and climate change in Northeast China: evidence of adaptation through land use shifts. Environ Res Lett 14:024014. https://doi.org/10.1088/1748-9326/aafa55

27. Sun S, Zhang C, Hu R (2020) Determinants and overuse of pesticides in grain production: a comparison of rice, maize and wheat in China. CAER 12:367–379. https://doi.org/10.1108/CAER-07-2018-0152
28. UNEP (2001) Diagnostic analysis of the Lake Xingkai/Khanka Basin (People's Republic of China and Russian Federation)
29. Urbaniak M (2007) Polychlorinated biphenyls: sources, distribution and transformation in the environment—a literature review. Acta Toxicologica 15:83–93
30. Fung F, Wang H-S, Menon S (2018) Food safety in the 21st century. Biomed J 41:88–95. https://doi.org/10.1016/j.bj.2018.03.003
31. Berbert AA, Kondo CRM, Almendra CL et al (2005) Supplementation of fish oil and olive oil in patients with rheumatoid arthritis. Nutrition 21:131–136. https://doi.org/10.1016/j.nut.2004.03.023
32. Calò L, Bianconi L, Colivicchi F et al (2005) N-3 fatty acids for the prevention of atrial fibrillation after coronary artery bypass surgery. J Am Coll Cardiol 45:1723–1728. https://doi.org/10.1016/j.jacc.2005.02.079
33. SanPin 2.3.2.1078-01 (2001) Hygienic requirements of safety and nutritional value of food products. Ministry of Health of Russia, Moscow
34. TR CU 021/2011 (2011) Technical regulations of the customs Union "on food safety"
35. TR EAEU 040/2016 (2016) Technical regulations of the Eurasian Economic Union "on the safety of fish and fish products"
36. Donets MM, Tsygankov VYu, Gumovskiy AN et al (2021) Organochlorine pesticides (OCPs) and polychlorinated biphenyls (PCBs) in Pacific salmon from the Kamchatka Peninsula and Sakhalin Island, Northwest Pacific. Marine Pollution Bulletin 169:112498. https://doi.org/10.1016/j.marpolbul.2021.112498

Chapter 10
Semi-aquatic Migratory Birds in the System of Persistent Organic Pollutant Monitoring (Exemplified by Avifauna of Lake Khanka, Sea of Okhotsk Basin)

Abstract Being subject to transboundary transport by water and air masses, persistent organic pollutants (POP) are precipitated at a great distance from sources of pollution and accumulated in aquatic and terrestrial ecosystems. Birds of semi-aquatic and aquatic ecosystems represent a widespread and highly mobile group of homeothermic animals comprising numerous species that make regular migrations over long distances. Migratory birds can be considered indicators of transboundary transfer of various pollutants. The present study provides data on POP levels in bodies of anatid birds (Anseriformes: Anatidae) from Lake Khanka (Sea of Okhotsk basin).

Keywords Migratory birds · Anatidae · Lake Khanka · POPs · Organochlorine pesticides · HCH isomers · DDT and metabolites · PCBs

10.1 Introduction

Persistent organic pollutants (POPs) are toxic lipophilic substances of anthropogenic origin, resistant to photolytic, chemical, and biological degradation. Such pollutants as hexachlorobenzene (HCB), organochlorine pesticides (OCPs), polychlorinated biphenyls (PCBs), and polycyclic aromatic hydrocarbons (PAHs) are usually considered to be globally important POPs. Recently, a variety of POPs have been recorded in abundance from terrestrial environments. This situation has worsened with the development of densely populated industrial centers, because both terrestrial and aquatic environmental components are important for the economy of the region [1].

Birds of semi-aquatic and aquatic ecosystems are a group of highly mobile homeothermic animals distributed all over the world. They are extremely sensitive to environmental changes and usually occupy high trophic levels. One of the factors of anthropogenic impact on them is the entry of persistent organic pollutants (POPs) to the breeding and wintering grounds, and also to the sites of concentration (stopover hotspots) on the routes of their seasonal migrations. Therefore, the level of POPs in birds is a marker of various types of toxicants in the environment and, more broadly, an indicator of trends in biosphere pollution [2–6]. Thus, migratory birds

© The Author(s), under exclusive license to Springer Nature Switzerland AG 2023 219
V. Tsygankov, *Persistent Organic Pollutants in the Ecosystems of the North Pacific*, Earth and Environmental Sciences Library, https://doi.org/10.1007/978-3-031-44896-6_10

indicate not only local, but also integral pollution along migration routes which are covered rather through a series of long stopovers at hotspot sites, well known as sites of intensive exchange of various viruses, than by a single flight [7–9].

Regular monitoring studies of POP levels in migratory birds of semi-aquatic and aquatic ecosystems in the south of the Russian Far East are currently under development [10]. In this paper, such data are presented for anatid birds (Anseriformes: Anatidae) from Lake Khanka, Sea of Okhotsk basin.

10.2   Pollution and Exposure of Birds to Organic Toxicants

Birds, like many other living organisms, are capable of accumulating organochlorine compounds (OCC). These substances build up in adipose tissues and, since the half-life of POPs is about 30 years, their concentration in the body usually increases throughout the animal's life. Different groups of birds occupy different trophic levels and, therefore, *upper* predators accumulate pollutants to the highest concentrations [4, 11].

One of the first reports on a mass mortality of birds caused by active use of pesticides dates back to 1954: the birds fed on dead fish whose tissues contained high levels of POPs, lethal for both fish and birds [12].

In the 1960s and 1970s, in the period of wide use of organochlorine pesticides (OCP), populations of many bird species showed a tendency to decline in areas where toxicants were applied. According to the results of avifauna studies conducted in the early 1980s, more than 10% of the analyzed individuals from different regions of the globe and belonging to different taxonomic and ecological groups contained not only dichlorodiphenyltrichloroethane (DDT), hexachlorocyclohexane (HCH), and polychlorinated biphenyls (PCB), but also such compounds as heptachlor, oxychlorodane, dieldrin, etc. As a rule, DDT and PCB dominated [2]. A decreasing trend of mortality among birds was observed after the restriction and ban on the use of a number of pesticides toxic to living organisms such as DDT, endrin, dieldrin, etc. [11].

Organic xenobiotics have a wide range of side effects on birds, affecting their endocrine, immune, and nervous systems, and also causing impairment of reproduction, development, and growth. However, the controversial results obtained indicate that the potential impact of POPs on the physiological parameters of birds consists of a multitude of factors, including many biological processes, species differences, sex, age, and types of compounds [4, 13]. Thus, level of toxicants varies markedly depending on fat reserves. When the general physiological condition of bird deteriorates and fat components are depleted, POPs are redistributed over the body, which leads to an increase in their content in the brain and liver [11]. The level of pollution is largely related also to the feeding specialization. As a rule, herbivorous bird species are characterized by lower concentrations of DDT and PCBs than those in omnivores and insectivores [2].

10.3   Use of Migratory Birds as Bioindicators

The use of birds of different trophic levels for biomonitoring purposes is a very promising approach. They respond rapidly and directly to the dynamics in the natural environment and ecological factors, in particular, anthropogenic ones [2, 14].

Analysis of the internal organs of birds such as muscle tissues, liver, heart, brain, etc. can show entry of organic xenobiotics with food ingested. Due to their lipophilic properties, organochlorine pesticides are accumulated and deposited primarily in adipose tissue. However, under stress, starvation, and during migrations, when birds consume large amounts of fat reserves, pollutants pass into internal organs and tissues, which leads to chronic or acute poisoning. In homeotherms, the level of organic xenobiotics in the blood and internal organs is inversely proportional to the amount of fat reserves [15].

Birds of prey, as the most representative group of organisms that indicate accumulation of organic toxicants at the upper links of the food web, have always remained an important subject of observation in the biomonitoring system. However, many authors prefer to overlook the challenges of studying small samples and consider more abundant bird species, which are usually located lower down the food chain. For example, typical shore birds such as the Common Tern (*Sterna hirundo*) and the European Herring Gull were used for monitoring the North Sea coast (*Larus argentatus*) [16, 17]. In turn, the pollutant research conducted by the US Fisheries and Wildlife Service (FWS) is based on the Bald Eagle (*Haliaeetus leucocephalus*), Mallard (*Anas platyrhynchos*), and American Black Duck (*Anas rubripes*). The Mallard and American Black Duck have vast ranges, together covering most of the United States. The Bald Eagle occupies the uppermost link in the food web and, therefore, is exposed to the most significant pressure from xenobiotics [11].

Not only bird tissues but also eggs are used in bioindication studies and long-term background monitoring of ecosystems pollution by OCCs [13, 15]. The toxic effect of POPs on birds was first documented by an analysis of bird eggs [4]. Bird eggs are used in numerous studies conducted to control organic xenobiotics, as eggs have a number of advantages over internal tissues. Eggs have a very constant composition, unlike tissues. They are produced by a clearly known part of population, adult females, which, however, may be a disadvantage, since it excludes sampling of other members. Egg sampling is strictly limited to certain time periods and is possible only during the egg-laying season which is specific and depends not only on the bird species, but also on the nesting ground. However, such samples are easy to handle. It is assumed that levels of pollutants in spoiled eggs differ insignificantly from recently fertilized eggs [11].

Besides the evident advantages of using eggs, there are also certain disadvantages. The presence of pollutants in eggs usually characterizes their entry within a short period before oviposition and, therefore, cannot be used to study pollutant's long-term pressure due to the lack of data for other time periods of the year [4, 11].

The use of wild bird eggs in a monitoring system has become traditional in world's practice. Thus, researchers from England have identified a noteworthy relationship for assessing the level of DDT contamination in the region. The weight of the eggshell in birds of prey decreased by 18.9% since the time of first DDT application, which in some cases resulted in fragility of eggs during incubation. Subsequently, an inverse relationship was revealed between the sedimentary dichlorodiphenyldichloroethylene (DDE) in eggs and the shell thickness in osprey, bald eagle, and peregrine falcon. As has been reported for all these species, the degree of reduction in the eggshell thickness is in a linear relationship with the concentration of DDE alone [4]. This metabolite is the most hazardous to birds, as it shows embryotoxicity and can negatively affect the reproductive success [5].

Due to the measures to ban the use of POPs that are currently underway, the condition of some populations of fish-eating birds and birds of prey in the Northern Hemisphere shows a positive trend [11]. Nevertheless, monitoring studies of bird organs contamination by organic toxicants remain highly relevant.

10.4 Monitoring of POPs in Birds Exemplified by the Anatid Population of Lake Khanka

We used waterfowl of the family Anatidae (Anseriformes), including many species that represent important hunting resources, as study objects. Waterfowl species dominate predatory members of the class in abundance, which makes it possible to collect extensive material for the study within a short time. Availability of information on levels of organic xenobiotics is extremely crucial both for monitoring the overall contribution of birds to the global background level and circulation of POPs and for estimating environmental risks to humans posed by the use of game birds in food.

Biological material for the study (with a total of 29 individuals) was collected by licensed scientific catch from the wetlands of Lake Khanka (Fig. 10.1) in the period from April to May 2021. Samples of pectoral muscles, liver, and plumage from birds were analyzed for levels of OCCs.

We detected organochlorine compounds (ΣOCP + ΣPCB) in all the samples analyzed. The highest average concentration, found in plumage, ranged from 100.0 to 3831.6 ng/g lipid weight (l.w.). In the liver and pectoral muscles, the levels of OCCs were in ranges of 12.1–1117.1 and 9.0–460.7 ng/g, respectively (Fig. 10.2). The high concentration in the liver compared to that in muscles confirms the depositing function of the organ.

We found organochlorine pesticides (ΣHCH + ΣDDT) in all the studied individuals but exclusively in plumage samples. In the other cases, the frequency of detection was 76% for pectoral muscles and 93% for the liver. The highest average OCP levels were recorded from plumage in a range from 38.3 to 3758.2 ng/g l.w. In the liver and pectoral muscles, the total of concentrations varied from 2.9 to 1058.7 and from 4.6 to 418.3 ng/g, respectively. The liver showed a higher average ΣOCP concentration than pectoral muscles.

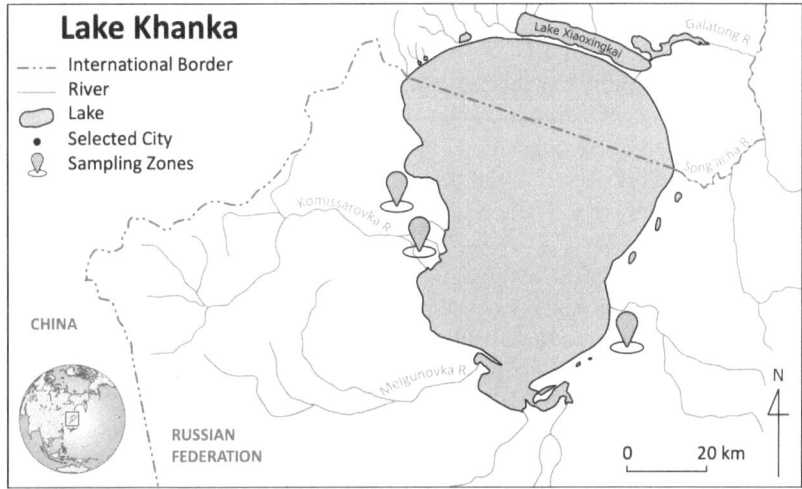

Fig. 10.1 Map of the sampling area

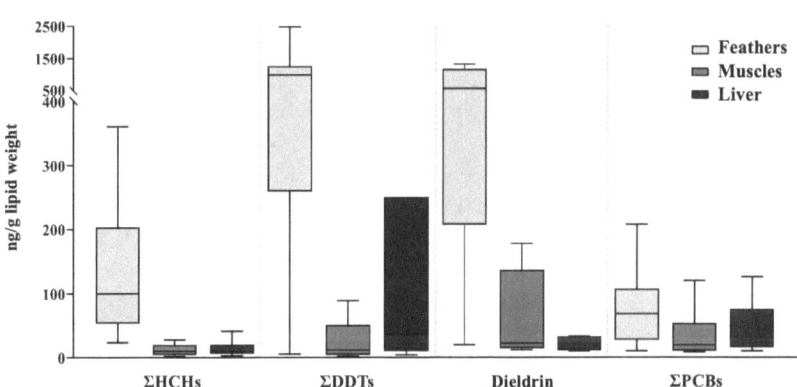

Fig. 10.2 Total range of ΣHCH, ΣDDT, dieldrin, and ΣPCB levels in birds' organs

In turn, HCH isomers in plumage also showed the highest concentrations and were detected in all samples. In muscles and liver, the detection frequency for them constituted 62 and 79%, respectively. Among the HCH isomers, the β- and δ-forms most frequently occurred in all organs, while α-HCH was represented only in a single case, in the liver. The concentration ranges in the liver and muscles were as follows: β-HCH, 1.6–15.3 (with a detection frequency of 79%) and 2.3–18.2 ng/g l.w. (59%); δ-HCH, 1.2–39.6 (48%) and 1.4–24.5 ng/g l.w. (24%), respectively. The isomeric composition of plumage was represented by all the forms under study: α-, β-, δ-, and γ-HCH. The first three forms were found in all the samples within the following ranges: 1.7–30.8, 7.1–111.6, and 12.5–275.2 ng/g l.w., respectively; the latter form had a detection frequency of 76% and ranged within 1.1–60.4 ng/g.

In the biological material analyzed, plumage had the highest average concentration and detection frequency of ΣDDT (DDT and its metabolites): 1035.0 ± 995.5 ng/g l.w. and 86%. In turn, the liver and pectoral muscles had similar detection frequencies, 38 and 41%, with concentration ranges of 2.1–240.0 and 4.1–1034.7 ng/g, respectively. The most diverse composition of DDT metabolites, which represented all the studied forms, was observed in the pectoral muscle samples: *p,p'*-DDT, *o,p'*-DDT, *p,p'*-DDD, *o,p'*-DDD, *p,p'*-DDE, and *o,p'*-DDE whose levels ranged within 7.6 (at a detection frequency of 3%), 13.3–59.0 (10%), 2.3–28.5 (14%), 10.4–240.0 (7%), 2.1–68.7 (10%), and 2.9–4.9 ng/g, respectively. DDT metabolites were recorded from plumage with the highest concentrations compared to other organs. Among these compounds, the most detectable form was *p,p'*-DDE that ranged within 3.7–3668.0 (1091.8 ± 1009.5) ng/g l.w. at a detection frequency of 76%. *o,p'*-DDD was represented in a single case (895.9 ng/g l.w.). *o,p'*-DDE and *o,p'*-DDT were also found within ranges of 2.1–476.5 and 39.2–346.2 ng/g l.w. at detection frequencies of 24 and 10%, respectively. The following metabolites were found in the liver samples: *p,p'*-DDT (3%), *p,p'*-DDD (7%), *o,p'*-DDD (21%), and *p,p'*-DDE (14%) that had concentration ranges of 7.0, 17.6–36.4, 10.4–1034.7, and 4.1–38.1 ng/g, respectively.

The levels of dieldrin in the biological material from birds were recorded at the following frequencies of detection: for liver, 14% (22.4 ± 11.8 ng/g l.w.); for pectoral muscles, 24% (59.3 ± 68.4 ng/g); for plumage, 17% (657.3 ± 509.6 ng/g).

We detected polychlorinated biphenyls in all the bird samples analyzed. Among the organs, the liver showed the greatest variety of PCB congeners, including PCB 28, PCB 52, PCB 101, PCB 118, PCB 153, PCB 138, and PCB 180. Absolutely all samples contained PCB 28 (46.7 ± 44.6 ng/g l.w.). In the liver and pectoral muscles, despite relatively equivalent average levels of 15.8 ± 10.4 and 15.9 ± 12.4 ng/g, respectively, the median values for the liver (13.3 ng/g) exceeded those for pectoral muscles (10.7 ng/g). PCB 52 had the highest detection frequency in the liver (31%) and was found in single samples of plumage (31.4 ng/g) and pectoral muscles (2.0 ng/g), both from the same individual. PCB 101 was recorded only from the liver at a detection frequency of 10% and within a range from 3.4 to 11.0 ng/g; in muscles and plumage, it was below the detection limits.

Among the other studied higher chlorinated PCB congeners, PCBs 118, 153, 138, and 180 showed the levels that ranged as follows: in pectoral muscles, 6.2–19.9, 4.5–56.4, 4.9–30.1, and 16.9–25.3 ng/g l.w.; in the liver, 3.8–20.2, 2.7–55.0, 7.7–48.6, and 6.5–14.9 ng/g; in plumage, 6.9–560.7, 1302.2–13.5, 15.5–615.1, and 32.4–242.9 ng/g, respectively.

We observed the following sequence of OCC dominance in muscles: HCH < DDT < PCB ≈ dieldrin. In the liver samples, the sequence was HCH < dieldrin < PCB < DDT; in the plumage samples, PCB < HCH < dieldrin < DDT. The pattern of distribution over organs in the spring season was as follows: muscles < liver < plumage.

Details of concentrations in the species under study are presented in Table 10.1.

The predominance of γ-HCH in plumage indicates the exposure to fresh pollution in the hydrosphere and atmosphere. The secretion of the uropygial gland can protect

Table 10.1 Concentrations of organochlorine compounds in the organs of birds of the family Anatidae, ng/g l.w.

Species	N	Organ	Lipid, %	ΣHCH[4]	ΣDDT[5]	Dieldrin	ΣPCB[6]
Mallard (*Anas platyrhynchos*)	3	Muscles	1.86–2.06[1] / 1.99 ± 0.11[2]	<DL[3]–3.8	<DL–2.1	12.9–137.2 / 55.1 ± 71.1	9.6–35.4 / 18.4 ± 14.7
		Liver	2.54–3.45 / 3.08 ± 0.48	<DL–7.0 / 5.6	<DL–926.2 / 468.3	<DL–11.0	13.9–30.0 / 20.1 ± 8.7
		Plumage	0.07–0.24 / 0.16 ± 0.08	151.2–233.0 / 193.6 ± 41.0	868.1–2894.5 / 2081.7 ± 1071.0	<DL	94.8–2766.8 / 1023.0 ± 1511.3
Falcated duck (*Anas falcata*)	6	Muscles	0.69–1.44 / 1.05 ± 0.37	<DL–21.8 / 14.4 ± 7.8	<DL–11.3	<DL	8.6–59.4 / 21.5 ± 19.3
		Liver	0.63–3.27 / 1.67 ± 0.94	<DL–11.3 / 8.6 ± 2.0	<DL	<DL	11.7–26.4 / 16.2 ± 5.5
		Plumage	0.31–1.60 / 0.96 ± 0.54	23.7–155.3 / 64.5 ± 50.0	<DL–274.6 / 579.5 ± 439.1	<DL–1311.1	10.9–38.3 / 16.2 ± 10.0
Eurasian teal (*Anas crecca*)	4	Muscles	0.85–1.96 / 1.50 ± 0.48	<DL–2.8	<DL–240.0	26.7–178.3 / 102.5	9.0–42.4 / 17.7 ± 16.5
		Liver	1.44–3.38 / 2.62 ± 0.83	<DL–24.0 / 12.1 ± 10.8	<DL–1034.7 / 527.2	<DL–33.7	13.2–58.4 / 31.1 ± 19.4
		Plumage	0.14–0.86 / 0.51 ± 0.39	46.4–245.2 / 170.1 ± 137.3	8.3–1242.1 / 524.0 ± 517.2	<DL–549.0	13.7–121.2 / 67.5 ± 44.4
Eastern spot-billed duck (*Anas zonorhyncha*)	1	Muscles	1.87	<DL	<DL	<DL	11.1
		Liver	2.35	<DL	<DL	31.3	76.9
		Plumage	0.23	245.2	<DL	394.8	54.2
Pintail (*Anas acuta*)	2	Muscles	0.74–0.93	8.9–28.4	<DL–6.2	<DL	10.9–100.8

(continued)

Table 10.1 (continued)

Species	N	Organ	Lipid, %	ΣHCH[4]	ΣDDT[5]	Dieldrin	ΣPCB[6]
			0.84	18.7			55.9
		Liver	0.56–1.14	28.0–53.0	4.1–250.6	<DL	22.7–81.2
			0.85	40.5	127.4		52.0
		Plumage	0.16–0.49	83.8–210.3	28.0–3674.4	<DL	28.3–108.1
			0.32	147.1	1851.2		68.2
Tufted duck (*Ayhya fuligula*)	3	Muscles	1.17–2.47	<DL–11.0	<DL	<DL–21.7	28.3–108.1
			1.61 ± 0.75				37.3 ± 43.4
		Liver	1.75–2.09	<DL–5.0	<DL	<DL–13.5	15.9–120.2
			1.90 ± 0.17				52.5 ± 58.7
		Plumage	0.15–0.93	53.9–447.3	1073.6–1325.3	<DL	70.5–107.6
			0.41 ± 0.45	224.4 ± 201.9	1193.2 ± 126.3		87.8 ± 18.7
Greater scaup (*Ayhya marila*)	7	Muscles	0.46–3.26	<DL–16.7	<DL–89.0	<DL–23.2	9.5–120.4
			1.85 ± 0.93	6.4 ± 5.9	26.0 ± 36.8		70.9 ± 41.9
		Liver	0.79–1.84	3.3–17.3	<DL–80.6	<DL	16.9–126.3
			1.43 ± 0.35	10.3 ± 5.7	41.1 ± 31.5		83.1 ± 37.7
		Plumage	0.25–2.25	32.5–215.9	<DL–2920.1	<DL–1011.0	26.7–283.5
			0.74 ± 0.72	115.7 ± 71.5	785.5 ± 1132.74	515.8	102.3 ± 90.6
Mandarin duck (*Aix galericulata*)	3	Muscles	0.57–1.41	18.6–27.0	13.3–59.0	<DL	14.2–35.0
			1.01 ± 0.42	22.2 ± 4.3	33.7 ± 23.2		23.8 ± 10.5
		Liver	0.57–1.41	20.6–41.6	<DL	<DL	10.9–21.1
			0.61 ± 0.42	33.0 ± 11.0			16.8 ± 5.3

(continued)

Table 10.1 (continued)

Species	N	Organ	Lipid, %	ΣHCH[4]	ΣDDT[5]	Dieldrin	ΣPCB[6]
		Plumage	0.38–0.94	38.8–166.6	968.7–1281.4	< DL	29.6–257.2
			0.61 ± 0.29	97.6 ± 64.5	1073.5 ± 180.1		121.4 ± 120.0

Note [1]Range, min–max (numerator, above line). [2]Arithmetic mean ± standard deviation (denominator, below line). [3]Below the detection limits. [4]Total of α-HCH, β-HCH, γ-HCH, and δ-HCH. [5]Total of p,p'-DDT, o,p'-DDT, p,p'-DDE, o,p'-DDE, p,p'-DDD, and o,p'-DDD. [6]Total of PCB 28, PCB 52, PCB 101, PCB 118, PCB 153, PCB 138, and PCB 180

waterfowl from soaking wet and other adverse effects of moisture, and also facilitates their movements in water [18]. Due to the activity of such lipid-producing organs, feathers can accumulate organic pollutants, possibly, because this uropygial oil acts as an adhesive for external contamination originating from the air [19]. In turn, the β-and α-forms that are more resistant to degradation in the environment were found in muscles and liver. The dominance of these isomers indicates a long period since the application of toxicants at birds' habitats and migration stopovers.

Despite the generally long-ago pollution, as evidenced by the predominance of DDE and DDD metabolites over DDT, the presence of *p,p'*-DDT and *o,p'*-DDT metabolites was also recorded, which probably indicates a recent entry of toxicants into birds' bodies. This suggests that many migratory species feed in areas with a high level of DDT use on their migration routes.

The homological composition of PCBs is characterized by a large number of higher chlorinated congeners (which are distinguished by a greater resistance and half-life due to their high solubility in fats) and a lower proportion of volatile, lower chlorinated analogues.

Interspecific differences in POP levels are well consistent with potential differences in trophic level between species due to the changes in diet structure and feeding grounds. Wintering grounds are also an important factor. The pattern of OCC accumulation reflects differences in migration routes. Different species even within the same family show different migration preferences [9, 20–22].

Various types of migrations in members of the family Anatidae cause certain organic xenobiotics to accumulate [11]. Thus, a characteristic accumulation of higher chlorinated PCB congeners such as 118, 153, 138, and 180 was identified for mallards, pintail, tufted duck, and greater scaup; a general predominance of ΣPCB over ΣOCP was also observed in the liver and muscles of the latter two species and falcated duck (*Anas falcata*) (Fig. 10.3).

Such a composition of toxicants in birds' bodies is explained by the specifics of migration routes that are confined to sea coasts and open sea waters. Thus, migration routes of large flocks may extend at a distance of up to 5–7 km offshore. Wintering grounds are located in Japan, South Korea, and China [9, 15, 21]. These oversea routes are areas of high shipping activities, which is very likely a source of PCBs in these waters.

Eurasian teal and mandarin duck showed a dominance of DDT, HCH, and dieldrin groups, i.e. OCP compounds, in the overall structure of toxicants (Fig. 10.3). A study by Kunisue and co-authors [3] identified a pattern of build-up of OCC concentrations in local birds, in particular PCBs in Japan, PCBs and chlordane in the Philippines, HCHs and DDT in India, DDT in Vietnam, and PCB and DDT in Lake Baikal. In this regard, the assumption can be made that the collected birds from Lake Khanka, which have their wintering grounds in the region of tropical Asia and migration routes that run over the territory of China, accumulate organic pollutants there.

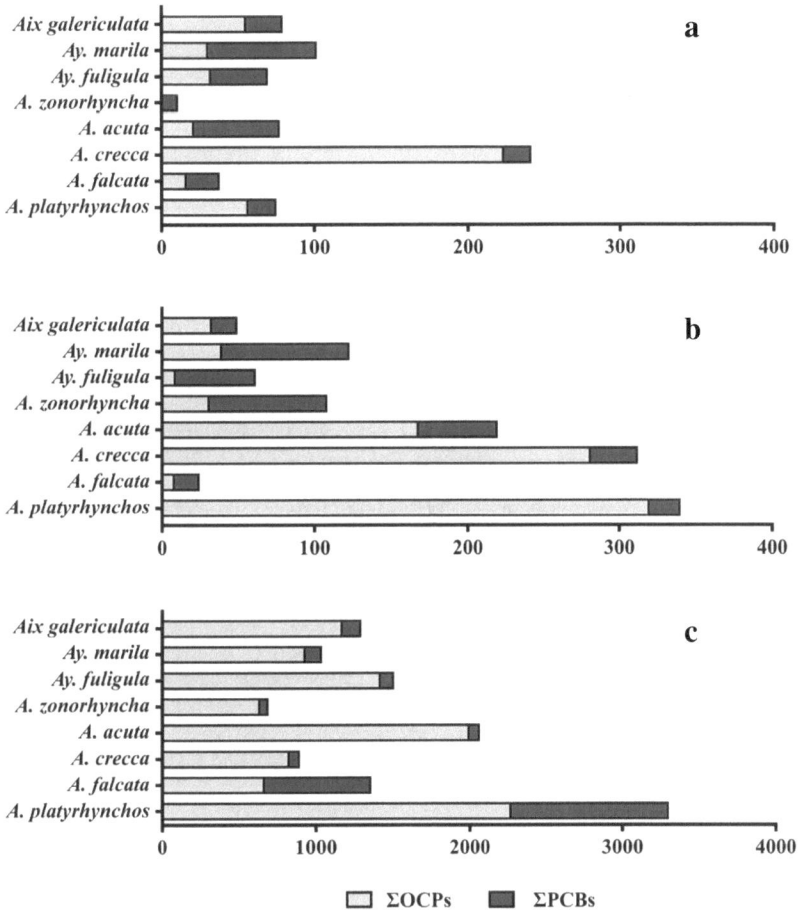

Fig. 10.3 Levels of ΣOCP and ΣPCB in muscles (**a**), liver (**b**), and plumage (**c**) of birds of the family Anatidae, ng/g l.w

10.5 Conclusions

Persistent organic pollutants have carcinogenic, cytotoxic, teratogenic and genotoxic effects. Therefore, the accumulation of toxicants in birds' bodies exerts a negative impact both on the health of adult individuals and on the development of embryos in eggs. Further monitoring of POP accumulation by migratory birds will provide new data on the contribution of their migrations to long-range transboundary transport of pollutants. Despite the predominance of forms that indicate a long period since the entry of xenobiotics into the environment, the detection of *fresh* pollutants necessitates continuing the study of birds to elucidate levels of OCPs in them and ways of transformation and accumulation of these substances in the environment.

References

1. Iatrou EI, Tsygankov V, Seryodkin I et al (2019) Monitoring of environmental persistent organic pollutants in hair samples collected from wild terrestrial mammals of Primorsky Krai, Russia. Environ Sci Pollut Res 26:7640–7650. https://doi.org/10.1007/s11356-019-04171-9
2. Rovinsky FYa, Voronova LD, Afanasev MI et al (1990) Background monitoring of ground ecosystems contamination by organochlorine compounds. Gidrometeoizdat, Leningrad
3. Kunisue T, Watanabe M, Subramanian A et al (2003) Accumulation features of persistent organochlorines in resident and migratory birds from Asia. Environ Pollut 125:157–172. https://doi.org/10.1016/S0269-7491(03)00074-5
4. Tanabe S, Subramanian A (2006) Bioindicators of POPs: monitoring in developing countries. Kyoto University Press; Trans Pacific Press, Kyoto, Japan; Melbourne
5. Tsygankov VYu, Lukyanova ON, Boyarova MD (2018) Organochlorine pesticide accumulation in seabirds and marine mammals from the Northwest Pacific. Mar Pollut Bull 128:208–213. https://doi.org/10.1016/j.marpolbul.2018.01.027
6. Tsygankov VYu, Lukyanova ON (2019) Current levels of organochlorine pesticides in marine ecosystems of the Russian far eastern seas. Contemp Probl Ecol 12:562–574. https://doi.org/10.1134/S199542551906009X
7. Lvov DK, Prilipov AG, Shchelkanov MYu, et al (2006) Molecular genetic analysis of the biological properties of highly pathogenic influenza a/H5n1 virus strains isolated from wild birds and poultry during epizooty in western Siberia (July 2005). Probl Virol 51
8. Lvov DK, Shchelkanov MYu, Prilipov AG et al (2008) Interpretation of the epizootic outbreak among wild and domestic birds in the south of the european part of Russia in December 2007. Probl Virol 53
9. Shchelkanov MYu (2010) Evolution of highly pathogenic avian influenza virus (H5N1) in ecosystems of Northern Eurasia (2005–2009). Abstract of the dissertation for the degree of doctor of biological sciences, D.I. Ivanovsky Institute of Virology
10. Dunaeva MN, Pankratov DV, Surovyi AL et al (2022) Reconstruction of epizootic outbreak provoked the largescale death of Rhinoceros auklet on the coast of the Japan Sea in the southern part of Primorsky Krai (July, 2021). Acta Biomedica Scientifica 7:90–97. https://doi.org/10.29413/ABS.2022-7.3.10
11. Furness RW (1993) Birds as monitors of pollutants. In: Furness RW, Greenwood JJD (eds) Birds as monitors of environmental change. Springer, Netherlands, Dordrecht, pp 86–143
12. Fedorov LA, Yablokov AV (1999) Pesticides—a toxic blow to the biosphere and human. Nauka, Moscow, Russia
13. Hao Y, Zheng S, Wang P et al (2021) Ecotoxicology of persistent organic pollutants in birds. Environ Sci Proc Impacts 23:400–416. https://doi.org/10.1039/D0EM00451K
14. Bezel VS (2006) Ecological toxicology: population and biocenotic aspects, scientific book. Goshchitsky Publishing House, Ekaterinburg, Russia
15. Ivanter EV, Medvedev NV (2007) Ecology toxicology of natural populations bird and mammals in the North. Nauka, Moscow
16. Becker PH (1989) Seabirds as monitor organisms of contaminants along the German North Sea coast. Helgolander Meeresunters 43:395–403. https://doi.org/10.1007/BF02365899
17. Perrins CM, Lebreton JD, Hirons GJM (1991) Bird population studies: relevance to conservation and management. Oxford University Press, Oxford, New York
18. Vystavnoy AL, Lider MG, Ryabikov AY (2006) Biochemical content and some features of coccygeal gland fluid in hens and ducks. Omsk Scientific Bulletin
19. Jaspers VLB, Voorspoels S, Covaci A et al (2007) Evaluation of the usefulness of bird feathers as a non-destructive biomonitoring tool for organic pollutants: a comparative and meta-analytical approach. Environ Int 33:328–337. https://doi.org/10.1016/j.envint.2006.11.011
20. Polivanova NN (1971) Birds of the Khanka Lake (hunting and commercial waterfowl and colonial). Far Eastern Scientific Center of the Academy of Sciences of the USSR, Vladivostok, Russia

21. Shchelkanov MYu, Usachev YV, Fedyaklna IT et al (2006) Newcastle disease virus in the populations of wild birds in the south of the Primorye territory in the period of autumn migrations in 2001–2004. Probl Virol 51
22. Nechaev VA, Gamova TV (2009) Birds of the Russian Far East (annotated catalog). Dal′nauka, Vladivostok

Chapter 11
Organochlorine Pesticides in Seabirds and Marine Mammals from the Sea of Okhotsk and Bering Sea

Abstract The results of measurements of HCH isomers and DDT and its metabolites in the organs of seabirds (*Larus schistisagus, Aethia cristatella, A. pusilla, Fulmarus glacialis*, and *Oceanodroma furcata*) and marine mammals (*Eschrichtius robustus* and *Odobenus rosmarus divergens*) from the Sea of Okhotsk and Bering Sea are considered in this chapter. It has been shown that the species-specific features of accumulation of lipophilic xenobiotics can largely be explained by the type of food consumed and the total fat content of the organs.

Keywords HCHs · DDTs · Seabirds · Marine mammals · Sea of Okhotsk · Bering Sea

11.1 Introduction

In the 1960s–1970s, cases of mass mortality of wild bird populations were reported from various parts of the world. One of the documented causes of such mortalities was the use of persistent organic pollutants (POPs), in particular, organochlorine pesticides (OCPs), which were widely applied as insecticides to control malaria vectors and also pests in agriculture and residential areas in the past. These anthropogenic substances, being lipophilic, are capable of accumulating in environmental components and exerting negative effects on living organisms. As was reported, partridges and pheasants died from endrin poisoning; bald eagles and geese, from dieldrin poisoning; cormorants, pelicans, and gulls, from dichlorodiphenyltrichloroethane (DDT) poisoning; starlings, from lindane (hexachlorocyclohexane, HCH) poisoning [1, 2].

After the ban on the use and production of a number of pesticides, bird mortalities have reduced substantially, but cases of poisoning continued. It has turned out that in areas contaminated by OCPs, birds become particularly sensitive to other groups of chemical and biological damaging agents. Thus, the presence of dichlorodiphenyldichloroethylene (DDE) masks the negative effect of mercury on birds' reproduction [1, 3].

V. Tsygankov, *Persistent Organic Pollutants in the Ecosystems of the North Pacific*,
Earth and Environmental Sciences Library,
https://doi.org/10.1007/978-3-031-44896-6_11

233

Lethal doses of pesticides are species-specific and range from 4 (for the bobolink *Dolichonyx oryzivorus* and Japanese quail *Coturnix japonica*) to 65 mg/kg (for sparrows). A number of experiments have revealed critical OCP concentrations in the bird brain [4].

In the Great Britain, a correlation was found, for the first time, between the thinning of eggshell and reproductive failure in the peregrine falcon and Eurasian sparrowhawk populations in areas where persistent organochlorine insecticides were used. The eggshell thickness decreased by 5–19% by the late 1960s, compared to 1940, in nine out of 17 individuals examined [5]. Studies in the USA and Canada also showed the degree of eggshell thinning: the thickness and weight of the eggshell decreased by 20% in many bird species [4, 6].

Birds are widely used as bioindicators for monitoring environmental pollution by organochlorine compounds. Birds can be both an intermediate and upper link in the food chain. Feeding on living organisms, they accumulate higher concentrations of toxic substances in their organs due to the biomagnification effect [7]. Depending on feeding specialization (plant eaters, plankton feeders, fish eaters, etc.) and types of migrations (migratory and residential), the POP level in birds' organs and tissues also varies. In the absence of local sources of pollution, POP accumulation indicates global pollution as a result of transboundary transport with air masses and sea currents [5, 8].

Researchers studying issues of global pollution of the natural environment and effects of this process on wildlife have long considered the Arctic as a target region. The Arctic almost completely lacks sources of anthropogenic pollution, but is exposed to the continuous effect of intensive transport from more southerly latitudes. Various pollutants easily reach the Arctic region by air or water, as well as with migrating organisms (the so-called biotransport), immediately become involved in the exchange of matter and energy, and, thus, have a negative impact on ecosystems and their inhabitants, including mammals [6, 9–15].

Marine mammals can be regarded as convenient species for long-term monitoring of marine pollution by POPs. They can be used as indicators of global pollution and biomonitors of trends in biosphere pollution variations [13, 15, 16].

11.2 Seabirds

We analyzed a total of 41 samples of organs from five seabird species collected in June and October 2012 from the coast of western Kamchatka and the Kuril Islands in the Sea of Okhotsk: the Slaty-Backed Gull (*Larus schistisagus*), Crested Auklet (*Aethia cristatella*), Least Auklet (*Aethia pusilla*), Northern Fulmar (*Fulmarus glacialis*) (light and dark color morphs), and Fork-Tailed Storm Petrel (*Oceanodroma furcata*). Depending on the body size of birds, different organs were examined: plumage, plumage with skin, muscles, liver, and whole carcasses (internal organs with plumage).

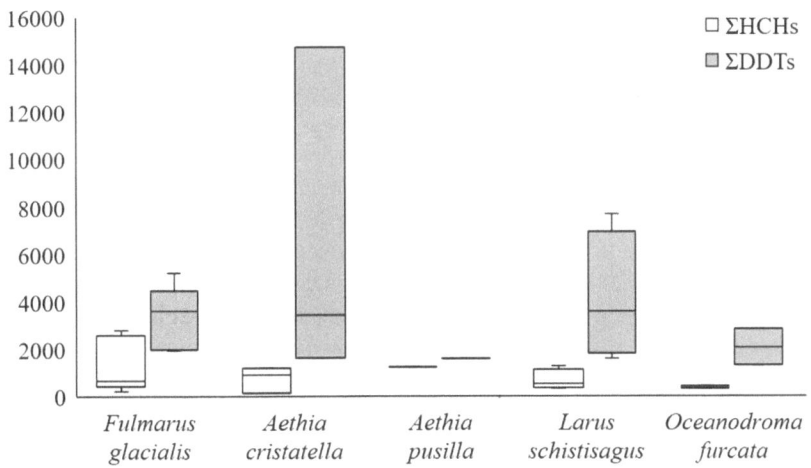

Fig. 11.1 Average concentration (median) of HCH isomers and DDT and its metabolites in birds' organs and tissues, ng/g l.w.

Pesticides were found in all samples. The total level in different organs varied from 28 to 16,095 ng/g lipid weight (l.w.). In plumage, the range of values was 28–8289 ng/g; in plumage with skin, 1567–16,095 ng/g; in liver, 1679–2478 ng/g, in muscles, 2230–3000 ng/g; in the homogenate of organs, 12.5–15,112.0 ng/g. The average total concentrations of pesticides in certain organs are shown in Fig. 11.1.

DDT was found within a range of 975–1978 ng/g only in plumage from fulmars captured in June. DDE was present in all the samples within a range from 27 to 15,276 ng/g. DDD was not detected in the samples.

γ-HCH was found only in some of the samples: in plumage and plumage with skin of fulmars in June, within 177–467 ng/g; in liver of slaty-backed gulls, at 160 ng/g. β-HCH was detected in single samples: in plumage of a crested auklet and in the carcass of a least auklet in June and in the liver of a fulmar in October. The concentration of β-HCH was 555–1151 ng/g. α-HCH was found in almost all the samples in a range from 160 to 3024 ng/g, except for fulmars captured in October (in one individual, it was absent from plumage, and in another one, it was present only in muscles) and slaty-backed gulls (below the detection limits in plumage of one individual). The average value of total OCP concentration in June (6580 ng/g) was almost twofold higher than in October (3442 ng/g).

Slaty-Backed Gull, Larus schistisagus. The total OCP level in all organs of slaty-backed gulls was within a range of 1567–10,357 ng/g l.w. The α-HCH concentrations ranged from 200 to 2017 ng/g; the amount of β-HCH was below the detection limits; γ-HCH was found in the liver of one of the two individuals, where the concentration was 160 ng/g; the α/γ-HCH ratio was 2.7, which indicates a long-term circulation of HCH isomers in the marine environment. The DDT and DDD concentrations were below the detection limits; the DDE level ranged from 1016 to 8339 ng/g.

Northern Fulmar, Fulmarus glacialis. The range of total OCP concentrations was from 28 to 8135 ng/g l.w. The total of HCH isomers ranged from 200 to 3040 ng/g: α-HCH, from 200 to 3024 ng/g; β-HCH was found at a concentration of 555 ng/g in the liver from one out of five birds; γ-HCH was found within a range from 177 to 467 ng/g in plumage and plumage with skin from three out of five individuals. In all the samples where γ-HCH was detected, the α/γ-HCH ratio was greater than 1.

The values of the total of DDT and its metabolites ranged from 28 to 5625 ng/g l.w.: DDT was found within a range of 1440–1978 ng/g l.w. in plumage from three individuals; DDD was below the detection limit; DDE ranged from 28 to 5608 ng/g l.w. The ratio of the average DDT and DDE concentrations was less than 1, which indicates a *fresh* entry of DDT into the marine environment and bodies of birds.

Crested Auklet, Aethia cristatella. The minimum total OCP concentration was 1842, and the maximum 16 095 ng/g l.w. The range of total HCH values ranged within 160–2214 ng/g; the range of α-HCH was from 160 to 1062 ng/g. β-HCH was detected at a concentration of 115 ng/g in plumage from one out of two individuals; γ-HCH was below the detection limit. DDT and DDD were also below the detection limits; the range of DDE was from 1604 to 15,276 ng/g.

Least Auklet, Aethia pusilla. The total of OCPs ranged from 2247 to 3499 ng/g. The total concentration of HCH isomers ranged from 626 to 1819 ng/g l.w.; α-HCH, from 406 to 1033 ng/g l.w.; β-HCH, from 786 to 810 ng/g; γ-HCH was not detected. DDT and DDD were below the detection limits; DDE ranged from 1450 to 1679 ng/g.

Fork-Tailed Storm Petrel, Oceanodroma furcata. One individual was analyzed. The concentration range of the total of OCPs was from 1705 to 3128 ng/g. Of the HCH isomers, only the α-HCH was detected within a range from 303 to 415 ng/g. DDT and DDD were below the detection limits; DDE ranged from 1290 to 2825 ng/g.

Total level of pesticides in carcasses (organs and tissues) of birds. The maximum OCP value was recorded from fulmars (5816 ng/g); the minimum, from the fork-tailed storm petrel (1705 ng/g). DDT and its metabolites showed similar pattern: the maximum levels were found in fulmars (5608 ng/g) and the minimum in the fork-tailed storm petrel (1209 ng/g). The highest level of HCH isomers was found in a least auklet (1819 ng/g) and the lowest level in a fulmar (208 ng/g l.w.). The average concentrations of HCH and DDT in certain organs are shown in Fig. 11.2.

Level of pesticides in plumage. The highest total OCP concentration was detected in plumage from a slaty-backed gull (8289 ng/g l.w.). The concentrations in the three fulmars captured in June were high and close (7119, 7916, and 8135 ng/g l.w.), but the OCP concentration in October decreased sharply (only 28 ng/g). The highest total concentration of DDE (the only identified DDT metabolite) was also found in a slaty-backed gull (766 ng/g); the lowest concentration, in the fulmar captured in October (28 ng/g). The total of HCH isomers reached the maximum in the fulmars captured in June (3040 ng/g); the minimum, in a crested auklet (160 ng/g).

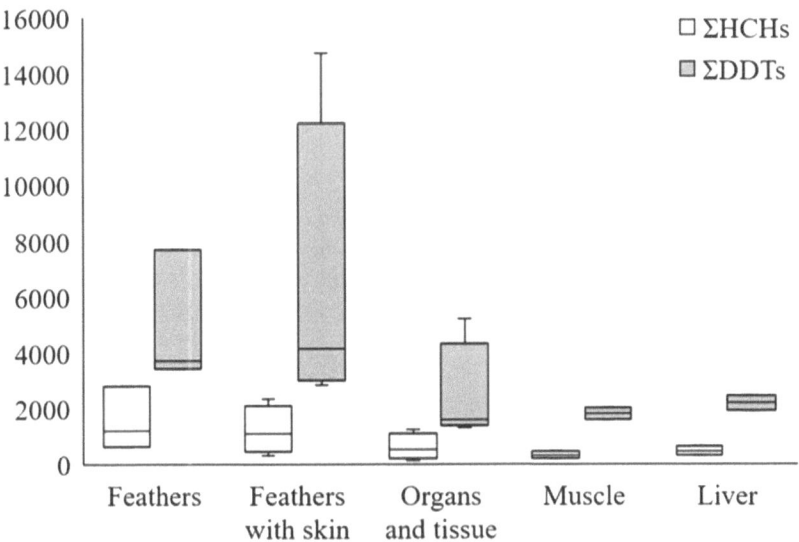

Fig. 11.2 The average content (median) of HCH isomers and DDT and its metabolites in seabirds' organs, ng/g l.w.

Level of pesticides in plumage with skin. The maximum total amount of OCPs and DDT metabolites was recorded from a crested auklet (16,095 and 15,276 ng/g, respectively); the minimum, from a slaty-backed gull (1567 and 1016 ng/g, respectively). The highest total level of HCH isomers was found in a fulmar (3047 ng/g); the minimum, in the fork-tailed storm petrel (303 ng/g l.w.).

OCPs in **liver** and **muscles** were recorded from two species of birds captured in October: the Northern Fulmar and the Slaty-Backed Gull.

Level of pesticides in liver. In the livers of fulmars, the total amounts of OCPs (2478 ng/g) and DDT metabolites (1923 ng/g) exceeded those in slaty-backed gulls (1679 and 1376 ng/g, respectively). The maximum level of HCH isomers was found in the liver of a slaty-backed gull (590 ng/g).

Level of pesticides in muscles. In muscles of slaty-backed gulls, the total amounts of OCPs (3000 ng/g) and DDT metabolites (2775 ng/g) were higher than in fulmars (2230 and 2030 ng/g, respectively). The total of HCH isomers (410 ng/g) was also higher than that in fulmars (200 ng/g).

The bird species analyzed differ in size, lifestyle, and amount of subcutaneous fat. OCPs are known to build up mainly in subcutaneous fat and, therefore, larger birds have a higher OCP content. Slaty-backed gulls (with their average body length of 64 cm) and northern fulmars (47 cm) had similar OCP concentrations in plumage with skin (5962 and 5949 ng/g, respectively). In the fork-tailed storm petrel (22 cm), which is much smaller than the bird species above, the pesticide concentration was 3128 ng/g. However, the accumulation and distribution of pesticides in the organs of birds can be regulated by other additional factors. For example, in a crested auklet of 26 cm in size, the OCP level in plumage with skin reached 15,604 ng/g. Further research is required to identify causes of such high OCP concentrations in this species.

The level of pesticides in the internal organs of birds is also influenced by their diet spectra. The diet items of fulmars include organisms characterized by a high coefficient of OCP accumulation (fish, fish eggs, mollusks, crustaceans and other invertebrates, carrion, whale entrails, various fatty waste, etc.), which results in a high pesticide content birds' carcasses (5874 ng/g). Crested auklets, least auklets, and fork-tailed storm petrels feed on small crustaceans, marine invertebrates, amphipods, etc. that accumulate pesticides to a lesser extent than fulmar's diet items. This is evidenced by a lower OCP level in the internal organs of these bird species: 1730, 2804, and 1705 ng/g l.w., respectively.

Data on HCH and DDT levels in birds from the Sea of Okhotsk and other areas of the world's oceans are presented in Table 11.1.

The total OCP level in the slaty-backed gulls, fork-tailed storm petrel, and least auklets from the Sea of Okhotsk was higher than in the Little Auk (*Alle alle*), Thick-Billed Murre (*Uria lomvia*), Black-Legged Kittiwake (*Rissa tridactyla*), and Black Guillemot (*Cepphus grylle*) from Baffin Bay, and lower than in the Northern Fulmar (*F. glacialis*) from Baffin Bay, the Lesser Black-Backed Gull (*L. fuscus*) and Great Black-Backed Gull (*L. marinus*) from Iceland, the Glaucous Gull (*L. hyperboreus*) from Baffin Bay, the Ivory Gull (*Pagophila eburnea*) and Great Skua (*Stercorarius skua*) from Iceland, and the Glaucous Gull (*L. hyperboreus*) from the Barents Sea.

In the organs and tissues of crested auklets, the overall level of OCPs was lower than that in great skuas from Iceland and glaucous gulls from the Barents Sea. The fulmars from the Sea of Okhotsk had a higher OCP level than fulmars from Baffin Bay, but lower than other bird species from Baffin Bay.

The total level of DDT metabolites in the fulmars, slaty-backed gulls, fork-tailed storm petrel, and least auklets from the Sea of Okhotsk was higher than that in little auks, thick-billed murre, black-legged kittiwakes, and black guillemots from Baffin Bay and lower than that in fulmars from Baffin Bay, black-backed gulls from Iceland, black-backed gulls from Iceland, glaucous gulls from Baffin Bay, ivory gulls from Iceland, great skuas from Iceland, and glaucous gulls from the Barents Sea. In all the birds analyzed, the total of DDT metabolites was lower than in great skuas from Iceland and glaucous gulls from the Barents Sea.

The concentration of the total of HCH isomers in the fork-tailed storm petrel was higher than that in black-legged kittiwakes, fulmars, black-backed gulls, thick-billed murres, ivory gulls, little auks, and black guillemots and lower than in the other species. The total of HCH isomers in the slaty-backed gulls and crested auklets from the Sea of Okhotsk was higher than in black-legged kittiwakes, fulmars, black-backed gulls, thick-billed murres, little auks, ivory gulls, black guillemots, and glaucous gulls from Baffin Bay and lower than in the other species. In the least auklets from the Sea of Okhotsk, the total of HCH isomers was higher than in great skuas from Iceland and lower than in the other species. The total of HCH isomers was higher in fulmars from the Sea of Okhotsk than in all the other species.

Table 11.1 The total level of organochlorine pesticides (DDT and HCH) in seabirds from various regions of the world's oceans, ng/g l.w.

Species	Region	$\sum$DDT	$\sum$HCH	References
Little Auk (*Alle alle*)	Arctic Ocean, Baffin Bay	571.4	233.9	[3]
Thick-Billed Murre (*Uria lomvia*)		1096.7	84.3	
Black Guillemot (*Cepphus grylle*)		1003.3	268.3	
Black-Legged Kittiwake (*Rissa tridactyla*)		1277.1	45.3	
Ivory Gull (*Pagophila eburnea*)		7049.3	138.1	
Glaucous Gull (*Larus hyperboreus*)		5956.1	449.2	
Northern Fulmar (*Fulmarus glacialis*)		4870.7	62.9	
Great Black-Backed Gull (*Larus marinus*)	Iceland	3420	0	[2]
Lesser Black-Backed Gull (*Larus fuscus*)		3370	36.7	
Great Skua (*Stercorarius skua*)		54,200	1400	
Glaucous Gull (*Larus hyperboreus*)	Barents Sea	215,109	1038	[8]
Northern Fulmar (*Fulmarus glacialis*)	Sea of Okhotsk	3335.3	1630.47	Present study
Crested Auklet (*Aethia cristatella*)		7575.36	884.78	
Slaty-Backed Gull (*Larus schistisagus*)		3576.41	674.66	
Fork-Tailed Storm Petrel (*Oceanodroma furcata*)		2017.05	377.1	
Least Auklet (*Aethia pusilla*)		1583.5	1220.97	

In all the samples analyzed and in the data provided by various authors (Table 11.1), a common pattern is observed: the total of HCH isomers is lower than the total of DDT metabolites.

All the birds under study are largely sedentary species and, accordingly, the level of bioaccumulation of organochlorine compounds by birds from the Sea of Okhotsk reflects the level of pollution of this region. The species listed in Table 11.1 were sampled at similar latitudes of the Northern Hemisphere. The ranges of POP concentrations in all birds were almost similar, which indicates similar degrees of pollution of the marine environment at these latitudes.

Species-specific patterns of OCP accumulation may primarily be related to feeding specialization. For example, glaucous gulls from the Barents Sea, feeding on fish, carrion, small mammals, and eggs of other birds, contain 216,157 ng OCP/g l.w. in their bodies [8]. Bird eggs are rich in fat essential for embryo development and, accordingly, contain OCPs at high concentrations, because during ovogenesis all toxic substances along with fat from the female enter her eggs that may become food for other birds [17, 18]. Bird eggs and, specifically, eggshell are often used as indicators of OCP contamination due to the maximum accumulation of pesticides compared to other marine organisms [4].

Great skuas, whose food consists mainly of fish that they take away from other birds, accumulate pollutants up to 55,600 ng/g [2]. Fish, unlike birds and mammals, contain usually lower levels of pesticides in their bodies, since they represent a lower link in the food chain. Thus, the OCP content in skuas is significantly lower than in glaucous gulls [19].

Inhabitants of Baffin Bay (Arctic Ocean) accumulate less OCPs due to the remoteness of the region from potential sources of pollution. For example, northern fulmars that feed on crustaceans, fish, squid, plankton, and, occasionally, carrion contain up to 4933.5 ng/g of toxicants in their bodies [3].

The accumulation of pesticides in birds affects various aspects of their physiology, e.g., causing serious deterioration of the reproductive function and thinning of eggshells, which results in impaired embryonic development and loss of offspring [4]. The different levels of OCP accumulation in some of the species reflect the different degrees of pollution of their habitats. The species-specific patterns of accumulation of lipophilic xenobiotics are largely determined by both the diet spectrum and the total fat content of certain organs. The detection of OCPs at marked concentrations in seabirds from the Sea of Okhotsk, which is isolated from intensive agricultural activities, indicates the general global background of POPs that has formed on the planet to date.

11.3   Marine Mammals

We studied samples of organs (muscles and liver) from seven gray whales landed in September 2010 and eight Pacific walruses landed in September 2011 by local hunters and whalers from the coastal waters off the village of Lorino (Mechigmen Bay, Chukchi Autonomous Okrug, Bering Sea).

The International Whaling Commission (IWC) granted the exclusive right to hunt gray whales of the eastern (California-Chukchi) population to the indigenous peoples of Chukotka (Russia) and Alaska (USA) for subsistence and maintaining their traditional way of living.

11.3.1 Gray Whale (Eschrichtius Robustus)

Pesticides were found in all samples. The range of OCP concentrations in whale organs was wide. Concentrations in the liver were significantly higher than in muscles (Fig. 11.3). This can probably be explained by the detoxifying properties of the liver which provides degradation of the major part of toxicants in the body.

In general, we found a tendency of the level of pollutants in muscles and liver to increase in larger individuals compared to smaller ones (Fig. 11.4). We also observed the same relationship with age, as mammals accumulate toxicants throughout their lives.

In females, pesticide concentrations show a tendency to increase until maturity. After the onset of reproductive age, OCP concentrations level off or even decrease [20]. Gray whales reach maturity at age 5–10 yr (on average, 8 yr) [21]. The organ and tissue samples that we analyzed were taken from immature females and, therefore, we only observed an increase in OCP concentration with age.

The concentration of pesticides in muscles ranged within 297–3581 ng/g l.w.; in the liver, within 769–13,808 ng/g l.w. (Table 11.2 and Fig. 11.5).

We did not find statistically significant sex-related differences in the HCH and DDT levels in muscles; however, the concentration of xenobiotics tended to increase in males (except DDT and DDD that were not found). In the livers of males, only the α-HCH concentration ($p = 0.05$) was significantly higher compared to that in females; however, as can be seen, the concentrations of all other xenobiotics increased in males. The low statistical significance of the results obtained is probably explained by the small sample size.

The total levels of various organochlorine pesticides in the muscles and liver samples from gray whales are listed in Table 11.2.

Statistically significant differences between muscles and liver were found only for β-HCH ($p = 0.002$). As regards other pollutants, DDT and its metabolites tended to increase in the liver and HCH in muscles.

When comparing the median values of $\sum$HCH, $\sum$DDT, and $\sum$OCP, we found that the organs differed significantly only in the level of $\sum$OCP (which was higher in

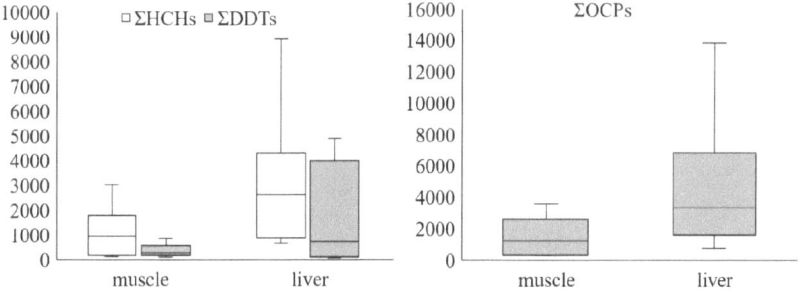

Fig. 11.3 The total content of OCPs, HCH isomers, and DDT and its metabolites in gray whale's muscles and liver, ng/g l.w.

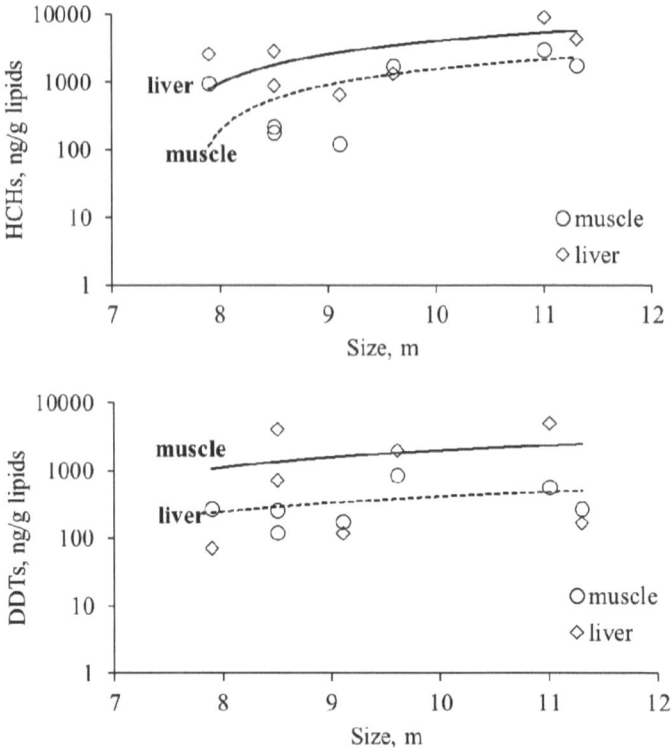

Fig. 11.4 The relationship of HCH and DDT concentrations in gray whale's organs with body size

Table 11.2 Mean values (numerator) and ranges of (denominator) of OCP concentrations in gray whales' organs, ng/g l.w.

Organs	α-HCH	β-HCH	γ-HCH	DDT	DDD	DDE
Muscles	728 ± 733 70–1974	131 ± 257 DL–676	285 ± 356 14–1043	44 ± 117 DL–310	< DL	317 ± 175 121–565
Liver	1047 ± 2257 72–6160	1499 ± 909 392–2754	535 ± 881 DL–2161	481 ± 729 DL–1552	34 ± 91 DL–240	1192 ± 1243 71–3132

Note DL is detection limit

the liver than in muscles) ($p = 0.048$). $\sum$HCH and $\sum$DDT only tended to increase in the liver.

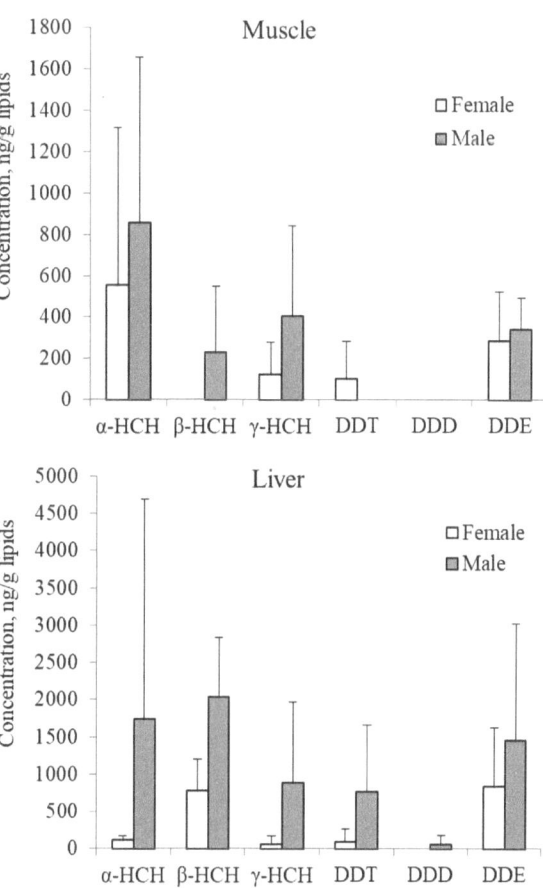

Fig. 11.5 OCP concentrations in muscles and livers of female and male gray whales, ng/g l.w.

11.3.2   *Pacific Walrus (Odobenus Rosmarus Divergens)*

Pesticides were found in all walrus samples. The OCP concentrations in the animals' organs had wide ranges. The concentrations of pesticides in the liver were significantly higher than in muscles. In male and female muscles, the total OCP concentrations were similar (Fig. 11.6). However, the average concentration in the female liver was higher than in the male liver. Males showed a tendency to accumulate pesticides with increasing body size, both in muscles and in the liver.

The total OCP concentration (ΣHCH + ΣDDT) in the liver was within a range of 4900–90,300 ng/g l.w. It was significantly higher than the range in muscles, 200–5700 ng/g l.w. All HCH isomers and a DDT metabolite were found in muscles; HCH isomers, DDT, and DDE were detected in the liver (Table 11.3 and Fig. 11.7).

There were no statistically significant differences between the levels of toxicants in male and female muscles; however, the concentration of all pollutants increased in females, except for α-HCH that was higher in males (Fig. 11.7). The levels of

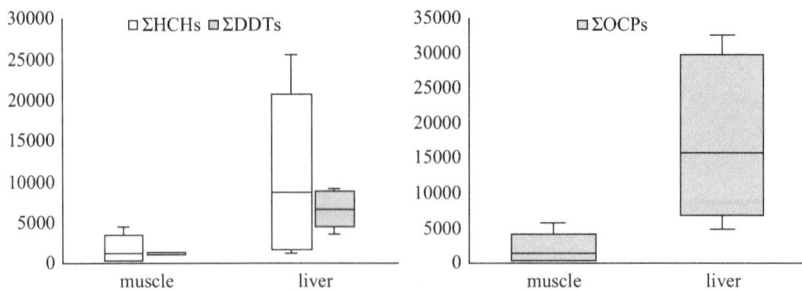

Fig. 11.6 The total levels of OCPs, HCH isomers, and DDT and its metabolites in Pacific walrus' muscles and liver, ng/g l.w.

Table 11.3 Mean values (numerator) and ranges (denominator) of OCP concentrations in Pacific walrus' organs, ng/g l.w.

	α-HCH	β-HCH	γ-HCH	DDT	DDE
Muscles	490 ± 679	405 ± 402	728 ± 986	405 ± 56	< DL
	DL–1826	DL–916	DL–2482	DL–1287	
Liver	8045 ± 10,647	440 ± 87	2056 ± 4085	4474 ± 3302	9282 ± 23,546
	216–25,307	DL–2536	DL–12,070	DL–9161	DL–67,238

Note DL is detection limit

toxicants in the liver, as well as in muscles, did not have any statistically significant differences, but the concentrations of all the pesticides detected increased in females, except for β-HCH that showed a higher concentration in males (Fig. 11.7).

The total levels of various OCPs in muscles and liver of Pacific walruses are provided in Table 11.3. When comparing organs, we obtained statistically significant results only for two substances, α-HCH ($p = 0.016$) and DDT ($p = 0.021$), with their concentrations being higher in the liver. γ-HCH and DDE tended to increase in the liver, and β-HCH did in muscles.

When comparing the median values of $\sum$HCH, $\sum$DDT, and $\sum$OCP in muscles and liver, we found that all results were statistically significant ($p = 0.001$–0.036) and indicated the predominance of pollutants in the liver.

11.3.3   Analysis of OCPs in the Organs of Gray Whales and Pacific Walruses

In the gray whale's organs, the lipid content reached 4% in muscles and 8% in the liver; in the Pacific walrus' organs, up to 6% in muscles and up to 10% in the liver. Calculations showed (Table 11.4) that the total OCP level in gray whale's fat was, on average, 6290 ng/g; in Pacific walrus' fat, 26,300 ng/g.

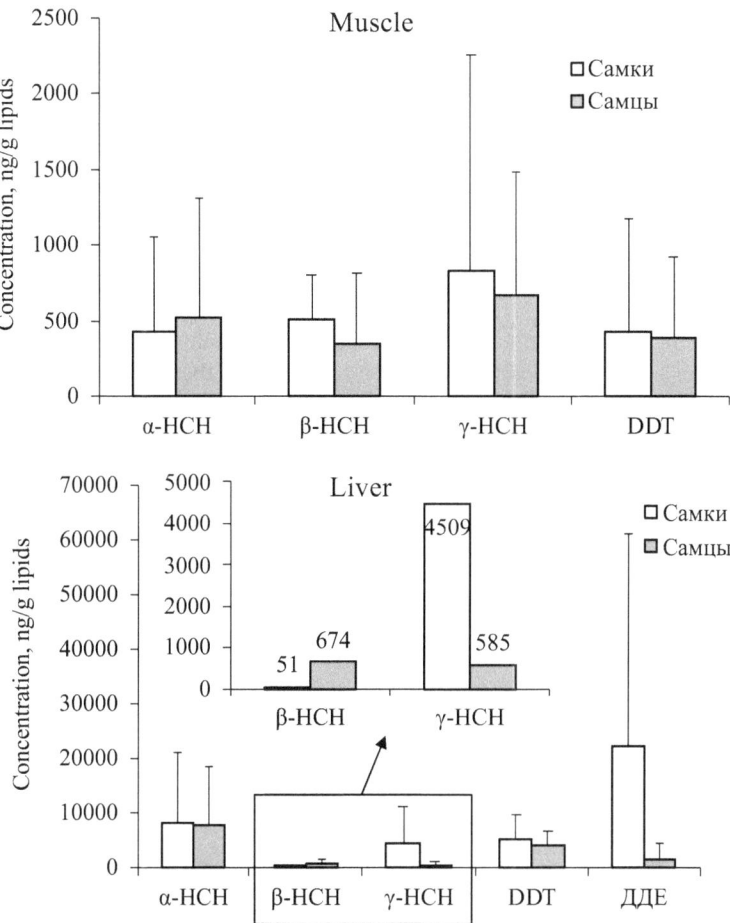

Fig. 11.7 OCP concentrations in muscles and liver of female and male Pacific walruses, ng/g l.w.

Table 11.4 Mean values (±standard deviation) of total OCP level in the lipid fraction from gray whale and Pacific walrus, ng/g l.w

Species analyzed	$\sum$HCH	$\sum$DDT	$\sum$OCP
Gray Whale (*Eschrichtius robustus*) (Mean ± SD, N = 7)	4220 ± 1428	2070 ± 427	6290 ± 1855
Pacific Walrus (*Odobenus rosmarus divergens*) (Mean ± SD, N = 8)	14,140 ± 4125	12,160 ± 2364	26,300 ± 6489

The species-specific patterns in the accumulation of lipophilic xenobiotics in marine mammals are determined largely by the total fat content of subcutaneous tissue and certain organs. The degree of maturity of individuals is also of great importance. The species that we studied have overlapping ranges of distribution, and the fat contents of their organs differ insignificantly, amounting to 8–10%. Consequently, the substantial differences in the pesticide level may be associated with the stage of reproductive cycle of the animals and their feeding habits. The diet of gray whales consists mainly of bottom crustaceans and other small benthic organisms that live both on the surface and in the subsurface layers of soft sediments (infauna). The major part of walrus' diet is made up of benthic invertebrates: bivalves, some species of shrimp, polychaete worms, priapulids, octopuses, holothurians, and also some fish species. Furthermore, walruses sometimes may prey on other seals: cases of walruses attacking ringed seals and harp seal pups have been documented (Fig. 11.8) [21].

Thus, transboundary atmospheric transport and sea currents may be considered possible sources of pesticides entering the marine environment in the Bering Sea.

The OCP concentrations in the marine mammals under study were significantly lower than those in the Striped Dolphin (*Stenella coeruleoalba*) from the northwestern coast of Japan, in the False Killer Whale (*Pseudorca crassidens*) from the Hawaii coast, in the Killer Whale (*Orcinus orca*) from the eastern coastal zone ranging from California to Alaska, in the Indo-Pacific Finless Porpoise (*Neophocaena phocaenoides*) from the coastal zone of South Korea, the Dugong (*Dugong dugon*) from off northeastern Australia, the California Sea Lion (*Zalophus californianus*), and also the Northern Elephant Seal (*Mirounga angustirostris*) and the Harbor Seal (*Phoca vitulina*) from off the urbanized areas of California. However, these

Fig. 11.8 Diet items of gray whale and Pacific walrus

values were higher than the OCP level in the Harbor Porpoise (*Phocoena phocoena relicta*) and the Bottlenose Dolphin (*Tursiops truncatus*) from the Black Sea, and the Harbor Seal (*Phoca vitulina*) from remote areas off California (Table 11.5).

Table 11.5 Total levels of organochlorine pesticides (DDT and HCH) in fat of marine mammals, ng/g l.w.

Species	Region	$\sum$DDT	$\sum$HCH	$\sum$OCP	Period	References
Striped Dolphin (*Stenella coeruleoalba*)	Northwestern coast of Japan	130,000	520	130,500	1978–2003	[22]
False Killer Whale (*Pseudorca crassidens*)	The Hawaii coasts	83,000	100	83,100	2008	[23]
Killer Whale (*Orcinus orca*)	Western coast of North America	99,000	1700	100,700	2007	[24]
Killer Whale (*Orcinus orca*)		160,000	1300	161,300	2004–2006	[25]
Indo-Pacific Finless Porpoise (*Neophocaena phocaenoides*)	Coastal waters of South Korea	45,000	3000	48,000	2003	[26]
Dugong (*Dugong dugon*)	Northeastern coast of Australia	52,000	0	52,000	1996–2000	[27]
Harbor Porpoise (*Phocaena phocaena relicta*)	Black Sea (Turkey)	8500	4900	13,400	2007	[28]
Bottlenose Dolphin (*Tursiops truncatus*)		5300	1000	6300	2007	
California Sea Lion (*Zalophus californianus*)	California, San Francisco	100,000	700	100,700	1997	[29]
Harbor Seal (*Phoca vitulina*)		350,000	100	350,100	1997	
Northern Elephant Seal (*Mirounga angustirostris*)	California, a remote polluted water area	59,000	400	59,400	1997	
Harbor Seal (*Phoca vitulina*)		21,000	100	21,100	1997	
California Sea Lion (*Zalophus californianus*)	California, Monterey Bay	570,000	2100	572,100	1997	
Harbor Seal (*Phoca vitulina*)	Northwest Atlantic	20,000	8500	208,500	2004	
Spotted Seal (*Phoca largha*)	Primorsky Krai (Sea of Japan)	37,600	345,000	382,600	2010	[30]

As the data in Table 11.5 show, the total level of DDT and its metabolites in the organs of marine mammals from other regions of the world's oceans is usually higher than the total of HCH isomers. For example, in striped dolphins from the coastal waters off Japan, the total amount of DDT reaches 130,000 ng/g, while HCH is only 520 ng/g. Gray whales and Pacific walruses have an opposite pattern: the HCH content markedly prevails over the DDT content. The quantitative prevalence of HCH over DDT was also reported for marine organisms from Peter the Great Bay, Sea of Japan [30, 31]. These facts are likely to indicate a more intensive use of lindane and technical HCH in agriculture in the Far Eastern region of Russia.

Thus, accumulation of pesticides is observed in the organs of gray whales of the eastern (California-Chukchi) population and Pacific walruses, but the overall level of their concentrations remains significantly lower than that in marine mammals from other regions of the world's oceans.

11.4   Conclusions

The different levels of pesticide accumulation in certain bird species primarily reflect their food preferences (Fig. 11.8), diet structures, migrations, and fat contents of some of the organs. Items of walrus' diet accumulate more pesticides in their bodies than gray whale's diet, because the coefficients of pollutant accumulation in mollusks and fish are higher than in crustaceans. A conclusion can be made from this fact that food is a determinative factor responsible for the differences in pesticide accumulation between gray whales and walruses. Consequently, seabirds and mammals are models of global biosphere pollution by organochlorine compounds that confirm the urgency of the worldwide problem of aquatic and air environment pollution.

References

1. Rovinsky FY, Voronova LD, Afanasev MI et al (1990) Background monitoring of ground ecosystems contamination by organochlorine compounds. Gidrometeoizdat, Leningrad
2. Jörundsdóttir H, Löfstrand K, Svavarsson J et al (2010) Organochlorine compounds and their metabolites in seven Icelandic seabird species—a comparative study. Environ Sci Technol 44:3252–3259. https://doi.org/10.1021/es902812x
3. Buckman AH, Norstrom RJ, Hobson KA et al (2004) Organochlorine contaminants in seven species of Arctic seabirds from northern Baffin Bay. Environ Pollut 128:327–338. https://doi.org/10.1016/j.envpol.2003.09.017
4. Tanabe S (2007) Chapter 18 Contamination by persistent toxic substances in the Asia-Pacific region. In: Li A, Tanabe S, Jiang G et al (eds) Developments in environmental science. Elsevier, pp 773–817
5. Kunisue T, Watanabe M, Subramanian A et al (2003) Accumulation features of persistent organochlorines in resident and migratory birds from Asia. Environ Pollut 125:157–172. https://doi.org/10.1016/S0269-7491(03)00074-5
6. Tanabe S, Subramanian A (2006) Bioindicators of POPs: monitoring in developing countries. Kyoto University Press, Trans Pacific Press, Kyoto, Japan, Melbourne

7. Tsygankov VY, Boyarova MD, Lukyanova ON (2016) Bioaccumulation of organochlorine pesticides (OCPs) in the northern fulmar (Fulmarus glacialis) from the Sea of Okhotsk. Mar Pollut Bull 110:82–85. https://doi.org/10.1016/j.marpolbul.2016.06.084

8. Knudsen LB, Sagerup K, Polder A et al (2007) Halogenated organic contaminants (HOCs) and mercury in dead or dying seabirds on Bjørnøya (Svalbard). Norwegian Polar Institute, Norway

9. Wania F, Mackay D (1995) A global distribution model for persistent organic chemicals. Sci Total Environ 160–161:211–232. https://doi.org/10.1016/0048-9697(95)04358-8

10. Wania F, MacKay D (1996) Peer reviewed: tracking the distribution of persistent organic pollutants. Environ Sci Technol 30:390A-396A. https://doi.org/10.1021/es962399q

11. Ivanter EV, Medvedev NV (2007) Ecology toxicology of natural populations bird and mammals in the North. Nauka, Moscow

12. Tsygankov VY, Boyarova MD, Lukyanova ON (2014) Persistent toxic substances in the muscles and liver of the pacific walrus Odobenus rosmarus divergens Illiger, 1815 from the Bering Sea. Russ J Mar Biol 40:147–151. https://doi.org/10.1134/S1063074014020102

13. Tsygankov VY, Lukyanova ON, Boyarova MD (2018) Organochlorine pesticide accumulation in seabirds and marine mammals from the Northwest Pacific. Mar Pollut Bull 128:208–213. https://doi.org/10.1016/j.marpolbul.2018.01.027

14. Lukyanova ON, Tsygankov VY, Boyarova MD, Khristoforova NK (2014) Pesticide biotransport by Pacific salmon in the northwestern Pacific Ocean. Dokl Biol Sci 456:188–190. https://doi.org/10.1134/S0012496614030089

15. Tsygankov VY (2019) Organochlorine pesticides in marine ecosystems of the Far Eastern Seas of Russia (2000–2017). Water Res 161:43–53. https://doi.org/10.1016/j.watres.2019.05.103

16. Tsygankov VY, Lukyanova ON (2019) Current levels of organochlorine pesticides in marine ecosystems of the Russian far eastern seas. Contemp Probl Ecol 12:562–574. https://doi.org/10.1134/S199542551906009X

17. Leat EHK, Bourgeon S, Borgå K et al (2011) Effects of environmental exposure and diet on levels of persistent organic pollutants (POPs) in eggs of a top predator in the North Atlantic in 1980 and 2008. Environ Pollut 159:1222–1228. https://doi.org/10.1016/j.envpol.2011.01.036

18. Morales L, Martrat MG, Olmos J et al (2012) Persistent organic pollutants in gull eggs of two species (Larus michahellis and Larus audouinii) from the Ebro delta natural park. Chemosphere 88:1306–1316. https://doi.org/10.1016/j.chemosphere.2012.03.106

19. Tsygankov VY, Boyarova MD, Lukyanova ON, Khristoforova NK (2017) Bioindicators of organochlorine pesticides in the Sea of Okhotsk and the Western Bering Sea. Arch Environ Contam Toxicol 73:176–184. https://doi.org/10.1007/s00244-017-0380-2

20. Hickie BE, Ross PS, Macdonald RW, Ford JKB (2007) Killer Whales (Orcinus orca) face protracted health risks associated with lifetime exposure to PCBs. Environ Sci Technol 41:6613–6619. https://doi.org/10.1021/es0702519

21. Burdin A, Filatova OA, Hoyt E (2009) Marine mammals of Russia. Kirov Regional Press, Kirov, Russia

22. Isobe T, Ochi Y, Ramu K et al (2009) Organohalogen contaminants in striped dolphins (Stenella coeruleoalba) from Japan: present contamination status, body distribution and temporal trends (1978–2003). Mar Pollut Bull 58:396–401. https://doi.org/10.1016/j.marpolbul.2008.10.008

23. Ylitalo GM, Baird RW, Yanagida GK et al (2009) High levels of persistent organic pollutants measured in blubber of island-associated false killer whales (Pseudorca crassidens) around the main Hawaiian Islands. Mar Pollut Bull 58:1932–1937. https://doi.org/10.1016/j.marpolbul.2009.08.029

24. Krahn MM, Bradley Hanson M, Schorr GS et al (2009) Effects of age, sex and reproductive status on persistent organic pollutant concentrations in "Southern Resident" killer whales. Mar Pollut Bull 58:1522–1529. https://doi.org/10.1016/j.marpolbul.2009.05.014

25. Krahn MM, Hanson MB, Baird RW et al (2007) Persistent organic pollutants and stable isotopes in biopsy samples (2004/2006) from Southern Resident killer whales. Mar Pollut Bull 54:1903–1911. https://doi.org/10.1016/j.marpolbul.2007.08.015

26. Park B-K, Park G-J, An Y-R et al (2010) Organohalogen contaminants in finless porpoises (Neophocaena phocaenoides) from Korean coastal waters: contamination status, maternal

transfer and ecotoxicological implications. Mar Pollut Bull 60:768–774. https://doi.org/10.1016/j.marpolbul.2010.03.023

27. Haynes D, Carter S, Gaus C et al (2005) Organochlorine and heavy metal concentrations in blubber and liver tissue collected from Queensland (Australia) dugong (Dugong dugon). Mar Pollut Bull 51:361–369. https://doi.org/10.1016/j.marpolbul.2004.10.020

28. Popa OM, Trif A, Marine N, Ursu N (2008) Organochlorine pesticides in the Black Sea dolphins. Lucrari stiintifice medicina veterinara V:768–773

29. Kajiwara N, Kannan K, Muraoka M et al (2001) Organochlorine pesticides, polychlorinated biphenyls, and butyltin compounds in blubber and livers of stranded California sea lions, elephant seals, and harbor seals from coastal California, USA. Arch Environ Contam Toxicol 41:90–99. https://doi.org/10.1007/s002440010224

30. Trukhin AM, Boyarova MD (2013) Chlorinated pesticides in tissues and organs of spotted seals (Phoca largha Pallas, 1811) from the Sea of Japan. Contemp Probl Ecol 6:336–342. https://doi.org/10.1134/S199542551303013X

31. Lukyanova O (2013) Persistent organic pollutants in marine ecosystems in Russian far east: sources, transport, biological effects. LAP LAMBERT Academic Publishing, Saarbrücken

Chapter 12
Biotransformation of Persistent Organic Pollutants in Ecosystems of the Russian North Pacific

Abstract Results of studies considering biotransformation of persistent organic pollutants (POPs) in marine organisms from the Far Eastern seas of Russia are provided in the chapter. A typical pattern of distribution of initial and degraded POPs in various organisms is shown.

Keywords POPs · Biotransformation · Degradation of initial compounds · Marine organisms · Far Eastern seas

12.1 Introduction

The major mechanisms of degradation of organic xenobiotics in the environment can be provisionally divided into abiotic (photochemical reactions) and biotic (processes of metabolic degradation involving living organisms).

The photochemical degradation of OCPs such as DDT and HCH, whose molecules contain aromatic groups and unsaturated bonds, occurs through absorption of solar energy in the ultraviolet and visible regions of the light spectrum [1]:

$$DDD \leftarrow DDT \rightarrow DDE \ldots \rightarrow PCB$$
$$\gamma\text{-}HCH \rightarrow \alpha\text{-}HCH \rightarrow \beta\text{-}HCH$$

After entering the body, many xenobiotics including persistent organic pollutants (POPs) are subject to biotransformation and excretion as metabolites. Biotransformation is based largely on enzymatic conversion of molecules. The biological meaning of this phenomenon is transformation of a chemical compound into a form convenient for excretion from the body and, thereby, reduction in the time of exposure to it.

Xenobiotics are metabolized through two stages (Fig. 12.1).

INITIAL COMPOUND

Oxidation

Reduction   Phase 1

Hydrolysis

INTERMEDIATE METABOLITES

Glucuronidation

Sulfation

Methylation   Phase 2

Binding to glutathione

FINAL METABOLITES

Fig. 12.1 Stages of xenobiotic metabolism (general view)

During the first stage of reductive/oxidative or hydrolytic transformation, the molecule acquires polar functional groups, which makes it capable of chemical reactions and more soluble in water. At the second stage, synthetic processes of conjugation of intermediate metabolites with endogenous molecules occur, resulting in the formation of polar compounds that are removed from the body by special excretion mechanisms.

The variety of catalytic properties of biotransformation enzymes and their low substrate specificity allows an organism to metabolize substances with various structures. Nevertheless, in animals of various species and humans, the pathways of xenobiotic metabolism are largely dissimilar, since the enzymes involved in transformations of these substances are often species-specific.

HCH in marine animals breaks down into an intermediate (α-HCH) and a more stable final metabolite (β-HCH). The presence of these two isomers indicates that these xenobiotics entered marine ecosystems and organisms long ago. DDT in marine organisms breaks down into DDD, DDE, and DDA (Fig. 12.2). The presence of DDE and DDD also indicates a long-term circulation of these pollutants in ecosystems and organisms.

Unlike OCPs, polychlorinated biphenyls (PCB) are transformed by other pathways (Fig. 12.3).

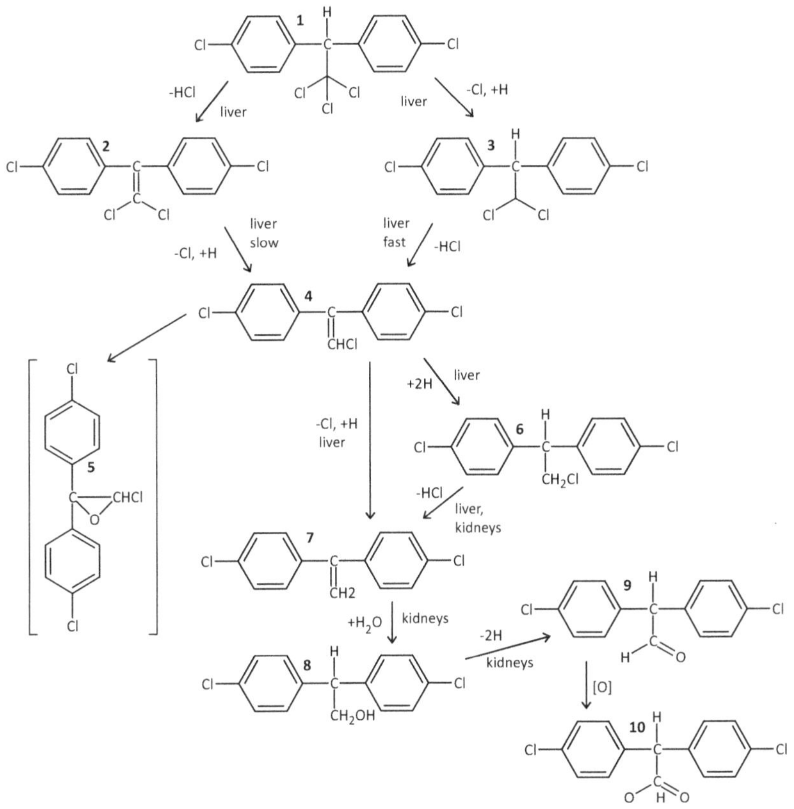

Fig. 12.2 Diagram of metabolic transformations of DDT in animals [2–4]: 1—DDT; 2—DDE; 3—DDD; 4—dichlorodiphenylchloroethylene; 5—dichlorodiphenylchloroethylene epoxide; 6—dichlorodiphenylchloroethane; 7—dichlorodiphenylethylene; 8—dichlorodiphenylethanol; 9—dichlorodiphenylethanal; 10—dichlorodiphenylethane (dichlorodiphenylacetic) acid

The pattern and dynamics of distribution of PCBs in the environment are largely determined by their physical properties such as chemical inertia, sufficiently high vapor density, and the ability to be adsorbed on particles. These are stable in the environment, poorly soluble in water, concentrated in bottom sediments of water bodies, and a minor part of them are subject to biotransformation by microorganisms and algae. Higher chlorinated compounds are particularly persistent. As PCBs become involved in biological food chains, lower chlorinated components are progressively lost due to their selective biotransformation. Therefore, the most hazardous higher chlorinated PCBs are accumulated in organisms.

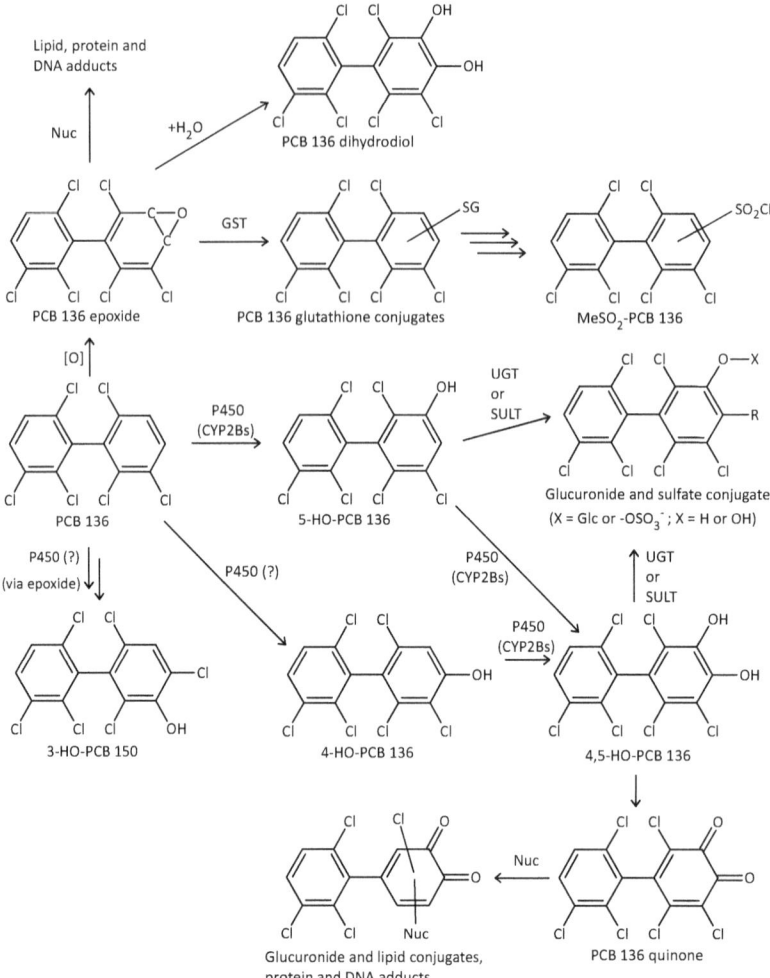

Fig. 12.3 Simplified diagram of metabolism of PCB 136 atropisomers [5]: Glc—glucuronide; GS—glutathione residue; GST—glutathione transferase; Nuc—cellular N- or S-nucleophile; P450—cytochrome P450 enzyme; SULT—sulfotransferase; UGT—UDP glucuronosyltransferase

12.2   Biotransformation of POPs in Organisms from the Far Eastern Seas

The conducted studies of POP biotransformation in marine organisms from the Far Eastern seas of Russia indicate similar situations in these regions. As a rule, transformed β- and α-HCH, DDD, and DDE, and higher chlorinated PCBs are present in organisms.

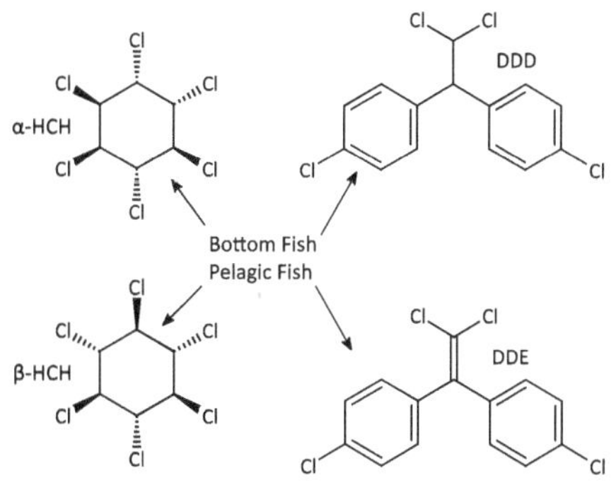

Fig. 12.4 Transformation of POPs in demersal and pelagic fish from the Far Eastern seas of Russia [6–9]

As is shown in the previous chapters, mainly HCH and DDT metabolites predominate in marine organisms. In turn, their emergence indicates a long-ago contamination event. However, judging where exactly the degradation of the initial compounds occurred is impossible. The substance could have entered the body several weeks or years ago, which might be caused by local, regional, or global pollution. The well-known monitoring technologies using demersal fish and mollusks allow assessment of local pollution. Knowing general features of their biology and ecology, one can estimate the actual accumulation and transformation of toxicants. Furthermore, the difference in the levels of POP accumulation by salmon and demersal fish can also be used for environmental assessments of local pollution.

The difference in the POP accumulation between demersal and pelagic fish is explained by different lipid contents, life cycles, and migration abilities, although the pathways of transformation of initial compounds are almost similar in them (Fig. 12.4).

Birds and mammals can be used as indicators for global and long-term POP monitoring. These organisms accumulate xenobiotics throughout their lives. The POP transformation in these organisms, recorded in our studies, indicates that the contamination occurred long ago and the initial compounds degraded. This is also evidenced by the ratio of HCH isomers and DDT and its metabolites (Figs. 12.5, 12.6).

A conclusion as to what kind of POP mixture was used can be made by detecting metabolites of initial compounds. For example, by detecting an increased concentration of γ-HCH in organisms and the environment, one may assume that a fresh contamination has occurred, or that a mixture of HCHs with the highest proportion of γ-HCH has entered the environment. Detection of β- and α-HCH in organisms and the environment indicates the use of a technical formulation: various HCH isomers differing in the conformation of the cycle that are represented by α-HCH (55–70%), β-HCH (5–14%), and γ-HCH (9–13%) in a technical mixture obtained through photochemical chlorination [14].

Detection of DDE and DDD at concentrations higher than that of DDT (or even in the absence of DDT) more accurately indicates a long-ago contamination event and a transformation of the initial compound, since a usual composition of technical DDT

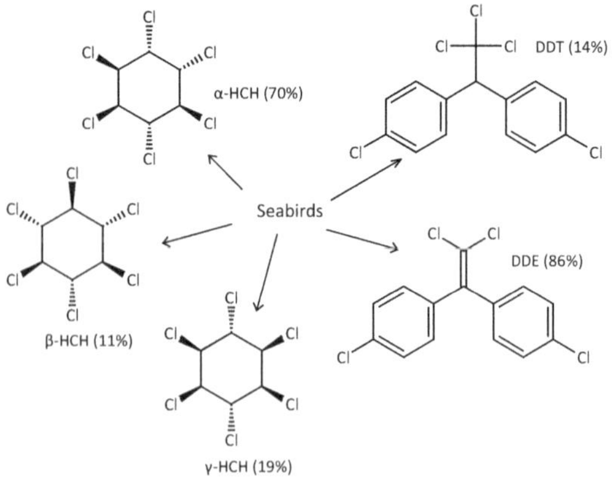

Fig. 12.5 Transformation of POPs in seabirds from the Sea of Okhotsk: 70% α- and 11% β-HCH indicate the breakdown of lindane (γ-HCH, 19%); 86% DDE indicates the breakdown of the initial DDT (14%) [10–12]

is as follows: **77.1% p,p′-DDT, 14.9% o,p′-DDT**, 0.3% p,p′-DDD, 0.1% o,p′-DDD, 4.0% p,p′-DDE, 0.1% o,p′-DDE, and trace concentrations of other compounds [15].

As for PCBs, their composition in organisms indicates selective biotransformation and accumulation of the most hazardous higher chlorinated PCBs (Fig. 12.7).

12.3 Biotransformation of POPs in the World

In the second half of the twentieth century, the global demand for food substantially increased with the growth of the world's population. This caused intensification of agriculture with inevitably much wider application of agricultural chemicals, including mainly pesticides [16]. Today, the total consumption of POPs worldwide is about two million tons a year, of which 69% is consumed by Europe and the USA, and only 31% by the rest of the countries. In South Asia, agriculture is the major sector of pesticide consumption, with 14% of world's farmlands extensively treated with pesticides [17].

The use of pesticides in agriculture and healthcare, mainly to control various pests and causative agents of human diseases, began in India. To achieve this goal, the production of basic pesticides was launched in 1952, and already more than 5000 t of pesticides, in particular DDT and HCH, were produced in 1958. In the mid-1990s, about 145 types of pesticides were registered, and the production reached 85,000 t. Recently, the pesticide consumption has shown a slight downward trend, probably due to the farmers' tendency to use biopesticides, natural plant sources, and other alternative methods [18]. Today, India is considered the Asia's largest and the world's 12th pesticide-producing country that provides 90,000 t of pesticides a year [19]. Furthermore, India participates in the production, use, and export of OCPs such as DDT on a large scale [20]. Currently, there are more than 125 major large- and medium-sized pesticide manufacturers and more than 500 pesticide formulations

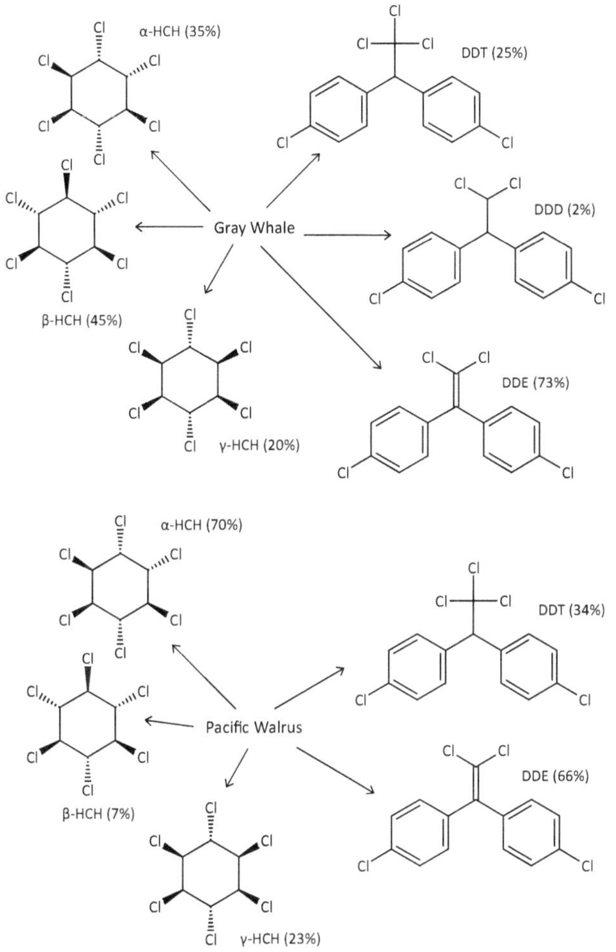

Fig. 12.6 Transformation of POPs in marine mammals from the Bering Sea. Gray whale: 35% α- and 45% β-HCH indicate the breakdown of lindane (γ-HCH, 20%); 75% DDE + DDD indicates the breakdown of initial DDT (25%). Pacific walrus: 70% α- and 7% β-HCH indicate the breakdown of lindane (γ-HCH, 23%); 66% DDE indicates the breakdown of initial DDT (34%) [11–13]

registered in India [21]. The world's countries establish safety standards for the initial and transformed compounds in products depending on the pesticide formulations used.

Table 12.1 shows, for example, transformations of the initial compounds in mollusks and fish from various regions of the planet depending on composition of pesticides and PCBs used.

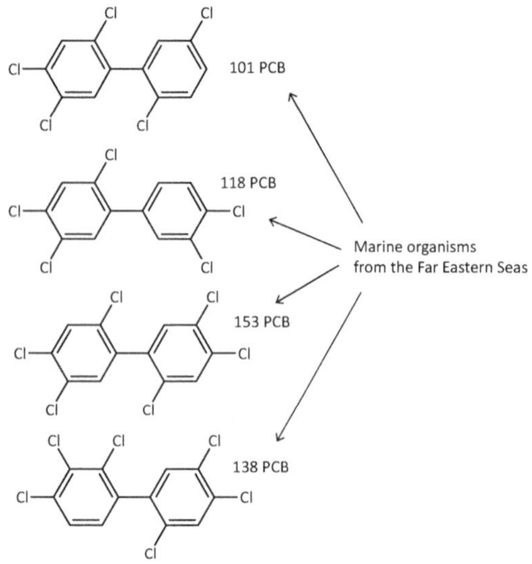

Fig. 12.7 Major higher chlorinated PCBs found in marine organisms from the Far Eastern seas of Russia [6–8, 12]

Thus, different pesticide compositions can be used in various regions of the planet, as evidenced by their different biotransformations in the environment and organisms. In the Far Eastern seas of Russia, mainly α-, β-HCH, DDE, DDD, and higher chlorinated PCBs (101, 118, 153, 138 PCB) are detected.

Table 12.1 Biotransformation of OCPs and PCBs in mollusks and fish from different regions of the world's oceans

Species	Region	Detected compounds	Reference
Mollusks			
Mytilus galloprovincialis	Adriatic coast (Croatia)	α-, β-, γ-HCH, *p,p'*-DDT, *p,p'*-DDD, *p,p'*-DDE	[22]
		105 PCB, 114 PCB, 118 PCB, 123 PCB, 156 PCB, 157 PCB, 167 PCB, 170 PCB, 189 PCB	
Crassostrea gigas	Atlantic coast of France	Lindane, *o,p'*-DDE, *p,p'*-DDE, *o,p'*-DDD, *p,p'*-DDD, *o,p'*-DDT, *p,p'*-DDT	[23]
		101 PCB, 118 PCB, 138 PCB, 153 PCB, 180 PCB	
Mytilus galloprovincialis	Tunisia	*o,p'*-DDE, *p,p'*-DDE, *p,p'*-DDD, *o,p'*-DDT	[24]
		101 PCB, 118 PCB, 138 PCB, 153 PCB, 180 PCB	
Adamussium colbecki	Ross Sea, Antarctica	Tetra-CB, penta-CB, hexa-CB, hepta-CB	[25]
Crassostrea virginica, Mytilus galloprovincialis, Mytilus trossulus, and Mytilus californianus	USA Pacific coast	*o,p'*-DDE, *p,p'*-DDE, *p,p'*-DDD, *o,p'*-DDT	[26]
Perna viridis	Indonesia	*p,p'*-DDE, *p,p'*-DDD, *p,p'*-DDT	[27]
Crassostrea rivularis	China	α-, β-, γ-HCH, *p,p'*-DDE, *p,p'*-DDD, *o,p'*-DDT, *p,p'*-DDT	[28]
Fish			
Freshwater fish species	India	β-HCH, *p,p'*-DDE, *p,p'*-DDD, *p,p'*-DDT	[29]
Pseudotolithus senegalensis, Mugil cephalus, Sphyraena piscatorum	Nigeria	β-HCH, *p,p'*-DDD	[30]
Oreochromis mossambicus, Clarias gariepinus	South Africa	α-, β-HCH, *p,p'*-DDE, *p,p'*-DDD, *p,p'*-DDT	[31]
Commercial fish species	China	β-HCH, *p,p'*-DDT	[32]

(continued)

Table 12.1 (continued)

Species	Region	Detected compounds	Reference
		101 PCB	
Chitala chitala, Channa striata, Rita rita,	Pakistan	α-, β-HCH, *p,p′*-DDT, *p,p′*-DDE	[33]
Clupisoma gaura, Sperata seenghala, Wallago attu, Cirrhinus mrigala, Catla catla, Cyprinus carpio, Labeo rohita, Labeo dyocheilus, Cirrhinus reba		Hexa-CB, tetra-CB, tri-CB	
Salmo salar	Norway	α-, γ-HCH	[34]
		28 PCB, 52 PCB, 101 PCB, 138 PCB, 153 PCB, 180 PCB	
Oncorhynchus tshawytscha, O. kisutch, O. nerka, O. keta, Salmo salar	Canada	β-, γ-HCH, *p,p′*-DDE, *p,p′*-DDD	[35]
		Penta-CB, tetra-CB	

12.4 Conclusion

Organochlorine compounds are detected mainly in the degraded forms, which indicates the removal of these toxicants from the industrial and agricultural sectors. The presence of PCBs in organisms from the Far Eastern seas can probably be explained by the fact that PCB-containing equipment (transformers and capacitors) is still used in Russia, but active disposal of such equipment has begun in the country recently. In the power systems of the fuel and energy industries of Russia, a total of about 7200 transformers and 360,000 capacitors are currently operated, where PCB is present as a dielectric. According to the requirements of the Stockholm Convention, the use of PCB-containing equipment in Russia will continue until 2025, and by 2028 it should be completely stopped.

Acknowledgements The work was supported by the Russian Science Foundation (no. 23-74-10032).

References

1. Tanabe S (2007) Chapter 18 Contamination by persistent toxic substances in the Asia-Pacific Region. In: Li A, Tanabe S, Jiang G et al (eds) Developments in environmental science. Elsevier, pp 773–817
2. Walker CH (1975) Variation in the intake and elimination of pollutants. In: Moriarty F (ed) Organochlorine insecticides: persistent organic pollutants. Academic Press, London, pp 73–131

3. Gold B, Leuschen T, Brunk G, Gingell R (1981) Metabolism of a DDT metabolite via a chloroepoxide. Chem Biol Interact 35:159–176. https://doi.org/10.1016/0009-2797(81)901 40-X

4. ATSDR (2002) Toxicological Profile for DDT, DDE, and DDD. U.S. Department of Health and Human Services, Public Health Service, Atlanta, GA

5. Kania-Korwel I, Lehmler H-J (2016) Chiral polychlorinated biphenyls: absorption, metabolism and excretion—a review. Environ Sci Pollut Res 23:2042–2057. https://doi.org/10.1007/s11 356-015-4150-2

6. Lukyanova ON, Tsygankov VY, Boyarova MD (2018) Organochlorine pesticides and polychlorinated biphenyls in the Bering flounder (Hippoglossoides robustus) from the Sea of Okhotsk. Mar Pollut Bull 137:152–156. https://doi.org/10.1016/j.marpolbul.2018.10.017

7. Donets MM, Tsygankov VY, Boyarova MD et al (2021) Flounders as indicators of environmental contamination by persistent organic pollutants and health risk. Mar Pollut Bull 164:111977. https://doi.org/10.1016/j.marpolbul.2021.111977

8. Donets MM, Tsygankov VY, Gumovskiy AN et al (2021) Organochlorine pesticides (OCPs) and polychlorinated biphenyls (PCBs) in Pacific salmon from the Kamchatka Peninsula and Sakhalin Island, Noethwest Pacific. Mar Pollut Bull 169:112498. https://doi.org/10.1016/j.mar polbul.2021.112498

9. Tsygankov VY, Donets MM, Gumovskiy AN, Khristoforova NK (2022) Temporal trends of persistent organic pollutants biotransport by Pacific salmon in the Northwest Pacific (2008–2018). Mar Pollut Bull 185:114256. https://doi.org/10.1016/j.marpolbul.2022.114256

10. Tsygankov VY, Boyarova MD, Lukyanova ON (2016) Bioaccumulation of organochlorine pesticides (OCPs) in the northern fulmar (Fulmarus glacialis) from the Sea of Okhotsk. Mar Pollut Bull 110:82–85. https://doi.org/10.1016/j.marpolbul.2016.06.084

11. Tsygankov VY, Lukyanova ON, Boyarova MD (2018) Organochlorine pesticide accumulation in seabirds and marine mammals from the Northwest Pacific. Mar Pollut Bull 128:208–213. https://doi.org/10.1016/j.marpolbul.2018.01.027

12. Tsygankov VY, Lukyanova ON (2019) Current levels of organochlorine pesticides in marine ecosystems of the Russian far eastern seas. Contemp Probl Ecol 12:562–574. https://doi.org/ 10.1134/S199542551906009X

13. Tsygankov VY, Boyarova MD, Lukyanova ON (2015) Bioaccumulation of persistent organochlorine pesticides (OCPs) by gray whale and Pacific walrus from the western part of the Bering Sea. Mar Pollut Bull 99:235–239. https://doi.org/10.1016/j.marpolbul.2015.07.020

14. Lobov VP, Efimov GA (1963) Pesticides. Gostekhizdat, Kyiv

15. Grung M, Lin Y, Zhang H et al (2015) Pesticide levels and environmental risk in aquatic environments in China—a review. Environ Int 81:87–97. https://doi.org/10.1016/j.envint.2015. 04.013

16. Merrington G (2002) Agricultural pollution: environmental problems and practical solutions. Spon Press, London, New York

17. Atapattu SS, Kodituwakku DC (2009) Agriculture in South Asia and its implications on downstream health and sustainability: a review. Agric Water Manag 96:361–373. https://doi.org/10. 1016/j.agwat.2008.09.028

18. Gupta PK (2004) Pesticide exposure—Indian scene. Toxicology 198:83–90. https://doi.org/10. 1016/j.tox.2004.01.021

19. Khan MJ, Zia MS, Qasim M (2010) Use of pesticides and their role in environmental pollution. World Acad Sci Eng Technol 72:122–128

20. Pozo K, Harner T, Lee SC et al (2011) Assessing seasonal and spatial trends of persistent organic pollutants (POPs) in Indian agricultural regions using PUF disk passive air samplers. Environ Pollut 159:646–653. https://doi.org/10.1016/j.envpol.2010.09.025

21. Abhilash PC, Singh N (2009) Pesticide use and application: an Indian scenario. J Hazard Mater 165:1–12. https://doi.org/10.1016/j.jhazmat.2008.10.061

22. Kljaković-Gašpić Z, Herceg-Romanić S, Kožul D, Veža J (2010) Biomonitoring of organochlorine compounds and trace metals along the Eastern Adriatic coast (Croatia) using Mytilus galloprovincialis. Mar Pollut Bull 60:1879–1889. https://doi.org/10.1016/j.marpolbul.2010. 07.019

23. Luna-Acosta A, Budzinski H, Le Menach K et al (2015) Persistent organic pollutants in a marine bivalve on the Marennes-Oléron Bay and the Gironde Estuary (French Atlantic Coast)—part 1: Bioaccumulation. Sci Total Environ 514:500–510. https://doi.org/10.1016/j.scitotenv.2014.08.071

24. Barhoumi B, Menach KL, Clérandeau C et al (2014) Assessment of pollution in the Bizerte lagoon (Tunisia) by the combined use of chemical and biochemical markers in mussels, Mytilus galloprovincialis. Mar Pollut Bull 84:379–390. https://doi.org/10.1016/j.marpolbul.2014.05.002

25. Grotti M, Pizzini S, Abelmoschi ML et al (2016) Retrospective biomonitoring of chemical contamination in the marine coastal environment of Terra Nova Bay (Ross Sea, Antarctica) by environmental specimen banking. Chemosphere 165:418–426. https://doi.org/10.1016/j.chemosphere.2016.09.049

26. Sericano JL, Wade TL, Sweet ST et al (2014) Temporal trends and spatial distribution of DDT in bivalves from the coastal marine environments of the continental United States, 1986–2009. Mar Pollut Bull 81:303–316. https://doi.org/10.1016/j.marpolbul.2013.12.049

27. Dwiyitno DL, Nordhaus I et al (2016) Accumulation patterns of lipophilic organic contaminants in surface sediments and in economic important mussel and fish species from Jakarta Bay, Indonesia. Mar Pollut Bull 110:767–777. https://doi.org/10.1016/j.marpolbul.2016.01.034

28. Gan JL, Jia XP, Jia T et al (2009) Distribution and change of DDT and HCH levels in oysters (Crassostrea rivularis) from coast of Guangdong, China between 2003 and 2007. J Environ Sci Health B 44:817–822. https://doi.org/10.1080/03601230903238657

29. Samidurai J, Subramanian M, Venugopal D (2018) Levels of organochlorine pesticide residues in fresh water fishes of three bird sanctuaries in Tamil Nadu, India. Environ Sci Pollut Res. https://doi.org/10.1007/s11356-018-3770-8

30. Unyimadu JP, Osibanjo O, Babayemi JO (2018) Levels of organochlorine pesticides in Brackish water fish from Niger River, Nigeria. J Environ Public Health 2018:1–9. https://doi.org/10.1155/2018/2658306

31. Buah-Kwofie A, Humphries MS, Pillay L (2018) Bioaccumulation and risk assessment of organochlorine pesticides in fish from a global biodiversity hotspot: iSimangaliso Wetland Park, South Africa. Sci Total Environ 621:273–281. https://doi.org/10.1016/j.scitotenv.2017.11.212

32. Qian Z, Luo F, Wu C et al (2017) Indicator polychlorinated biphenyls (PCBs) and organochlorine pesticides (OCPs) in seafood from Xiamen (China): levels, distributions, and risk assessment. Environ Sci Pollut Res 24:10443–10453. https://doi.org/10.1007/s11356-017-8659-4

33. Robinson T, Ali U, Mahmood A et al (2016) Concentrations and patterns of organochlorines (OCs) in various fish species from the Indus River, Pakistan: a human health risk assessment. Sci Total Environ 541:1232–1242. https://doi.org/10.1016/j.scitotenv.2015.10.002

34. Nøstbakken OJ, Hove HT, Duinker A et al (2015) Contaminant levels in Norwegian farmed Atlantic salmon (Salmo salar) in the 13-year period from 1999 to 2011. Environ Int 74:274–280. https://doi.org/10.1016/j.envint.2014.10.008

35. Kelly BC, Ikonomou MG, Higgs DA et al (2011) Flesh residue concentrations of organochlorine pesticides in farmed and wild salmon from British Columbia, Canada. Environ Toxicol Chem 30:2456–2464. https://doi.org/10.1002/etc.662

Chapter 13
Environmental Risks from Persistent Organic Pollutants in Marine Organisms of the Northwestern Pacific Ocean

Abstract In this chapter, we provide the results of calculations of the environmental risk to the health of Far East residents from eating Pacific salmon, flounder, and marine mammals containing persistent organic pollutants (POPs). The incremental lifetime cancer risk was found for all the organisms analyzed. There is an urgent need for monitoring of POPs and their impact on the health of residents of the Russian Far East.

Keywords POPs · DDTs · HCHs · PCBs · Environmental health risks · Pacific salmon · Flounder · Marine mammals

13.1 Introduction

Understanding the effects of environmental contaminants on public health requires knowledge of the continuum from contaminant source to public health outcomes [1]. It can be described as a conceptual framework for assessing human health risk, which combines knowledge and methods from chemistry, physiology, biology, mathematics, physics, medicine, and other relevant disciplines.

Assessment of environmental risk to human health is a complex multifactorial process that takes into account the physicochemical properties of the substance (identified experimentally or using models), routes of entry into the body, dose–effect relationship, bioaccumulation potential, and distribution and behavior in the environment. Despite numerous variables used in the risk assessment, it provides a methodological approach to determining safe levels of exposure to various contaminants, thus, preventing a decrease in the quality of life and the health of population [2].

All POPs are substances with high toxicity, capable of bioaccumulation and biomagnification, which causes them to build up in the human body to high concentrations, even if their levels in food are below permissible [3]. Despite the high relevance of such studies, these are almost not conducted in Russia.

Health risk assessment includes the following steps:

1. Hazard identification (qualitative identification of potential health threats that may be caused by certain physical, biological, or chemical agents).
2. Impact assessment (assessment of the effects that these agents have on humans).
3. Characteristics of the dose–effect relationship (estimation of activity and probability of a specific type of harm per unit of exposure to an agent or agents).
4. Risk characteristics (description of the expected health impacts of agents, quantitative and qualitative description of uncertainty).

The assessment is almost always followed by the risk management. The goal of the health risk assessment is to determine whether the assessed risk is acceptable and, in case of unacceptability, take measures to mitigate it.

13.2 Technique for Calculating Environmental Health Risk

To assess the environmental health risk, we calculated the Hazard Quotient (HQ), which shows the probability of acute poisoning to occur within one year, and the Incremental Lifetime Cancer Risk (ILCR) from pollutants entering the human body [4, 5].

HQ and ILCR were calculated without taking into account the entry of OCPs and PCBs from other sources (water, air, or skin contact) and, therefore, indicate only the possible adverse health effects that result from eating *contaminated* food. HQ was calculated by the following formula:

$$HQ = \frac{EDI}{TDI},$$

where *EDI* is the average estimated daily intake of the toxicant, mg/kg per day; *TDI* is the tolerable daily intake, an amount of toxic substances that does not cause poisoning in human, mg/kg per day. A HQ value > 0.2 indicates that the risk of acute poisoning from certain toxicant potentially exists for humans. Estimated daily intake (EDI) was calculated as follows:

$$EDI = \frac{C_{food} \cdot IR_{food} \cdot AF_{GIT} \cdot D_d \cdot D_y}{BW \cdot 365 \cdot LE},$$

where C_{food} is the concentration of the toxicant in fish, mg/kg; IR_{food} is the average food ingestion rate, kg/day; AF_{GIT} is the absorption factor for the gastrointestinal tract (assumed to be equal to 1 if other data unavailable); D_d is the number of days in the year when contaminated fish is eaten; D_y is the number of years when contaminated fish is eaten (not used for hazard quotient calculation and is assumed to be 1; for ILCR calculation, it is assumed to be 65 yr); *BW* is the average human body weight, kg (according to the official Russian statistics, the average body weight of Far East residents is 70 kg); *LE* is the life expectancy, yr (according to the official Russian statistics, it is 70 yr for the population of the Far East).

ILCR was calculated as follows:

$$ILCR = EDI \cdot SF_{oral},$$

where SF_{oral} is the oral slope factor, a carcinogenic potential showing how much the cancer risk increases with the intake of the toxicant, kg a day/mg. If an ILCR value $> 1 \times 10^{-5}$ is calculated, the toxicant under consideration can potentially increase the probability of cancer.

13.3   Environmental Risks to the Health of Far East Residents from POPs

13.3.1   Pacific Salmon

Aquatic ecosystems are considered the final *acceptors* of POPs [6] that, due to their lipophilicity, build up in tissues of aquatic organisms, in particular fish. In many countries such as Japan, Indonesia, etc., fish and seafood constitute the major part of the diet of local residents. In this regard, State's regulation of maximum permissible concentrations (MPC) of POPs (in particular OCPs and PCBs) in tissues of marine organisms is carried out almost worldwide (Table 13.1). In Russia, many sanitary and hygienic standards have been adopted to provide safety of food products. The most important documents establishing safety requirements for fish and non-fish objects of marine fisheries are SanPiN 2.3.2.1078-01, TR CU 021/2011, and TR EAEU 040/2016 [7–9].

Pacific salmon are one of the valuable national resources of Russia. These are the most popular fish species ranked 2nd of 3rd in terms of catch sizes after walleye pollock and herring [24]. However, we detected POPs in all salmon fish analyzed, which can potentially cause various negative effects on public health.

The Russian Far East is a region with the highest rate of consumption of fish and seafood in the country, which is by 9 kg/yr higher than the average value for other regions (29 kg/yr in the Far East vs. an average of 20 kg/yr across Russia) [24]. This suggests the need for a continuous assessment of the environmental health risk from ingesting such fish as, in particular, Pacific salmon and fish products derived from them due to their wide distribution and high value.

To calculate the environmental risk to the health of Far East residents from eating Pacific salmon, we assumed an absolute value of 29 kg of fillets per person per year, because there is currently no available information on the average ingestion rate for each fish organ. The data for calculating the risk from eating salmon are provided in Chap. 7. To calculate the risk, concentrations of toxicants were expressed in terms of mg/kg wet weight (w.w.) (Table 13.2).

Table 13.1 Hygienic requirements (permissible concentrations) to the safety of fish and seafood products as regards OCP and PCB content in Russia and other countries, ng/g wet weight

Country/Organization	Substance	MPC	Note	References
Russia	HCH[1]	200	Types of products derived from marine fish and meat of marine mammals (except liver and fish oil); roe, fish milt, and products derived from them; analogues of caviar	[8]
		100	Fish oil	
		1000	Fish liver and derived products	
	DDT[2]	2000	Sturgeon, salmonids, fatty herring—all types of products (except liver, roe, and milt), including dried, smoked, salted, spicy, pickled, semi-finished fish, and other ready-to-eat products	
		400	Fish roe and milt (all species) and products derived from them; analogues of caviar	
		3000	Fish liver and derived products	
	PCB[3]	2000	Types of fish products (except liver and fish oil) and meat of marine mammals, including dried foods	
		5000	Fish liver and derived products	
		3000	Fish oil	
	HCH	200	Live fish, raw, chilled, frozen, minced fish, fillets, meat of marine mammals	[7]
		200	Fish roe and milt (all species) and products derived from them; analogues of caviar	
		1000	Fish liver and derived products	
	DDT	2000	Live fish, raw, chilled, frozen, minced fish, fillets (sturgeon, salmonids, herring), meat of marine mammals	
		2000	Fish roe and milt (all species) and products derived from them; analogues of caviar	
		3000	Fish liver and derived products	
	PCB	2000	Types of fish products (except liver and fish oil) and meat of marine mammals, including dried foods	
		3000	Fish oil	

(continued)

Table 13.1 (continued)

Country/Organization	Substance	MPC	Note	References
		5000	Fish liver and derived products	
	PCB	2000	Types of food fish products (except liver and food oil from fish), including dried foods	[9]
United States	DDT[4]	5000	Edible part	[10]
	PCB	2000	Fish and shellfish	[11]
Canada	HCH[5]	100	All fish species	[12]
	DDT	5000	All fish species	
	PCB	2000	All fish species	
Australia, New Zealand	HCH[6]	10	All fish species	[13]
	Lindane	1000	All fish species	
	DDT[7]	1000	All fish species	
	PCB	500	All fish species	[14]
Thailand	HCH	500	All fish species	[15]
	DDT	5000	All fish species	
China	HCH[8]	100	Marine-derived products	[16]
	DDT[7]	500	Marine-derived products	
	PCB[9]	500	Marine animals and derived products	[17]
Hong Kong	DDT	18,000	Fish, seafood and derived products	[18]
WHO[10]	HCH	200		[19]
Germany	HCH	500		[20]
	DDT	5000		
FAO/WHO[11]	DDT	200	All fish species	[21]
CREM/CBI[12]	DDT	5000		[22]
EU[13]	DDT	100		[23]

[1] HCH (total of α-, β-, γ-isomers)

[2] DDT and its metabolites (DDD and DDE)

[3] Total of all PCB congeners

[4] The current level of DDT, DDD, and DDE is used both for certain pesticides and for their sum. However, DDT, DDD, and DDE concentrations below 200 ng/g are not taken into account when calculating the total amount

[5] In the table of permissible levels of pollutants in fish and fish products, HCH (total of isomers) is referred to as "Other agricultural chemicals or their derivatives"

[6] HCH (all isomers except γ-HCH, lindane)

[7] Total of p,p'-DDT, o,p'-DDT, p,p'-DDE, and p,p'-DDD

[8] Total of α-, β-, γ-, and δ-HCH

[9] Total of PCB 28, PCB 52, PCB 101, PCB 118, PCB 138, PCB 153, and PCB 180

[10] The World Health Organization

[11] The Food and Agriculture Organization of the United Nations (FAO) and WHO

[12] The Center for Resource and Environmental Management/The Centre for the Promotion of Imports from developing countries

[13] European Union

Table 13.2 Average concentrations of pesticides and PCBs in muscles of Pacific salmon caught in various years

Species	Sampling year	Concentration, mg/kg w.w.						
		α-HCH	β-HCH	γ-HCH	DDT[1]	DDD	DDE	ΣPCB
Pink salmon	2012	0.025	0.0049	0.0031	< DL[2]	< DL	0.0023	–[3]
	2017	0.002	0.0006	0.00008	0.00047	0.00027	0.0014	0.00164
Chum salmon	2013	0.021	0.014	0.0065	< DL	< DL	0.0047	–
	2018	0.0002	0.00002	0.00001	< DL	< DL	0.00002	0.0002
Masu salmon	2017	0.0006	0.00025	0.00012	0.00046	0.0005	0.00016	0.00205
Chinook salmon	2010	0.007	< DL	0.0076	< DL	< DL	0.0056	–
	2018	0.00003	0.00023	0.00005	0.00005	0.00012	0.00006	0.00142
Sockeye salmon	2011	0.017	< DL	0.00272	< DL	< DL	0.0016	–
	2017	< DL	0.00038	< DL	0.00004	0.00039	0.00002	0.0005
	2018	0.00003	0.00023	0.00004	0.00026	0.00024	0.00005	0.00108

[1] Total of o,p' and p,p' isomers;
[2] Below the detection limit;
[3] Not analyzed

The concentrations in pink salmon caught in 2017 represent the total levels of toxicants in Lake Azabachye and the Poronay River, because it is often impossible to know the exact place of catch when buying fish. It is also worth noting that, when concentrations were expressed in terms of mg/kg wet weight, none of the compounds exceeded the permissible levels of the safety standards.

The results of environmental risk calculations for pink, chum, masu, Chinook, and sockeye salmon are presented in Tables 13.3, 13.4, 13.5, 13.6 and 13.7.

As can be seen, the incremental lifetime cancer risk from ingestion of Pacific salmon was recorded for all fish caught in 2010–2012. It was found to be caused by the HCH isomers (mainly the α-form). The fish caught in 2017–2018 were mostly safe, which reflects a decrease in POP concentrations in the organs of Pacific salmon and in the environment in general. The exception was pink salmon caught in 2017 (cancer risk due to α-HCH).

As already mentioned, an absolute value of 29 kg of fish meat per year was assumed for calculations, and the risk is probably acceptable. Nevertheless, the results of calculations show a potentially possible increase in the risk of cancer diseases for the population of the Far East. This, in turn, indicates the necessity to introduce a risk assessment system into the regulatory framework of Russia and update the existing regulatory documents that establish requirements for food safety, in particular, for fish and seafood.

Table 13.3 Assessment of the environmental risk from POPs detected in muscles (fillets) of pink salmon for the local population of the Russian Far East

Compound	EDI for HQ	HQ	EDI for ILCR	ILCR
2012				
α-HCH	2.9×10^{-5}	9.6×10^{-2}	2.7×10^{-5}	*1.7×10^{-4}*
β-HCH	5.5×10^{-6}	1.8×10^{-2}	5.2×10^{-6}	9.4×10^{-6}
γ-HCH	3.6×10^{-6}	1.2×10^{-2}	3.4×10^{-6}	4.4×10^{-6}
DDE	2.6×10^{-6}	5.3×10^{-3}	2.5×10^{-6}	8.4×10^{-7}
2017				
α-HCH	2.2×10^{-6}	7.2×10^{-3}	2.0×10^{-6}	*1.3×10^{-5}*
β-HCH	6.8×10^{-7}	2.3×10^{-3}	6.4×10^{-7}	1.2×10^{-6}
γ-HCH	8.8×10^{-8}	2.9×10^{-4}	8.3×10^{-8}	1.1×10^{-7}
DDT	5.4×10^{-7}	1.1×10^{-3}	5.1×10^{-7}	1.7×10^{-7}
DDD	3.1×10^{-7}	6.2×10^{-4}	2.9×10^{-7}	7.0×10^{-8}
DDE	1.6×10^{-6}	3.2×10^{-3}	1.5×10^{-6}	5.1×10^{-7}
ΣPCB	1.9×10^{-6}	1.4×10^{-2}	1.8×10^{-6}	3.5×10^{-6}
Permissible value		**0.2**		**1×10^{-5}**

Italics are values indicating potential risk

Table 13.4 Assessment of the environmental risk from POPs detected in muscles (fillets) of chum salmon for the local population of the Russian Far East

Compound	EDI for HQ	HQ	EDI for ILCR	ILCR
2013				
α-HCH	2.4×10^{-5}	8.0×10^{-2}	2.2×10^{-5}	*1.4×10^{-4}*
β-HCH	1.6×10^{-5}	5.4×10^{-2}	1.5×10^{-5}	*2.7×10^{-5}*
γ-HCH	7.4×10^{-6}	2.5×10^{-2}	7.0×10^{-6}	9.1×10^{-6}
DDE	5.4×10^{-6}	1.1×10^{-2}	5.0×10^{-6}	1.7×10^{-6}
2018				
α-HCH	2.1×10^{-7}	6.9×10^{-4}	2.0×10^{-7}	1.2×10^{-6}
β-HCH	2.3×10^{-8}	7.6×10^{-5}	2.1×10^{-8}	3.9×10^{-8}
γ-HCH	6.1×10^{-9}	2.0×10^{-5}	5.7×10^{-9}	7.5×10^{-9}
DDE	2.6×10^{-8}	5.3×10^{-5}	2.5×10^{-8}	8.5×10^{-9}
ΣPCB	2.8×10^{-7}	2.2×10^{-3}	2.6×10^{-7}	5.3×10^{-7}
Permissible value		**0.2**		**1×10^{-5}**

Italics are values indicating potential risk

Table 13.5 Assessment of the environmental risk from POPs detected in muscles (fillets) of masu salmon caught in 2017 for the local population of the Russian Far East

Compound	EDI for HQ	HQ	EDI for ILCR	ILCR
α-HCH	7.2×10^{-7}	0.00241	6.8×10^{-7}	4.3×10^{-6}
β-HCH	2.9×10^{-7}	0.00097	2.7×10^{-7}	4.9×10^{-7}
γ-HCH	1.4×10^{-7}	0.00045	1.3×10^{-7}	1.7×10^{-7}
DDT	5.2×10^{-7}	0.00105	4.9×10^{-7}	1.7×10^{-7}
DDD	5.7×10^{-7}	0.00113	5.3×10^{-7}	1.3×10^{-7}
DDE	1.8×10^{-7}	0.00037	1.7×10^{-7}	5.9×10^{-8}
ΣPCB	2.3×10^{-6}	0.01799	2.2×10^{-6}	4.4×10^{-6}
Permissible value		**0.2**		**1×10^{-5}**

Table 13.6 Assessment of the environmental risk from POPs detected in muscles (fillets) of Chinook salmon for the local population of the Russian Far East

Compound	EDI for HQ	HQ	EDI for ILCR	ILCR
2010				
α-HCH	8.0×10^{-6}	2.7×10^{-2}	7.5×10^{-6}	*4.7×10^{-5}*
γ-HCH	8.7×10^{-6}	2.9×10^{-2}	8.2×10^{-6}	*1.1×10^{-5}*
DDE	6.4×10^{-6}	1.3×10^{-2}	6.0×10^{-6}	2.0×10^{-6}
2018				
α-HCH	3.3×10^{-8}	1.1×10^{-4}	3.1×10^{-8}	2.0×10^{-7}
β-HCH	2.6×10^{-7}	8.6×10^{-4}	2.4×10^{-7}	4.4×10^{-7}
γ-HCH	6.2×10^{-8}	2.1×10^{-4}	5.8×10^{-8}	7.6×10^{-8}
DDT	5.6×10^{-8}	1.1×10^{-4}	5.3×10^{-8}	1.8×10^{-8}
DDD	1.4×10^{-7}	2.8×10^{-4}	1.3×10^{-7}	3.2×10^{-8}
DDE	7.0×10^{-8}	1.4×10^{-4}	6.6×10^{-8}	2.2×10^{-8}
ΣPCB	1.6×10^{-6}	1.2×10^{-2}	1.5×10^{-6}	3.1×10^{-6}
Permissible value		**0.2**		**1×10^{-5}**

Italics are values indicating potential risk

13.3.2 Flounder

The Far Eastern Seas (the Sea of Japan, Sea of Okhotsk, and Bering Sea) are the major fisheries zones of the Russian Federation. Flounder harvested in the Far East are among the most important target species of commercial fisheries, accounting for 9.5% of the total size of fish catch in the region [24]. The large catch size, the variety of species, and the low price on the market explain the particular importance of flounder in the structure of the diet of the local population. Among the most important species are righteye flounders of the genus *Hippoglossoides* widely distributed in the Sea of Okhotsk and the Sea of Japan, including the Tatar Strait. We estimated POP levels in

Table 13.7 Assessment of the environmental risk from POPs detected in muscles (fillets) of sockeye salmon for the local population of the Russian Far East

Compound	EDI for HQ	HQ	EDI for ILCR	ILCR
2012				
α-HCH	1.9×10^{-5}	6.4×10^{-2}	1.8×10^{-5}	*1.1×10^{-4}*
γ-HCH	3.1×10^{-6}	1.0×10^{-2}	2.9×10^{-6}	3.8×10^{-6}
DDE	1.8×10^{-6}	3.6×10^{-3}	1.7×10^{-6}	5.7×10^{-7}
2017				
β-HCH	4.4×10^{-7}	1.5×10^{-3}	4.1×10^{-7}	7.4×10^{-7}
DDT	5.1×10^{-8}	1.0×10^{-4}	4.8×10^{-8}	1.6×10^{-8}
DDD	4.4×10^{-7}	8.9×10^{-4}	4.2×10^{-7}	1.0×10^{-7}
DDE	2.5×10^{-8}	5.0×10^{-5}	2.4×10^{-8}	8.0×10^{-9}
ΣPCB	5.2×10^{-7}	4.0×10^{-3}	4.9×10^{-7}	9.7×10^{-7}
2018				
α-HCH	3.8×10^{-8}	1.3×10^{-4}	3.6×10^{-8}	2.3×10^{-7}
β-HCH	2.6×10^{-7}	8.6×10^{-4}	2.4×10^{-7}	4.4×10^{-7}
γ-HCH	4.5×10^{-8}	1.5×10^{-4}	4.3×10^{-8}	5.5×10^{-8}
DDT	2.9×10^{-7}	5.9×10^{-4}	2.8×10^{-7}	9.4×10^{-8}
DDD	2.7×10^{-7}	5.5×10^{-4}	2.6×10^{-7}	6.2×10^{-8}
DDE	5.4×10^{-8}	1.1×10^{-4}	5.1×10^{-8}	1.7×10^{-8}
ΣPCB	1.2×10^{-6}	9.5×10^{-3}	1.2×10^{-6}	2.3×10^{-6}
Permissible value		**0.2**		**1×10^{-5}**

Italics are values indicating potential risk

muscles of flathead flounder (*H. dubius*) and Bering flounder (*H. robustus*) (Chap. 7). To calculate the environmental risk, we expressed the concentrations of toxicants in terms of mg/kg w.w. (Table 13.8), since only this organ is used for food.

According to the data obtained, the OCP and PCB concentrations in flounder from all the study areas do not exceed the safety standards established in Russia [7]. The average values of estimated daily intake of the toxicant (EDI), hazard quotient (HQ), and incremental lifetime cancer risk (ILCR) from eating meat of righteye flounders for all the study areas are presented in Table 13.9.

According to the data obtained, HQ was less than 0.2 in all cases. However, in flounder from Rifovaya Cove (Sea of Japan), we recorded a slightly more than twofold excess of the permissible ILCR levels due to high values of accumulated PCBs. This suggests the necessity to include the methodology of health risk assessment in the evidence-based methodological framework of regulatory documents in the Russian Federation. To date, there is an urgent need for environmental protection measures in this cove (reduction in the number of vacationers, replacement of old PCB-containing transformers with newer ones, and reduction in shipping activity in this area) and for a thorough analysis of fish caught here.

Table 13.8 Average concentrations of the studied POPs in flounder muscles, mg/kg w.w.

Region	α-HCH	β-HCH	γ-HCH	DDD	DDE	PCB
Eastern Sea of Okhotsk	$-^1$	1.0×10^{-4}	–	1.0×10^{-4}	–	2.5×10^{-4}
Southern Sea of Okhotsk	3.9×10^{-5}	2.9×10^{-4}	–	1.3×10^{-4}	7.2×10^{-5}	7.1×10^{-4}
Tatar Strait	–	3.8×10^{-4}	–		–	7.8×10^{-4}
Sea of Japan	1.2×10^{-5}	3.7×10^{-4}	1.4×10^{-5}	9.3×10^{-5}	6.7×10^{-5}	1.0×10^{-2}
MPC^2	0.2^3			2.0^4		2.0^5

^{1} Concentrations below the detection limits of the equipment;
^{2} Maximum permissible concentration [7];
^{3} For total of all isomers;
^{4} For total of all metabolites;
^{5} For total of all congeners

Table 13.9 Assessment of environmental risk for residents of the Russian Far East eating flounders from the Far Eastern seas of Russia

Toxicant	EDI for HQ	HQ	EDI for ILCR	ILCR
Eastern Sea of Okhotsk				
β-HCH	1.14×10^{-7}	0.0004	1.08×10^{-7}	1.94×10^{-7}
DDD	1.17×10^{-7}	0.0002	1.1×10^{-7}	2.64×10^{-8}
PCB	2.88×10^{-7}	0.0022	2.71×10^{-7}	5.43×10^{-7}
Southern Sea of Okhotsk				
α-HCH	4.46×10^{-8}	0.0001	4.2×10^{-8}	2.65×10^{-7}
β-HCH	3.33×10^{-7}	0.0011	3.13×10^{-7}	5.64×10^{-7}
DDD	1.45×10^{-7}	0.0003	1.37×10^{-7}	3.28×10^{-8}
DDE	8.28×10^{-8}	0.0002	7.8×10^{-8}	2.65×10^{-8}
PCB	8.13×10^{-7}	0.0063	7.66×10^{-7}	1.53×10^{-6}
Tatar Strait				
β-HCH	4.34×10^{-7}	0.0014	4.09×10^{-7}	7.36×10^{-7}
PCB	8.92×10^{-7}	0.0069	8.4×10^{-7}	1.68×10^{-6}
Sea of Japan				
α-HCH	1.4×10^{-8}	0.00005	1.32×10^{-8}	8.31×10^{-8}
β-HCH	4.22×10^{-7}	0.0014	3.97×10^{-7}	7.15×10^{-7}
γ-HCH	1.59×10^{-8}	0.0001	1.5×10^{-8}	1.94×10^{-8}
DDD	1.06×10^{-7}	0.0002	1.0×10^{-7}	2.4×10^{-8}
DDE	7.61×10^{-8}	0.0002	7.17×10^{-8}	2.44×10^{-8}
PCB	1.15×10^{-5}	0.0885	1.08×10^{-5}	2.17×10^{-5}
Permissible value		**HQ < 0.2**		**ILCR < 1 × 10^{-5}**

Italics are values indicating potential risk

13.3.3 Marine Mammals

Currently, a characteristic feature of the Russian North is two different types of economy carried out in parallel—one is traditional and the other is advanced—which affects the diet of the local population [25].

The traditional diet of indigenous people of the Far North has been formed for centuries and, due to the unavailability of easily digestible carbohydrates (vegetables, fruits, cereals, etc.), consisted mainly of fats and proteins. In the inner regions of the continent, people's need for essential nutrients was met primarily by reindeer husbandry and hunting. On the coast of the Russian North, marine organisms constituted the major food supply. The delivery of easily accessible foods (pasta, cereals, and vegetables) rich in carbohydrates has led to a decrease in the consumption of traditional, *barbaric* food (meat of whales, walruses, and seals). In contrast to the traditional food of the peoples of the Russian North, food products usual for modern people were delivered from more southerly regions of the country [26], which, in turn, resulted in a significant reduction in consumption of products from marine mammals (according to 2008 data, from 83.6 to 52 kg per year per person). The collapse of the USSR and the following economic crisis almost completely cut off the supply of foods to the Far North regions and forced indigenous people to rely on traditional food sources again.

To date, there are no available official statistics as regards the consumption of meat and derived products from marine mammals by indigenous people of Chukotka. However, according to the study of [26], up to 80% of local residents prefer traditional meals, including those based on products of marine hunting, whales and walruses. In this regard, we calculated the maximum permissible ingestion rates (kg per month per person) for the analyzed gray whale and Pacific walrus individuals, at which risks of incremental lifetime cancer (ILCR) or acute poisoning (HQ) exist. The HQ and ILCR were estimated separately, since these take into account different parameters and can be manifested at different ingestion rates.

The concentrations of the pollutants that we detected in muscles and liver of female and male gray whales and Pacific walruses (Chap. 11), expressed in terms of mg/kg w.w., are provided in Table 13.10. It is worth noting that the levels of pollutants in meat of the marine mammals under study did not exceed the sanitary and safety standards of the Russian Federation.

We estimated the maximum rates of ingestion of meat and liver from gray whales and Pacific walruses (separately for males and females) to assess the potential incremental lifetime cancer risk by selecting the average monthly ingestion rate at which the ILCR value for α-HCH (a substance with the highest cancer slope factor among all the compounds detected) would be 1.0×10^{-5} (a value indicating an individual risk). When assessing the risk of acute poisoning, we, based on the highest concentrations of certain HCH isomers, determined the average daily ingestion rate as forming the lowest permissible daily dose (for HCH isomers, 0.0003; for DDE, 0.0005). The results of the calculations are provided in Tables 13.11 and 13.12.

Table 13.10 Levels of pollutants found in muscles (numerator) and liver (denominator) of gray whale and Pacific walrus, mg/kg w.w.

Species	α-HCH	β-HCH	γ-HCH	DDT	DDE
Gray whale	0.025 ± 0.026 0.017 ± 0.037	0.016 ± 0.011 0.025 ± 0.015	0.010 ± 0.012 0.021 ± 0.016	$-^1$ 0.018 ± 0.012	0.011 ± 0.006 0.020 ± 0.020
Pacific walrus	0.014 ± 0.017 0.421 ± 0.557	0.016 ± 0.007 0.046 ± 0.059	0.029 ± 0.025 0.144 ± 0.241	0.027 ± 0.005 0.268 ± 0.156	$-$ 1.944 ± 2.229
MPC	0.2^2 1.0			0.2^3 3.0	

[1] Insufficient samples to assess risk;
[2] Total of α-, β-, and γ-HCH;
[3] Total of DDT and its metabolites

Table 13.11 Maximum rates of ingestion of marine mammal organs at which the poisoning risk arises

Species	POP	Permissible ingestion rate, kg/month[1]	EDI for HQ[1]	HQ[1]
Gray whale (*Eschrichtius robustus*)	α-HCH	5.0 5.1	6.0×10^{-5} 4.2×10^{-5}	**0.20** 0.14
	β-HCH		3.8×10^{-5} 6.0×10^{-5}	0.13 **0.20**
	γ-HCH		2.3×10^{-5} 5.0×10^{-5}	0.08 0.17
	DDT		$< DL^2$ 4.5×10^{-5}	$< DL$ 0.09
	DDE		2.6×10^{-5} 4.8×10^{-5}	0.05 0.10
Pacific walrus (*Odobenus rosmarus*)	α-HCH	4.4 0.11	2.9×10^{-5} 2.2×10^{-5}	0.10 0.07
	β-HCH		3.3×10^{-5} 2.4×10^{-6}	0.11 0.01
	γ-HCH		6.0×10^{-5} 7.4×10^{-6}	**0.20** 0.03
	DDT		5.6×10^{-5} 1.4×10^{-5}	0.11 0.03
	DDE		$< DL$ 1.0×10^{-4}	$< DL$ **0.20**

[1] The value in the numerator is for muscles; in the denominator, for liver;
[2] Below the detection limits
Bold are values indicating potential risk

Table 13.12 Maximum rates of ingestion of marine mammal organs at which cancer diseases become potentially probable

Species	POP	Permissible ingestion rate, kg/month[1]	EDI for ILCR[1]	ILCR[1]
Gray whale (*Eschrichtius robustus*)	α-HCH	0.14 0.21	1.6×10^{-6} 1.6×10^{-6}	**1.0×10^{-5}** **1.0×10^{-5}**
	β-HCH		1.0×10^{-6} 2.3×10^{-6}	1.8×10^{-6} 4.1×10^{-6}
	γ-HCH		6.2×10^{-7} 1.9×10^{-6}	8.1×10^{-7} 2.5×10^{-6}
	DDT		$< DL^2$ 1.7×10^{-6}	$< DL$ 5.8×10^{-7}
	DDE		6.9×10^{-7} 1.8×10^{-6}	2.4×10^{-7} 6.2×10^{-7}
Pacific walrus (*Odobenus rosmarus*)	α-HCH	0.26 0.01	1.6×10^{-6} 1.6×10^{-6}	**1.0×10^{-5}** **1.0×10^{-5}**
	β-HCH		1.8×10^{-6} 1.7×10^{-7}	3.3×10^{-6} 3.1×10^{-7}
	γ-HCH		3.3×10^{-6} 5.4×10^{-7}	4.3×10^{-6} 7.0×10^{-7}
	DDT		3.1×10^{-6} 1.0×10^{-6}	1.0×10^{-6} 3.4×10^{-7}
	DDE		$< DL$ 7.3×10^{-6}	$< DL$ 2.5×10^{-6}

[1] The value in the numerator is for muscles; in the denominator, for liver;
[2] Below the detection limits
Bold are values indicating potential risk

According to the calculations, the potential risk of poisoning was observed when residents ate meat and livers from gray whales (5.0 and 5.1 kg a month, respectively) and Pacific walruses (4.4 and 0.11 kg a month). The ingestion rates for the potential incremental lifetime cancer risk were as follows: for meat and liver from whales, 0.14 and 0.21 kg a month, respectively; from Pacific walrus, 0.26 and 0.01 kg a month. In all cases, the increased potential incremental lifetime cancer risk was due to α-HCH.

According to the study of [26], the average annual rate of whale meat ingestion is 4.3 kg per month per person. The results that we obtained show that the rates of ingestion of meat from marine mammals cannot cause acute poisoning in humans but, however, can potentially contribute to cancer prevalence. Acceptable rates of ingestion of whale and walrus liver should be very low to prevent both the potential risk of poisoning and the increased lifetime cancer risk. To date, there is an urgent need for continuous monitoring of POPs in organs of marine mammals (especially those used as food), assessment of health risk to indigenous residents of the Chukotka Autonomous Okrug, and also State's measures aimed to reduce this risk.

13.4 Conclusions

Environmental health risk assessment is a novel and advanced approach to evaluating the quality and safety of food products based on fundamental knowledge in medicine, toxicology, and ecology, which makes it particularly relevant to be included in the regulatory framework of the Russian Federation. Despite the significant decrease in POP levels in the world's oceans, we have recorded a potential cancer risk from all organisms consumed by residents of the Russian Far East that we analyzed. Meat and livers of the marine mammals used as food by the indigenous peoples of the Chukotka Autonomous Okrug pose the greatest hazard. Flounder caught in the Far Eastern seas of Russia have shown the lowest hazard: risk was recorded only for fish from the Sea of Japan where coastal waters are considered highly polluted. Pacific salmon, as migratory species, are less susceptible to local sources of marine pollution and reflect the general trend towards the removal of POPs from ecosystems of the world's oceans. However, despite the decrease in pollution, the risk is theoretically possible at an ingestion rate of 29 kg/yr. To reduce the potential threat to the health of the Far East population, government's intervention and risk management are recommended.

Acknowledgements The work was supported by the Russian Science Foundation (no. 23-74-10032)

References

1. Dong Z, Liu Y, Duan L et al (2015) Uncertainties in human health risk assessment of environmental contaminants: a review and perspective. Environ Int 85:120–132. https://doi.org/10.1016/j.envint.2015.09.008
2. Bourgeois M, Johnson G, Harbison R (2017) Human health risk assessment. In: International encyclopedia of public health. Elsevier, pp 84–94
3. Byun G-H, Moon H-B, Choi J-H et al (2013) Biomagnification of persistent chlorinated and brominated contaminants in food web components of the Yellow Sea. Mar Pollut Bull 73:210–219. https://doi.org/10.1016/j.marpolbul.2013.05.017
4. US EPA IRIS (2007) United States Environment Protection Agency (USEPA)'s integrated risk information system
5. Health Canada (2010) Federal contaminated site risk assessment in Canada, part II: health Canada toxicological reference values (TRVs) and chemical-specific factors, version 2.0
6. Lukyanova ON, Tsygankov VY, Boyarova MD (2018) Organochlorine pesticides and polychlorinated biphenyls in the Bering flounder (Hippoglossoides robustus) from the Sea of Okhotsk. Mar Pollut Bull 137:152–156. https://doi.org/10.1016/j.marpolbul.2018.10.017
7. SanPin 2.3.2.1078-01 (2002) Hygienic requirements of safety and nutritional value of food products. Ministry of Health of Russia, Moscow
8. TR CU 021/2011 (2011) Technical regulations of the Customs Union "On food safety." Customs Union Commission
9. TR EAEU 040/2016 (2016) Technical regulations of the Eurasian Economic Union "On safety of fish and fish products." Eurasian Economic Commission, Moscow
10. Food and Drug Administration (2018) Compliance Policy Guide. Sec. 575.100 pesticide residues in food and feed—enforcement Criteria (Compliance Policy Guide 7141.01). https://www.fda.gov/regulatory-information/search-fda-guidance-documents/compliance-policy-guide-sec-575100-pesticide-residues-food-and-feed-enforcement-criteria-compliance

11. Food and Drug Administration, Department of Health and Human Services (2018) Code of Federal Regulations. Title 21—Food and Drugs. Chapter I—Food and Drug Administration Department of Health and Human Services. Subchapter B—Food for Human Consumption. Section 109.30 Tolerances for polychlorinated biphenyls (PCB's). https://www.accessdata.fda.gov/scripts/cdrh/cfdocs/cfcfr/CFRSearch.cfm?fr=109.30

12. Government of Canada CFIA (2012) C.F.I.A. Standards and methods manual. https://www.inspection.gc.ca/food-safety-for-industry/archived-food-guidance/fish-and-seafood/manuals/standards-and-methods/eng/1348608971859/1348609209602?chap=7#s20c7. Accessed 15 Jan 2020

13. Federal Register of Legislation (2017) Australia New Zealand food standards code—Schedule 19—Maximum levels of contaminants and natural toxicants

14. Federal Register of Legislation (2017) Australia New Zealand food standards code—Schedule 21—Extraneous residue limits

15. Tanabe S, Kannan K, Tabucanon MS et al (1991) Organochlorine pesticide and polychlorinated biphenyl residues in foodstuffs from Bangkok, Thailand. Environ Pollut 72:191–203. https://doi.org/10.1016/0269-7491(91)90099-I

16. GB 2763-2016 (2016) National food safety standard for maximum residue limits for pesticides in foods. National Health and Family Planning Commission, Ministry of Agriculture, China Food and Drug Administration, Beijing

17. GB 2762-2017 (2017) National food safety standard for maximum levels of contaminants in foods. National Health and Family Planning Commission; China Food and Drug Administration, Beijing

18. Food and Environmental Hygiene Department (2014) First Hong Kong total diet study: organochlorine pesticide residues. Centre for Food Safety, Food and Environmental Hygiene Department, Hong Kong

19. Yahia D, Elsharkawy EE (2014) Multi pesticide and PCB residues in Nile tilapia and catfish in Assiut city, Egypt. Sci Total Environ 466–467:306–314. https://doi.org/10.1016/j.scitotenv.2013.07.002

20. Kasozi GN, Kiremire BT, Bugenyi FWB et al (2006) Organochlorine residues in fish and water samples from Lake Victoria, Uganda. J Environ Qual 35:584. https://doi.org/10.2134/jeq2005.0222

21. Mwevura H, Othman OC, Mhehe GL (2002) Organochlorine pesticide residues in sediments and biota from the coastal area of Dar es Salaam city, Tanzania. Mar Pollut Bull 45:262–267. https://doi.org/10.1016/S0025-326X(01)00331-9

22. Ogwok P, Muyonga JH, Sserunjogi ML (2009) Pesticide residues and heavy metals in Lake Victoria Nile Perch, Lates niloticus, Belly Flap Oil. Bull Environ Contam Toxicol 82:529–533. https://doi.org/10.1007/s00128-009-9668-x

23. Daba D, Hymete A, Bekhit AA et al (2011) Multi residue analysis of pesticides in wheat and khat collected from different regions of Ethiopia. Bull Environ Contam Toxicol 86:336–341. https://doi.org/10.1007/s00128-011-0207-1

24. Vorozhbit OY, Danilovskikh TE, Kuzmicheva IA et al (2016) The fishing industry of the Russian Far East: current state, problems and prospects for increasing competitiveness. Vladivostok State University of Economics and Service, Vladivostok

25. Dudarev AA (2009) Persistent polychlorinated hydrocarbons and heavy metals in Arctic biosphere: the main regularities of exposure and reproductive health of indigenous people. Biosfera 1:186–202

26. Kozlov AI, Nuvano V, Zdor A (2008) Diet of Chukotka. Chem Life 42–45

Chapter 14
Non-target Screening Analysis of "New" Persistent Organic Pollutants in the Far Eastern Seas

Abstract A non-target screening analysis has been conducted to search for persistent organic pollutants (POPs) "new" to the Far Eastern region for the first time. The detection of "nonconventional" POPs in the environment necessitates regular monitoring with methods of qualitative and quantitative analyses.

Keywords Non-target screening analysis · "New" POPs · "Conventional" POPs · Far Eastern seas

14.1 Introduction

Analysis of compound mixtures in environmental samples is a major challenge. Since sampling matrix is complex in most cases, traditional trace analytical methods were developed specifically for certain sample types and groups of substances. This traditional targeted approach provides high sensitivity, reliable identification, quantification of target compounds, and has been successfully used for decades. However, such a traditional approach has a substantial drawback: it will always skip all unknown compounds or other non-target substances even at high concentrations or with a pronounced toxic potential. There is every reason to assume that the concentration of unknown compounds is higher than that of known ones. In many cases, efficacy studies have shown that concentrations of known compounds are not high enough to explain some of the toxic potentials of samples [1, 2]. To fill this knowledge gap, non-target screening methods are very important tools for the study of environmental chemistry. Over the last decade, new analytical hardware and software tools have been developed that make the non-target approach to screening much more realistic and feasible than previously.

The number and variety of products manufactured by the chemical industry is increasing every year worldwide. Moreover, the amount of chemicals used for commercial purposes is immense. Europe alone uses more than 22,000 types of substances at a rate of more than 1 ton per year [3]. Many human-made compounds are harmful to the environment and have been proven to cause health problems in susceptible species [4–8]. After the ban on persistent organic pollutants (POPs), manufacturers began producing new chemical analogues.

In this regard, the POP research is currently focused, in addition to "conventional priority pollutants", also on "new pollutants" that are present in the environment at trace levels but exhibit high toxicity. There are few repots in literature on the use of the full scan mode for non-target screening of toxic compounds in living organisms [9–11].

Mass spectrometry (MS) based on the time-of-flight (TOF) technology has become more affordable, stable, and useful for environmental analysis over the past decade. TOF–MS provides full mass spectrum with much higher sensitivity than the standard quadrupole-trap MS, which makes it a versatile tool for both target and non-target analysis of environmental pollutants. In combination with gas or liquid chromatography (GC–MS or LC–MS), mass spectrometry allows a very wide range of chemical compounds to be isolated and detected.

In the non-target approach, also referred to as general unknown screening, there is no information about pollutants in sample. Data on substances is derived exclusively from chromatograms and mass spectra. Information obtained using a MS detector is usually huge and requires automatic processing of the data files recorded. NIST/EPA/NIH/PESTICIDES/AMDIS (automatic identification libraries) is software that not only identifies hidden compounds but also accelerates evaluation of full-scale analysis on the basis of GC–MS and, thus, is a reliable tool for environmental and toxicological studies.

14.2 "New" POPs in the Far Eastern Seas of Russia

In our study, we used GC–MS (Shimadzu GCMS-QP2010 Ultra). Examples of chromatograms for non-target screening analysis are shown in Figs. 14.1 and 14.2.

For 10 years of studying POPs in the Far Eastern seas, we have collected a sample bank containing about 10,000 specimens. We used the standard sample preparation protocol described in Chap. 2; for qualitative identification, instead of reference standards, we used NIST/EPA/NIH/**PESTICIDES**/AMDIS (automatic identification libraries). Samples were accessed from the bank selectively for matrices: fish, birds, and humans. Reliability of identification of compounds was evaluated by their match to the standard substances (at least 80%) from the library. The "conventional" POPs detected during non-targeted analysis are listed in Table 14.1. The spectrum of "new" POPs is presented in Table 14.2.

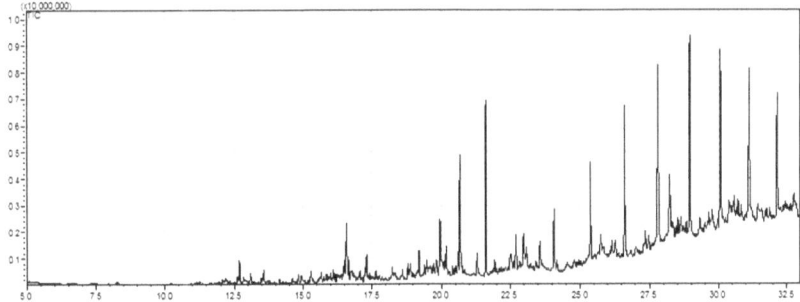

Fig. 14.1 Unlabeled chromatogram of a sample (example) in the case of non-target screening analysis on Shimadzu GCMS-QP2010 Ultra

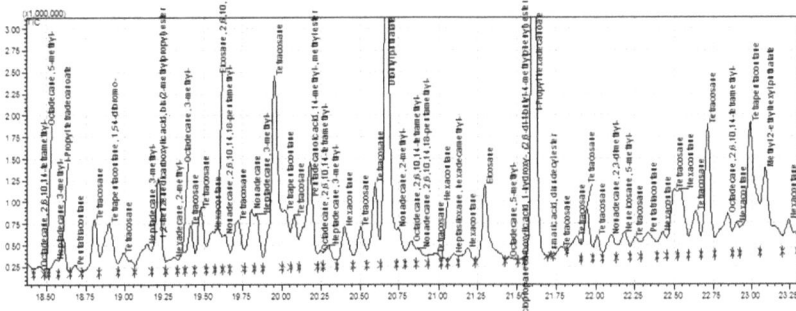

Fig. 14.2 Labeled chromatogram of a sample (example) in the case of non-target screening analysis on Shimadzu GCMS-QP2010 Ultra

Table 14.1 "Conventional" organic compounds most commonly found by non-target screening of samples from the Far Eastern seas

Name of compound, CAS ID No	Other names (according to CAS Common Chemistry and PubChem)
α-HCH 319–84-6	(1α,2α,3β,4α,5β,6β)-1,2,3,4,5,6-Hexachlorcyclohexan; (1α,2α,3β,4α,5β,6β)-1,2,3,4,5,6-Hexachlorocyclohexane; (1α,2α,3β,4α,5β,6β)-1,2,3,4,5,6-hexaclorociclohexano; ALPHA-1,2,3,4,5,6-HEXACHLORCYCLOHEXAN; Cyclohexane, 1,2,3,4,5,6-hexachloro-, (1α,2α,3β,4α,5β,6β)-; Cyclohexane, 1,2,3,4,5,6-hexachloro-, α-; Hexachlorocyclohexane; α-1,2,3,4,5,6-Hexachlorocyclohexane; α-666; α-Benzenehexachloride; α-Benzohexachloride; α-BHC; α-HCH; α-Hexachloran; α-Hexachlorane; α-Hexachlorcyclohexane; α-Hexachlorocyclohexane; α-Lindane
β-HCH 319–85-7	(1α,2β,3α,4β,5α,6β)-1,2,3,4,5,6-Hexachlorcyclohexan; (1α,2β,3α,4β,5α,6β)-1,2,3,4,5,6-Hexachlorocyclohexane; (1α,2β,3α,4β,5α,6β)-1,2,3,4,5,6-hexaclorociclohexano; BETA-1,2,3,4,5,6-HEXACHLORCYCLOHEXAN; Cyclohexane, 1,2,3,4,5,6-hexachloro-, (1α,2β,3α,4β,5α,6β)-; Cyclohexane, 1,2,3,4,5,6-hexachloro-, β-; Hexachlorocyclohexane; β-1,2,3,4,5,6-Hexachlorocyclohexane; β-666; β-Benzene hexachloride; β-BHC; β-HCH; β-Hexachloran; β-Hexachlorobenzene; β-Hexachlorocyclohexane; β-Lindane
γ-HCH 58–89-9	(1α,2α,3β,4α,5α,6β)-1,2,3,4,5,6-Hexachlorocyclohexane; 1,2,3,4,5,6-G-HEXACHLOROCYCLOHEXANE; 1,2,3,4,5,6-Hexachlorocyclohexane; 666; Aalindan; Aficide; Agrocide; Agrocide III; Agrocide WP; Ameisenmittel Merck; Aparasin; Aphtiria; Aplidal; BHC; BHC (insecticide); Celanex; Chloresene; Codechine; Cyclohexane, 1,2,3,4,5,6-hexachloro-, (1α,2α,3β,4α,5α,6β)-; Cyclohexane, 1,2,3,4,5,6-hexachloro-, γ-; DBH; Detmol Extract; Entomoxan; Esoderm; Fenoform forte; Gamacid; Gamacide; Gamacide 20; Gamene; Gamma benzene hexachloride; Gamma-HCH; Gammalin; Gexane; HCC; HCCH; HCH; Heclotox; Hexatin; Lidano (HCH); Lidenal; Lindafor; Lindane; Lindane (g-BHC or g-HCH); Lindane [cyclohexane, 1,2,3,4,5,6-hexachloro-(1α,2α,3β,4α,5α,6β)-]; γ-1,2,3,4,5,6-Hexachlorocyclohexane; γ-666; γ-Benzene hexachloride; γ-Benzohexachloride; γ-BHC; γ-HCH; γ-HCH or γ-BHC; etc
δ-HCH 319–86-8	(1α,2α,3α,4β,5α,6.β)-1,2,3,4,5,6-Hexachlorocyclohexane; (1α,2α,3α,4β,5α,6β)-1,2,3,4,5,6-Hexachlorcyclohexan; (1α,2α,3α,4β,5α,6β)-1,2,3,4,5,6-hexachlorocyclohexane; (1α,2α,3α,4β,5α,6β)-1,2,3,4,5,6-hexaclorociclohexano; Cyclohexane, 1,2,3,4,5,6-hexachloro-, (1α,2α,3α,4β,5α,6β)-; Cyclohexane, 1,2,3,4,5,6-hexachloro-, δ-; Delta-BHC; Hexachlorocyclohexane; δ-1,2,3,4,5,6-Hexachlorocyclohexane; δ-666; δ-Benzene hexachloride; δ-BHC; δ-HCH; δ-Hexachlorocyclohexane; δ-Lindane

(continued)

Table 14.1 (continued)

Name of compound, CAS ID No	Other names (according to CAS Common Chemistry and PubChem)
o,p′-DDT 789–02-6	(. + -.)-1-(o-Chlorophenyl)-1-(p-chlorophenyl)-2,2,2-trichloroethane; (. + -.)-o,p'-DDT; 1-(2-Chlorophenyl)-1-(4-chlorophenyl)-2,2,2-trichloroethane; 1-(o-Chlorophenyl)-1-(p-chlorophenyl)-2,2,2-trichloroethane; 1,1,1-Trichloro-2-(o-chlorophenyl)-2-(p-chlorophenyl)ethane; 2-(2-Chlorophenyl)-2-(4-chlorophenyl)-1,1,1-trichloroethane; 2,2,2,o,p′-Pentachlorethylidenbisbenzol 2,2,2,o,p′-pentachloroethylidenebisbenzene; 2,2,2,o,p′-pentacloroetilidenbisbenceno; 2,4′-DDT; 2,4′-Dichlorodiphenyltrichloroethane; Benzene, 1-chloro-2-[2,2,2-trichloro-1-(4-chlorophenyl)ethyl]-; Ethane, 1,1,1-trichloro-2-(o-chlorophenyl)-2-(p-chlorophenyl)-; NSC 33,446; NSC 57,644; o,p′-Chlorophenothane; o,p′-DDT
p,p′-DDT 50–29-3	1,1,1-Trichloro-2,2-bis(4-chlorophenyl) ethane; 1,1,1-Trichloro-2,2-bis(p-chlorophenyl)ethane; 1,1,1-Trichloro-2,2-di(p-chlorophenyl)ethane; 1,1-Bis(4-chlorophenyl)-2,2,2-trichloroethane 1,1-Bis(p-chlorophenyl)-2,2,2-trichloroethane; 2,2,2-Trichloro-1,1-bis(4-chlorophenyl)ethane; 2,2-bis(p-Chlorophenyl)-1,1,1-trichloroethane; 4,4′-DDT; 4,4′-Dichlorodiphenyltrichloroethane; 4,4-Dichlorodiphenyl-trichloroethane; Aavero-extra; Agritan; Arkotine; Benzene, 1,1′-(2,2,2-trichloroethylidene)bis[4-chloro-; Benzene, 1,1′-(2,2,2-trichloroethylidene)bis[4-chloro-; Benzene, 1,1′-(2,2,2-trichloroethylidene)bis[4-chloro-; Benzochloryl; Bovidermol; Chlorophenothane; Chlrphenothan; Chlorphenotoxum; Citox; D.D.T.; DDT; Deoval; Detox; Detox (pesticide); Detoxan; Dibovin; Dichlorodiphenyltrichloroethane; Dicophane; ENT-1506; Estonate; Ethane, 1,1,1-trichloro-2,2-bis(4-chlorophenyl)-; Ethane, 1,1,1-trichloro-2,2-bis(p-chlorophenyl)-; ETHANE, 1,1-BIS(4-CHLOROPHENYL)-2,2,2-TRICHLORO-; Gesafid; NSC 8939; p,p′-DDT; p,p′-Dichlorodiphenyltrichloroethane; p,p′-Dichlorodiphenyltrichloromethylmethane; etc
o,p′-DDD 53–19-0	1-(2-Chlorophenyl)-1-(4-chlorophenyl)-2,2-dichloroethane; 1-(o-Chlorophenyl)-1-(p-chlorophenyl)-2,2-dichloroethane; 1,1-Dichloro-2-(o-chlorophenyl)-2-(p-chlorophenyl)ethane; 2-(2-Chlorophenyl)-2-(4-chlorophenyl)-1,1-dichloroethane; 2-(p-Chlorophenyl)-2-(o-chlorophenyl)-1,1-dichloroethane; 2,4′-Dichlorodiphenyldichloroethane; Benzene, 1-chloro-2-[2,2-dichloro-1-(4-chlorophenyl)ethyl]-; Benzene, 1-chloro-2-[2,2-dichloro-1-(4-chlorophenyl)ethyl]-; CB 313; Chloditan; Chlodithane; Ethane, 1,1-dichloro-2-(o-chlorophenyl)-2-(p-chlorophenyl)-; Lysodren; Mitotan; Mitotan; mitotane; mitotano; NSC 38,721; o,p′-DDD; o,p′-Dichlorodiphenyldichloroethane; o,p′-TDE; Opeprim

(continued)

Table 14.1 (continued)

Name of compound, CAS ID No	Other names (according to CAS Common Chemistry and PubChem)
p,p'-DDD 72–54-8	1,1'-(2,2-Dichloroethylidene)bis[4-chlorobenzene]; 1,1-Bis(4-chlorophenyl)-2,2-dichloroethane; 1,1-Bis(p-chlorophenyl)-2,2-dichloroethane; 1,1-Dichloro-2,2-bis(4-chlorophenyl)ethane; 1,1-Dichloro-2,2-bis(p-chlorophenyl)ethane; 2,2-Bis(4-chlorophenyl)-1,1-dichloroethane; 2,2-bis(para-Chlorophenyl)-1,1-dichloroethane; 2,2-Bis(p-chlorophenyl)-1,1-dichloroethane; 4,4'-DDD; 4,4'-Dichlorodiphenyldichloroethane; 4,4'-TDE; Benzene, 1,1'-(2,2-dichloroethylidene)bis[4-chloro-; Benzene, 1,1'-(2,2-dichloroethylidene)bis[4-chloro-; Benzene, 1,1'-(2,2-dichloroethylidene)bis[4-chloro-; DDD; Dichlorodiphenyl dichloroethane; Dichlorodiphenyldichloroethane; DICHLORODIPHENYLDICHLOROETHANE (DDD); Dilene; Ethane, 1,1-dichloro-2,2-bis(p-chlorophenyl)-; ME 1700; NSC 8941; p,p'-DDD; p,p'-Dichlorodiphenyldichloroethane; p,p'-Dichlorodiphenylethylene dichloride; p,p'-TDE; Rhothane; TDE; TDE (1,1-DICHLORO-2,2-BIS(P-CHLOROPHENYL)ETHANE); UN 2761
o,p'-DDE 3424–82-6	1,1-Dichloro-2-(o-chlorophenyl)-2-(p-chlorophenyl)ethylene; 2-(2-Chlorophenyl)-2-(4-chlorophenyl)-1,1-dichloroethylene; 2,2,o,p'-tetrachlorovinylidenebisbenzene; 2,2,o,p'-Tetrachlorvinylidenbisbenzol; 2,2,o,p'-tetraclorovinilidenobisbenceno; 2,4'-DDE; 2,4'-Dichlorodiphenyldichloroethylene; Benzene, 1-chloro-2-[2,2-dichloro-1-(4-chlorophenyl)ethenyl]-; Ethylene, 1,1-dichloro-2-(o-chlorophenyl)-2-(p-chlorophenyl)-; NSC 59,908; o,p'-DDE

(continued)

Table 14.1 (continued)

Name of compound, CAS ID No	Other names (according to CAS Common Chemistry and PubChem)
p,p'-DDE 72–55-9	1,1'-(Dichloroethenylidene)bis(4-chlorobenzene); 1,1-Bis(4-chlorophenyl)-2,2-dichloroethene; 1,1-BIS-(4-CHLORPHENYL)-2,2-DICHLOR-AETHEN; 1,1-Bis(p-chlorophenyl)-2,2-dichloroethylene; 1,1-Dichloro-2,2-bis(p-chlorophenyl)ethylene; 1,1-Dichloro-2,2-di(p-chlorophenyl)ethylene; 2,2-bis(4-Chlorophenyl)-1,1-dichloroethylene; 2,2-bis(p-chlorophenyl)-1,1-dichloroethylene; 2,2-Bis(p-chlorphenyl)-1,1-dichlorethylen; 2,2-bis(p-clorofenil)-1,1-dicloroetileno; 2,2-Dichloro-1,1-bis(4-chlorophenyl)ethylene; 4,4'-DDE; 4,4'-Dichlorodiphenyldichloroethylene; Benzene, 1,1'-(2,2-dichloroethenylidene)bis[4-chloro-; Benzene, 1,1'-(dichloroethenylidene)bis(4-chloro-; Benzene, 1,1'-(dichloroethenylidene)bis[4-chloro-; Benzene, 1,1'-(dichloroethenylidene)bis[4-chloro-; Benzene, 1,1'-(dichloroethenylidene)bis[4-chloro-; DDE; Dichloro diphenyl dichloroethane; DICHLORODIPHENYLDICHLOROETHYLENE (DDE); Ethylene, 1,1-dichloro-2,2-bis(p-chlorophenyl)-; Ethylene, 1,1-dichloro-2,2-bis(p-chlorophenyl)-,; NSC 1153; *p,p'*-DDE; *p,p'*-Dichlorodiphenyldichloroethylene; UN 2761
PCB 28 7012–37-5	1,1'-Biphenyl, 2,4,4'-trichloro-; 2,4,4'-Trichlorbiphenyl; 2,4,4'-Trichloro-1,1'-biphenyl; 2',4,4'-Trichlorobiphenyl; 2,4,4'-trichlorobiphenyl; 2,4,4'-trichlorobiphenyle; 2,4,4'-triclorobifenilo; 4,2',4'-Trichloro-1,1'-biphenyl; 4,2',4'-Trichlorobiphenyl; Biphenyl, 2,4,4'-trichloro-; CB 28; K 28; PCB 28
PCB 52 35,693–99-3	1,1'-Biphenyl, 2,2',5,5'-tetrachloro-; 2,2',5,5'-TCB; 2,2',5,5'-Tetrachloro-1,1'-biphenyl; 2,2',5,5'-Tetrachlorobiphenyl; 2,5,2',5'-Tetrachlorobiphenyl; CB 52; K 52; PCB 52
PCB 155 33,979–03-2	2,2',4,4',6,6'-hexachlorobiphenyl; 2,4,6,2',4',6'-hexachlorobiphenyl; 246-HCB; PCB-155
PCB 101 37,680–73-2	1,1'-Biphenyl, 2,2',4,5,5'-pentachloro-; 2,2',4,5,5'-PCB; 2,2',4,5,5'-Pentachloro-1,1'-biphenyl; 2,2',4',5,5'-Pentachlorobiphenyl; 2,2',4,5,5'-Pentachlorobiphenyl; 2,4,5,2',5'-Pentachlorobiphenyl; 2,5,2',4',5'-Pentachlorobiphenyl; CB 101; K 101; PCB 101
PCB 118 31,508–00-6	1,1'-Biphenyl, 2,3',4,4',5-pentachloro-; 2,3',4,4',5-Pentachloro-1,1'-biphenyl; 2,3',4,4',5-Pentachlorobiphenyl; 2,4,5,3',4'-Pentachlorobiphenyl; 3,4,2',4',5'-Pentachlorobiphenyl; Biphenyl, 2,3',4,4',5-pentachloro-; CB 118; PCB 118
PCB 143 68,194–15-0	2,2',3,4,5,6'-Hexachlorobiphenyl; 1,1'-Biphenyl, 2,2',3,4,5,6'-hexachloro-; 1,1'-Biphenyl,2,2',3,4,5,6'-hexachloro-; 2,2',3,4,5,6'-Hexachloro-1,1'-biphenyl; 1,2,3,4-tetrachloro-5-(2,6-dichlorophenyl)benzene; PCB 143 (2,2',3,4,5,6'-Hexachlorobiphenyl)

(continued)

Table 14.1 (continued)

Name of compound, CAS ID No	Other names (according to CAS Common Chemistry and PubChem)
PCB 153 35,065–27-1	1,1′-Biphenyl, 2,2′,4,4′,5,5′-hexachloro-; 1,1′-Biphenyl, 2,2′,4,4′,5,5′-hexachloro-; 2,2′,4,4′,5,5′-HEXACHLORBIPHENYL; 2,2′,4,4′,5,5′-Hexachloro-1,1′-biphenyl; 2,2′,4,4′,5,5′-Hexachlorobiphenyl; 2,4,5,2′,4′,5′-Hexachlorobiphenyl; Biphenyl, 2,2′,4,4′,5,5′-hexachloro-; CB 153; K 153; PCB 153; UN 2315
PCB 138 35,065–28-2	1,1′-Biphenyl, 2,2′,3,4,4′,5′-hexachloro-; 2,2′,3,4,4′,5′-Hexachloro-1,1′-biphenyl; 2,2′,3′,4,4′,5-Hexachlorobiphenyl; 2,2′,3,4,4′,5′-Hexachlorobiphenyl; 2,3,4,2′,4′,5′-Hexachlorobiphenyl; 2,4,5,2′,3′,4′-Hexachlorobiphenyl; CB 138; K 138; PCB 138
PCB 180 35,065–29-3	1,1′-Biphenyl, 2,2′,3,4,4′,5,5′-heptachloro-; 2,2′,3,4,4′,5,5′-Heptachloro-1,1′-biphenyl; 2,2′,3,4,4′,5,5′-Heptachlorobiphenyl; 2,3,4,5,2′,4′,5′-Heptachlorobiphenyl; CB 180; K 180; PCB 180

Table 14.2 "New" organic compounds most commonly found by the non-targeted screening analysis of samples from the Far Eastern seas

Name of compound, CAS ID No	Other names (according to Pubchem)	Description
Aldicarb 116–06–3	2-Methyl-2-(methylthio)propanal O-((methylamino)carbonyl)oxime; 2-Methyl-2-(methylthio)propionaldehyde O(methylcarbamoyl)oxime; 2-Methyl-2-(methylthio)propionaldehyde O-(methylcarbamoyl)oxime; Propanal, 2-methyl-2-(methylthio)-, O-[(methylamino)carbonyl]oxime; Propionaldehyde, 2-methyl-2-(methylthio)-, O-(methylcarbamoyl)oxime; Aldicarb; Aldicarbe; NSC 379,586; OMS 771; Temik; Temik 10G; Temik G 10; UC 21,149; UN 2757; etc	A carbamate insecticide, which is the active substance in the pesticide Temik. It is mainly used as a nematicide. Aldicarb is a cholinesterase inhibitor preventing the breakdown of acetylcholine in the synapse. In case of severe poisoning, the victim dies from respiratory failure
Cyromazine 66,215–27–8	2-N-cyclopropyl-1,3,5-triazine-2,4,6-triamine; 2-cyclopropylamino-4,6-diamino-s-triazine; CGA 72,662; CGA-72662; cyromazin; cyromazine dihydrochloride; Larvadex; N-cyclopropylmelamine; Neporex; Vetrazin; etc	Cyromazine is a triazine insect growth regulator used as an insecticide and acaricide. It is a cyclopropyl derivative of melamine. Cyromazine affects the nervous system of immature larvae of some insects. In veterinary medicine, it is used as an ectoparasiticide
Metribuzin 21,087–64–9	4-amino-6-tert-butyl-3-(methylsulfanyl)-1,2,4-triazin-5(4H)-one; 4-amino-6-tert-butyl-4,5-dihydro-3-methylthio-1,2,4-triazin-5-one; Lexone; Sencor; Metribuzine; Zenkor; Sencorex; Senkor; Lexone DF; Sencor DF; Lexone 4L; Sencor 4F; Bayer 6159H; Bayer 6443H; Bayer 94,337; BAY dic 1468; Sencorex L.F.; Metribuzin [BSI:ISO]; etc	A herbicide from the triazinone group. Metribuzin is included as an active ingredient in many pesticide formulations. The mechanism of action is based on inhibition of photosystem II

(continued)

Table 14.2 (continued)

Name of compound, CAS ID No	Other names (according to Pubchem)	Description
Prohydrojasmon-1 158,474–72-7	Prohydrojasmon; Prohydrojasmon [ISO]; propyl 2-[(1R,2R)-3-oxo-2-pentylcyclopentyl]acetate; propyl [(1r,2r)-3-oxo-2-pentylcyclopentyl]acetate; etc	A growth inhibitor herbicide. Detailed information is not available
Oxamyl 23,135–22-0	DPX 1410; DPX 1410L; Du Pont 1410; NSC 379,588; Oxamyl; Oxamyl (pesticide); Thioxamyl; Vydate; Vydate CLV; Vydate G; Vydate L Ethanimidothioic acid, 2-(dimethylamino)-N-[[(methylamino)carbonyl]oxy]-2-oxo-, methyl ester; N',N'-dimethylcarbamoyl(methylthio)methylenamine N-methylcarbamate; N',N'-Dimethylcarbamoyl(methylthio)methylenamin-N-methylcarbamat; N-methylcarbamate de N',N'-dimethylcarbamoyl(methylthio)methylenamine; N-metilcarbamato de N',N'-dimetilcarbamoil(metiltio)metilenamina; Oxamimidic acid, N',N'-dimethyl-N-[(methylcarbamoyl)oxy]-1-thio-, methyl ester; etc	Oxamyl is a carbamate pesticide. It is produced in two forms: granular and liquid. The granular form has been banned in the United States. It is commonly sold under the trade name Vydate. It is classified as an extremely hazardous substance in the United States, as defined in Sect. 302 of the U.S. Emergency Planning and Community Right-to-Know Act (42 U.S.C. 11,002), and is subject to strict reporting requirements by facilities that produce, store, or use it in significant quantities
Thiocyclame 31,895–21-3	Thiocyclam; Thiocyclame; Sultamine; 5-(Dimethylamino)-1,2,3-trithiane; Thiocyclam [BSI:ISO]; N,N-dimethyl-1,2,3-trithian-5-amine; N,N-Dimethyl-1,2,3-trithian-5-ylamine; Thiocyclame [ISO-French]; 1,2,3-Trithian-5-amine, N,N-dimethyl-; N,N-Dimethyl-1,2,3-trithian-5-amine (9CI); N,N-dimethyltrithian-5-amine; etc	Thiocyclam is a nereistoxin analogue insecticide, a selective insecticide with contact action for lepidopterous, etc. pests. At low concentrations, thiocyclam acts as an acetylcholine receptor agonist; at higher concentrations, as a channel blocker

Table 14.2 (continued)

Name of compound, CAS ID No	Other names (according to Pubchem)	Description
Terbucarb 1918–11-2	Terbucarb; Terbutol; Azak; Terbucarb [ISO]; AZAC; AZAR; Phenol, 2,6-bis(1,1-dimethylethyl)-4-methyl-, methylcarbamate; UNII-17B625R8YT; Carbamic acid, methyl-, 2,6-di-tert-butyl-p-tolyl ester; 2,6-Di-tert-butyl-p-tolyl methylcarbamate; 2,6-Di-t-butyl-4-methylphenyl-N-methylcarbamate; (2,6-ditert-butyl-4-methylphenyl) N-methylcarbamate; Phenol, 2,6-bis(1,1-dimethylethyl)-4-methyl-, methylcarbamate (9CI); etc	It belongs to phenylcarbamate herbicides and is classified as a mitotic-inhibiting herbicide. The compound was found to disrupt the mitotic cycle in nodules with microorganisms in plants
Bis(2-ethylhexyl)phthalate 117–81-7	DEHP; Bis(2-ethylhexyl) phthalate; Di(2-ethylhexyl)phthalate Diethylhexyl phthalate; Di-sec-octyl phthalate; Octyl phthalate; Fleximel; Octoil; Ethylhexyl phthalate; Palatinol AH; Celluflex DOP; Vestinol AH; Bisoflex DOP; Kodaflex DOP, Staflex DOP; Truflex DOP; Flexol DOP	It is used in polyvinyl chloride (PVC) production. It exhibits low toxicity with both acute (short-term) and chronic (long-term) exposures. Acute exposure to large doses of DEHP can cause gastrointestinal disorders in humans. Information about reproductive disorders or carcinogenic effects of DEHP in humans is not available

(continued)

Table 14.2 (continued)

Name of compound, CAS ID No	Other names (according to Pubchem)	Description
Triadimefon 43,121–43–3	TRIADIMEFON; Bayleton; Azocene; Triadimefone; Amiral; Fenxiunin; Haleton; Acizol; Adifon; Mighty; Miltek; Strike; Typhon; Nurex; 1-(4-Chlorophenoxy)-3,3-dimethyl-1-(1,2,4-triazol-1-yl)-butan-2-one; 1-(4-Chlorophenoxy)-3,3-dimethyl-1-(1H-1,2,4-triazol-1-yl)-2-butanone; 1-(1,2,4-Triazoyl-1)-1-(4-chloro-phenoxy)-3,3-dimethylbutanone; 1-(4-Chlorophenoxy)-3,3-dimethyl-1-(1,2,4-triazol-1-yl)butanone; etc	Triadimefone is a systemic fungicide. The mechanism of action of the agent is to inhibit demethylation at the C-14 position during the ergosterol biosynthesis. Reactions leading to demethylation at the C-14 position are catalyzed by mixed-function oxygenases. One of these reactions involves the cytochrome P-450
Spirodiclofen 148,477–71–8	Spirodiclofen; 3-(2,4-dichlorophenyl)-2-oxo-1-oxaspiro[4.5]dec-3-en-4-yl 2,2-dimethylbutanoate; Envidor; [3-(2,4-dichlorophenyl)-2-oxo-1-oxaspiro[4.5]dec-3-en-4-yl] 2,2-dimethylbutanoate; Spirodiclofen [ISO]; [2-(2,4-dichlorophenyl)-3-oxo-4-oxaspiro[4.5]dec-1-en-1-yl] 2,2-dimethylbutanoate; etc	An acaricide and insecticide applied to control mites. Spirodiclofen belongs to the tetronic acid class and acts by inhibiting lipid biosynthesis
Triadimenol 55,219–65–3	TRIADIMENOL; Baytan; Bayfidan; Spinmaker; Triafol; Baytan 15; Triaphol; Triadimenol [BSI:ISO]; alpha-tert-Butyl-beta-(4-chlorophenoxy)-1H-1,2,4-triazole-1-ethanol; 1-(4-Chlorophenoxy)-3,3-dimethyl-1-(1H-1,2,4-triazol-1-yl)butan-2-ol; 1-(4-chlorophenoxy)-3,3-dimethyl1-1-(1,2,4-triazol-1-yl)butan-2-ol; 2-(4-Chlorophenoxy)-1-tert-butyl-2-(1H-1,2,4-triazole-1-yl)ethanol; etc	A pesticide and a systemic fungicide from the triazole class. With exposure to triadimenol, the balance of terpenoid biosynthesis products shifts toward phytohormones: abscisic acid, gibberellin, and cytokinin

(continued)

Table 14.2 (continued)

Name of compound, CAS ID No	Other names (according to Pubchem)	Description
Flutriafol 76,674–21-0	Flutriafol; Flutriafol [BSI:ISO]; alpha-(2-Fluorophenyl)-alpha-(4-fluorophenyl)-1H-1,2,4-triazole-1-ethanol; 1-(2-fluorophenyl)-1-(4-fluorophenyl)-2-((1,2,4-triazol-1-yl)ethanol; (RS)-2,4′-Difluoro-alpha-(1H-1,2,4-triazol-1-ylmethyl)benzhydryl alcohol; flutriafen; etc	A pesticide and a fungicide from the class of triazole derivatives. The agent blocks the ergosterol biosynthesis, thereby disrupting the cell wall formation and preventing the growth of hyphae in mycelium. It is also used as a fumigant especially against powdery mildew
Di-n-butyl Phthalate (DBP) 84–74-2	Di-n-butyl phthalate; dibutyl phthalate; n-Butyl phthalate; Butyl phthalate; Elaol; Genoplast B; Palatinol C; Polycizer DBP; Unimoll DB; Dibutyl-o-phthalate; Witcizer 300; Dibutyl 1,2-benzenedicarboxylate; etc	It is used as a plasticizer of polymer materials based on PVC, rubbers, epoxy resins, some cellulose esters, and as a high-boiling solvent. A component of the BF-6 glue. Dibutyl phthalate is also used as a repellent

(continued)

Table 14.2 (continued)

Name of compound, CAS ID No	Other names (according to Pubchem)	Description
Dimethylvinphos 2274–67-1	Dimethylvinphos; (Z)-DIMETHYLVINPHOS; Dimethylvinphos (Z type); 2-Chloro-1-(2,4-dichlorophenyl)vinyldimethyl phosphate; [(Z)-2-chloro-1-(2,4-dichlorophenyl)ethenyl] dimethyl phosphate; 2-Chloro-1-(2,4-dichlorophenyl)ethenyl dimethyl phosphate; Phosphoric acid, 2-chloro-1-(2,4-dichlorophenyl)ethenyl dimethyl ester; Phosphoric acid, 2-chloro-1-(2,4-dichlorophenyl)vinyl dimethyl ester; etc	An organophosphate pesticide. These compounds inhibit the activity of acetylcholinesterase (a nervous system enzyme) and other esterases, which leads to the accumulation of acetylcholine in nerve synapses, thus, causing symptoms of poisoning. The effect on the CNS is accompanied by dystrophic changes and death of nerve cells due to hypoxia
Hexazinone 51,235–04-2	HEXAZINONE; Velpar; Brushkiller; Velpar L; 3-Cyclohexyl-6-(dimethylamino)-1-methyl-1,3,5-triazine-2,4(1H,3H)-dione; Gridball 1,3,5-Triazine-2,4(1H,3H)-dione, 3-cyclohexyl-6-(dimethylamino)-1-methyl-; Hexazinoc; Hexazinone [ANSI:BSI:ISO]; 3-Cyclohexyl-1-methyl-6-(dimethylamino)-s-trazine-2,4(1H,3H)-dione; 3-cyclohexyl-6-(dimethylamino)-1-methyl-1,3,5-triazine-2,4-dione; 3-Cyclohexyl-6-dimethylamino-1-methyl-1,2,3,4-tetrahydro-1,3,5-triazine-2,4-dione; etc	A non-selective broad-spectrum herbicide from the class of triazines and diketones (triazinones). It acts by inhibiting photosynthesis at the photosystem II level and causes a tendency to increase the activity of serum alkaline phosphatase (SAP) and serum glutamic-pyruvic transaminase (SGPT)

(continued)

Table 14.2 (continued)

Name of compound, CAS ID No	Other names (according to Pubchem)	Description
Tebuconazole 107,534–96–3	Tebuconazole; Folicur; Fenetrazole; Ethyltrianol; Terbuconazole; Terbutrazole; Etiltrianol; Raxil; Preventol A 8; Tebuconazol; LYNX; Tebuconazole [ISO]; 3-((1H-1,2,4-Triazol-1-yl)methyl)-1-(4-chlorophenyl)-4,4-dimethylpentan-3-ol; 1-(4-Chlorophenyl)-4,4-dimethyl-3-(1,2,4-triazol-1-ylmethyl)pentan-3-ol; 1-(4-Chlorophenyl)-4,4-dimethyl-3-(1H-1,2,4-triazol-1-ylmethyl)-3-pentanol; (RS)-1-(4-Chlorophenyl)-4,4-dimethyl-3-(1H-1,2,4-triazol-1-ylmethyl)pentan-3-ol; (alpha-(2-(4-Chlorophenyl)ethyl)-alpha-(1,1-dimethylethyl)-1H-1,2,4-triazole-1-ethanol; etc	A pesticide and an effective systemic fungicide belonging to the third-generation triazoles. It suppresses the ergosterol biosynthesis in membranes of phytopathogens' cells, inhibiting demethylation at the C-14 position. The resulting D5-sterols also affect metabolism, which distinguishes tebuconazole from other triazoles
Piperophos 24,151–93–7	Rilof; Avirosan; Piperofos; Piperophos [BSI:ISO]; 1-(Di-N-propoxyphosphinothioylthiomethylcarbonyl-2-methylpiperidine); O,O-Dipropyl S-2-methyl-piperidinocarbonyl-methyl phosphorodithioate; Phosphorodithioic acid, O,O-dipropyl S-(2-pipecolinocarbonylmethyl) ester; S-2-Methylpiperidinocarbonylmethyl O,O-dipropyl phosphorodithioate; 2-dipropoxyphosphinothioylsulfanyl-1-(2-methylpiperidin-1-yl)ethenone; Phosphorodithioic acid, S-[2-(2-methyl-1-piperidinyl)-2-oxoethyl] O,O-dipropyl ester; etc	An organophosphate pesticide. These compounds inhibit the activity of acetylcholinesterase (a nervous system enzyme) and other esterases, which leads to the accumulation of acetylcholine in nerve synapses, thus, causing symptoms of poisoning. The effect on the CNS is accompanied by dystrophic changes and death of nerve cells due to hypoxia

(continued)

Table 14.2 (continued)

Name of compound, CAS ID No	Other names (according to Pubchem)	Description
Imibenconazole 86,598–92–7	Imibenconazole; Imibenconazole [ISO]; 4-Chlorobenzyl N-2,4-dichlorophenyl-2-(1H-1,2,4-triazol-1-yl)thioacetamidate; 1H-1,2,4-Triazole-1-ethanamidothioic acid, N-(2,4-dichlorophenyl)-, (4-chlorophenyl)methyl ester; (4-chlorophenyl)methyl (1E)-N-(2,4-dichlorophenyl)-2-(1,2,4-triazol-1-yl)ethanimidothioate; etc	A fungicide and seed protectant active against a wide range of pathogens including ascomycetes, basidiomycetes, and deuteromycetes. It is persistent and resistant to rain washout
Molinate 2212–67–1	Ordram; S-Ethyl azepane-1-carbothioate; Jalan; Higalnate; Molmate; Hydram; Felan; Yalan; Yulan; Sakkimol; S-Ethyl hexahydro-1H-azepine-1-carbothioate; Molinate [BSI:ISO]; 1H-Azepine-1-carbothioic acid, hexahydro-, S-ethyl ester; S-Ethyl-N-hexamethylenethiocarbamate; Ethyl 1-hexamethyleneiminecarbothiolate; S-Ethyl N,N-hexamethylenethiocarbamate; etc	A herbicide. It is very toxic and can be fatal if inhaled, ingested, or absorbed through the skin. It causes cancer, has a potential to bioaccumulate and biomagnify, and is an agent that destroys spermatozoa in the male testes and blocks spermatogenesis
Chloroneb 2675–77–6	1,4-Dichloro-2,5-dimethoxybenzene; Demosan; Tersan SP; Demasan; Tersan-SP; Chloronebe; Demosan 65W; Terraneb; Terraneb SP; Benzene, 1,4-dichloro-2,5-dimethoxy-; Nuflo D; Chloroneb [ANSI:BSI:ISO]; Dichloro-2,5-dimethoxybenzene; etc	The fungitoxicity is due to specific inhibition of DNA polymerization. Available data is insufficient to assess the carcinogenic potential

(continued)

Table 14.2 (continued)

Name of compound, CAS ID No	Other names (according to Pubchem)	Description
Jasmolin I 4466-14-2	Jasmolin I; Jasmoline I; (1S)-2-methyl-4-oxo-3-[(2Z)-pent-2-en-1-yl]cyclopent-2-en-1-yl (1R,3R)-2,2-dimethyl-3-(2-methylprop-1-en-1-yl)cyclopropanecarboxylate; Cyclopropanecarboxylic acid, 2,2-dimethyl-3-(2-methyl-1-propenyl)-, (1S)-2-methyl-4-oxo-3-(2Z)-2-pentenyl-2-cyclopenten-1-yl ester, (1R,3R)-; (1R-(1alpha(S*(Z)),3beta))-2-Methyl-4-oxo-3-(2-pentenyl)-2-cyclopenten-1-yl 2,2-dimethyl-3-(2-methyl-1-propenyl)cyclopropanecarboxylate; (1R,3R)-2,2-Dimethyl-3-(2-methyl-1-propen-1-yl)cyclopropanecarboxylic Acid (1S)-2-Methyl-4-oxo-3-(2Z)-2-penten-1-yl-2-cyclopenten-1-yl Ester; etc	Pyrethrins are contact insecticides that are very quickly absorbed into the insect's body and affect the nervous system, disrupting the transmission of nerve impulses along axons
Cinerin I 97-12-1	Cinerin; Cinerin I [BSI:ISO]; 3-(2-butenyl)-2-methyl-4-oxo-2-cyclopenten-1-yl 2,2-dimethyl-3-(2-methyl-1-propenyl)cyclopentanecarboxylate; Cinerine I; (1S)-3-[(2Z)-but-2-en-1-yl]-2-methyl-4-oxocyclopent-2-en-1-yl (1R,3R)-2,2-dimethyl-3-(2-methylprop-1-en-1-yl)cyclopropanecarboxylate; 2,2-dimethyl-3-(2-methyl-1-propenyl)cyclopropanecarboxylic acid 3-(2-butenyl)-2-methyl-4-oxo-2-cyclopenten-1-yl ester; Chrysanthemummonocarboxylic acid, cinerolone ester; Cinerolone, chrysanthemummonocarboxylic acid ester; etc	Pyrethrins are contact insecticides that are very quickly absorbed into the insect's body and affect the nervous system, disrupting the transmission of nerve impulses along axons
Cycloate 1134-23-2	Etsan; Ronit; Ro-Neet; Hexylthiocarbam; Eurex; Sabet; S-Ethyl N-cyclohexylthiocarbamate; Ro-Neet E; Cycloate [BSI:ISO]; Carbamothioic acid, cyclohexylethyl-, S-ethyl ester; S-Ethylethylcyclohexylthiocarbamate; etc	A thiocarbamate. It is used as a selective herbicide against annual dicotyledonous and some monocotyledonous weeds. It is categorized as a low-toxic substance and considered non-carcinogenic to humans

14.3   Conclusion

A non-target screening analysis of lipophilic xenobiotics in the Far Eastern region has been carried out for the first time. The detection of all these compounds in the samples indicates the presence of various POPs in the environment and living organisms and, therefore, necessitates regular monitoring with methods of qualitative and quantitative analyses. Since most of these pollutants are not banned, their amounts in the environment should be strictly regulated to avoid negative consequences for living organisms, including humans.

References

1. Bader T, Schulz W, Kümmerer K, Winzenbacher R (2016) General strategies to increase the repeatability in non-target screening by liquid chromatography-high resolution mass spectrometry. Anal Chim Acta 935:173–186. https://doi.org/10.1016/j.aca.2016.06.030
2. Kunzelmann M, Winter M, Åberg M et al (2018) Non-targeted analysis of unexpected food contaminants using LC-HRMS. Anal Bioanal Chem 410:5593–5602. https://doi.org/10.1007/s00216-018-1028-4
3. Rebryk A, Gallampois C, Haglund P (2022) A time-trend guided non-target screening study of organic contaminants in Baltic Sea harbor porpoise (1988–2019), guillemot (1986–2019), and white-tailed sea eagle (1965–2017) using gas chromatography–high-resolution mass spectrometry. Sci Total Environ 829:154620. https://doi.org/10.1016/j.scitotenv.2022.154620
4. Tsygankov VY, Boyarova MD, Kiku PF, Yarygina MV (2015) Hexachlorocyclohexane (HCH) in human blood in the south of the Russian Far East. Environ Sci Pollut Res 22:14379–14382. https://doi.org/10.1007/s11356-015-4951-3
5. Tsygankov VY, Khristoforova NK, Lukyanova ON et al (2017) Selected organochlorines in human blood and urine in the South of the Russian far east. Bull Environ Contam Toxicol 99:460–464. https://doi.org/10.1007/s00128-017-2152-0
6. Lukyanova ON, Tsygankov VY, Boyarova MD (2018) Organochlorine pesticides and polychlorinated biphenyls in the Bering flounder (Hippoglossoides robustus) from the Sea of Okhotsk. Mar Pollut Bull 137:152–156. https://doi.org/10.1016/j.marpolbul.2018.10.017
7. Iatrou EI, Tsygankov V, Seryodkin I et al (2019) Monitoring of environmental persistent organic pollutants in hair samples collected from wild terrestrial mammals of Primorsky Krai, Russia. Environ Sci Pollut Res 26:7640–7650. https://doi.org/10.1007/s11356-019-04171-9
8. Donets MM, Tsygankov VY, Polevschikov AV et al (2022) Organochlorine compounds in commercial bivalves from the Mekong and Saigon-Dong Nai River Deltas (South Vietnam). Water Air Soil Pollut 233:64. https://doi.org/10.1007/s11270-022-05540-w
9. Rager JE, Strynar MJ, Liang S et al (2016) Linking high resolution mass spectrometry data with exposure and toxicity forecasts to advance high-throughput environmental monitoring. Environ Int 88:269–280. https://doi.org/10.1016/j.envint.2015.12.008
10. Singer HP, Wössner AE, McArdell CS, Fenner K (2016) Rapid screening for exposure to "Non-Target" pharmaceuticals from wastewater effluents by combining HRMS-based suspect screening and exposure modeling. Environ Sci Technol 50:6698–6707. https://doi.org/10.1021/acs.est.5b03332
11. Blum KM, Andersson PL, Renman G et al (2017) Non-target screening and prioritization of potentially persistent, bioaccumulating and toxic domestic wastewater contaminants and their removal in on-site and large-scale sewage treatment plants. Sci Total Environ 575:265–275. https://doi.org/10.1016/j.scitotenv.2016.09.135

Chapter 15
Persistent Organic Pollutants in the Human Body: The Experience of Russia and the Former Soviet Republics

Abstract This review summarizes data on the distribution and use of persistent organic pollutants (POPs) in Russia. Recent biomonitoring studies of POPs in human organs and tissues are described. Results of Russian and Soviet studies on the effect of xenobiotics on the human body are considered. The research on acute effects that toxicants exert on the human health reached its culmination in the 1960s and 1970s. In the 1980s–1990s, first publications on chronic effects were published.

Keywords POPs · Biomonitoring · Human health · Impact on the human body · USSR · Russia

List of Abbreviation

CA   Congenital anomalies;
HCH   Hexachlorocyclohexane;
DDE   Dichlorodiphenyldichloroethylene;
DDT   Dichlorodiphenyltrichloroethane;
GIT   Gastrointestinal tract;
LDH   Lactate dehydrogenase;
ENT   Otorhinolaryngological (ear, nose, throat);
MPC   Maximum permissible concentration;
PCB   Polychlorinated biphenyls;
PCDD   Polychlorinated dibenzo-p-dioxins;
POP   Persistent organic pollutants;
OCP   Organochlorine pesticides;
OCC   Organochlorine compounds;
CNS   Central nervous system

V. Tsygankov, *Persistent Organic Pollutants in the Ecosystems of the North Pacific*,
Earth and Environmental Sciences Library,
https://doi.org/10.1007/978-3-031-44896-6_15

15.1   Introduction

In conditions of modern civilization, environmental factors account for 40–50% of the total external impact on the human health. Environmental pollution is one of the major causes of the increase in diseases that lead to weakening and modification of the protective functions and adaptive potential of the human body [1, 2]. The uncontrolled pollution of the biosphere by products of organochlorine pesticide degradation was recorded in the second half of the twentieth century [3].

Pesticides are persistent organic pollutants (POPs) that have a damaging effect on all levels of living matter organization. Their accumulation in the environment has raised a global health concern. Due to their toxicity, persistence in the environment, and lipophilicity, these are categorized as general metabolic toxins with pronounced side mutagenic and carcinogenic effects [4, 5].

The world community has adopted a number of documents that provide a legal basis for the development and implementation of global measures aimed at reducing the accumulation of POPs in the environment and the associated risks. Within the framework of the Stockholm Convention on Persistent Organic Pollutants, global monitoring of POP levels in various habitats and also animal, plant, and human organisms has become an objective of extensive and integrated scientific projects to address these issues [6].

In compliance with the Stockholm Convention, Russia conducts biomonitoring of POPs all over its territory. To date, all links of marine and terrestrial food webs have been investigated [7–10]. Nevertheless, the question as to the impact that POPs have on the health of human as the highest link in the food chain still remains open. Despite the substantial research effort in the country [9, 11–19], the array of data collected on POPs at different food chain links and in the human body requires serious systematization and generalization. Further studies are still needed to assess the implication of detected POP concentrations for the health care management. In this regard, the aim of the review was to systematize the results of monitoring studies in Russia involving human as a bioindicator of POPs, as well as to summarize data on pathologies presumably associated with effects of these xenobiotics on the human body.

15.2   Distribution and Use of POPs in Russia

A total of approximately 1500 names of pesticides representing mono-constituent substances, their various mixtures, and also preparations having biological activities based on phytopathogenic strains of microorganisms, fungi, etc. have been registered in Russia [20]. Information about the major groups of pesticides is summarized in Table 15.1.

In Russia, there is a Catalog of Pesticides and Agrochemicals approved for the use by the Ministry of Agriculture of the Russian Federation. However, as can be

Table 15.1 Classification of pesticides by destination, routes of entry, and major effect [21]

Group	Destination, route of entry, and major effect
Insecticides	
Contact action	Cause death of harmful insects upon contact;
Intestinal action	Cause death of harmful insects after entering their guts;
Systemic action	Can move through the vascular system of plant and poison insects foraging on it;
Fumigants	Act in a gaseous state through the respiratory organs of insects
Herbicides	
Contact action	Cause death of weeds upon contact;
Systemic action	Can move through the vascular system of plant and cause its death;
Soil action	Act on the root system of plants or on germinating seeds;
Selective action	Affect only certain plant species;
Total action	Destroy all vegetation
Fungicides	
Contact action	Are applied to control pathogenic fungi;
Systemic action	Can move through the vascular system of plants and kill pathogenic fungi;
Protective action	Can protect against the effects of pathogenic fungi;
Therapeutic action	Can have a therapeutic effect against pathogenic fungi
Other groups	
Larvicides	Kill insect larvae and caterpillars;
Acaricides	Kill herbivorous ticks;
Ovicides	Destroy eggs of harmful insects and ticks;
Nematocides	Kill roundworms;
Zoocides, or rodenticides	Kill rodents;
Molluscicides	Kill mollusks;
Bactericides	Kill pathogenic bacteria

seen in Table 15.1, the concept of *pesticides* is interpreted quite broadly. These include both protective agents and growth stimulants. All the substances listed on the Catalog are registered in accordance with Federal Law No. 109 of June 19, 1997 "On Safe Handling of Pesticides and Agrochemicals". In 2018, new hygienic standards (GN 1.2.3539-18) for pesticides were introduced by the order of the Chief Sanitary Inspector. Until 2016, the use of pesticides and chemicals was regulated by GN 1.2.3111-13.

Many pesticides are no longer used due to their low efficiency or high toxicity, as indicated in the respective international treaties. Nevertheless, an inventory of 2003 showed that 24,000 t of banned pesticides and/or expired agents are stored within the territory of Russia. Furthermore, the fact that 60% of warehouses do not meet sanitary standards and safety requirements raises even greater concerns. The large-scale and often uncontrolled and unjustified use of pesticides has resulted in the situation where the maximum permissible concentrations (MPC) for pesticides in different environments are exceeded in 60% of the territories surveyed. Among the most problematic regions are those with the highest agricultural activities that make a substantial contribution to the total agricultural production in the country, including Krasnodar Krai (2700 t of non-disposed banned or obsolete pesticides stored in warehouses), Rostov, Voronezh, and Kurgan oblasts, and Altai Krai (up to 1000 t per each of the regions) [22].

According to published data, the most polluted environment in the vast majority of cases is the soil of farmlands [23, 24]. However, areas of forest fires, burning municipal solid waste landfills, and even small bonfires at household plots and summer settlements can also be important additional sources of POPs [24]. Thus, according to the report of the Ministry of Natural Resources and Ecology of the Russian Federation "The Status of Pesticide Pollution of the Natural Environment Objects in the Russian Federation", soils of various types from 39 constituent entities of the Russian Federation were studied in 2017. The excess of MPCs for pesticides was recorded from 11 Russia's federal constituent entities (from 12 entities in 2016) [25].

Another important example of distribution of POPs in Russia is polychlorinated biphenyls (PCBs), one of the most common POP classes hazardous to the environment and humans. PCB mixtures have unique physical and chemical properties that provide convenience of storage and use (non-flammability, resistance to acids and alkalis, low solubility in water, and wide dielectric characteristics). Such properties explain the extremely wide range of applications of these chemicals in industries. Although PCBs are resistant to hydrolysis and biotransformation in water, they may become a source of even more toxic compounds such as polychlorinated dibenzo-p-dioxins (PCDD) over time, as a result of photolysis through a series of successive reactions [26].

The major sources of PCB contamination are numerous hydropower plants, railway facilities, and industrial enterprises that produce and use capacitors and transformers. A portion of PCBs are formed as a by-product of garbage incineration. An inventory of PCBs and PCB-containing equipment in the Russian Federation in 2000 revealed approximately 7500 transformers and 340 000 capacitors containing about 21,000 t of PCBs. However, according to expert estimates, the actual amount of PCBs is much higher and can reach at least 28,000 t [27, 28].

In the Soviet Union, the Commonwealth of Independent States (CIS), and the Russian Federation until 2011, PCBs were widely applied in the industrial manufacture of transformers. The major production facilities were located in the cities of Serpukhov (Russia), Ust-Kamenogorsk (Kazakhstan), Gyumri (Armenia), the paint and varnish factories in Yaroslavl, Chelyabinsk, Zagorsk, and Kotovsk (all in Russia); factories for the production of lubricants were in Nizhny Novgorod, St. Petersburg, Orenburg, Ufa, Perm (i.e., in Russia) [29]. Currently, the largest amounts of PCBs are found in the Volga and Ural regions, and the second largest values were recorded from the Central, East Siberian, North Caucasian, and Volga-Vyatka economic regions [30].

Since the Federal Law No. 164 of the Russian Federation dated June 27, 2011 "On Ratification of the Stockholm Convention on Persistent Organic Pollutants (POPs)" was put into effect, Russia assumed obligations to destroy the existing PCB stockpiles. Electrical equipment containing PCBs must be decommissioned by 2025, and PCB-containing waste disposed by 2028. In Russia, emissions of pollutants into the atmosphere are controlled by special programs. Collected samples are analyzed at the laboratories accredited for dioxin and PCB analyses in Moscow, St. Petersburg, Obninsk, and Ufa [31].

An important manifestation of the toxic effects of PCBs is the immunity suppression. The entry of PCBs into the human body also induces cancers of various localizations, damage to the liver, kidneys, nervous system, and skin lesions (neurodermatitis, eczema, and rash). Upon entering the body of a fetus or child, PCBs exert teratogenic effects and contribute to congenital anomalies, developmental delays, immune dysfunction, and impaired hematopoiesis. PCBs also have a mutagenic effect, which negatively affects the health of subsequent generations.

15.3 Biomonitoring of POPs in Human Body

In many countries, most POPs are banned or their use is strictly regulated. In 2006, the World Health Organization (WHO) has decided to allow further use of DDT, the most notorious organochlorine pesticide, in 12 countries, including India, North Korea, and some South African states [6]. These POPs can come to Russia via transboundary transport, with sea currents, migrating organisms, and foods [32, 33].

POPs can enter the human body by alimentary, respiratory, and contact routes, through damaged and even intact skin. Both rural residents living near POP-treated lands and a large part of human population are exposed to these substances due to their wide application for household purposes [34]. A typical example that can be used for analyzing the distribution of POPs in the human body is organochlorine pesticides (OCP).

Biological monitoring, which provides an integral indicator of OCP contamination and accumulation in the human body, is currently conducted in four of the largest agricultural entities of the Southern Federal District: Krasnodar Krai, Astrakhan, Volgograd, and Rostov oblasts. Unfortunately, there is a lack of available data on the Republics of Adygea and Kalmykia, whose lands are subject to the natural and anthropogenic desertification processes that have turned these regions into recognized zones of ecological disaster [35–37].

A short-term, one- or two-fold excess of the MPC for pesticides contributes to the emergence of local foci of environmentally caused diseases [3, 38, 39]. A significant direct relationship between the prevalence of diseases and the total pesticide's pressure on the human body has been reported [40, 41].

Since 1998, biomonitoring of organochlorine compounds has been carried out in the former republics of the Soviet Union: Kazakhstan [39], Belarus [12, 42], Armenia [43], Ukraine [44], and Kyrgyzstan [16, 18, 45]. There are data collected through biological monitoring in Altai Krai [46], Arkhangelsk Oblast [47, 48], Baikal Region, Irkutsk Oblast [13, 49], Far East [8, 10, 50, 51], Murmansk Oblast (Russian Lapland) [52–54], Oryol Oblast [55], Samara Oblast [56–59], and Chukotka [2, 60].

The data of the currently conducted observations are summarized in Table 15.2.

In the 1960s–1980s, numerous studies were carried out in the Soviet Union, where effects of pesticides on the following human organs and systems were considered: brain [68], liver [67, 71], eyes [72], skin [73], blood [74], immune system [75, 76], reproductive system [77, 78], nervous system [79, 80], vestibular system [81], and also on carbohydrate metabolism and oxidative processes [82].

The high solubility in fats and low solubility in water cause DDT to be retained in adipose tissue. In general, consumers of high trophic levels tend to accumulate large amounts of DDT compared to producers and consumers of lower trophic levels [76]. It has been reported that plant foods contain significantly less dioxins and PCBs (plants poorly metabolize lipophilic substances), which can also be observed in analysis of biomaterial samples from vegetarians who do not eat animal foods [83].

In 2005, the I.M. Sechenov First Moscow State Medical University used gas–liquid chromatography and gas chromatography–mass spectrometry to determine the level of DDT and its metabolite, DDE, in 49 official phytopreparations (liquid extracts and tinctures) from yarrow grass, thyme, nettle leaves, and viburnum and hawthorn fruits. It was found that the absolute concentrations of DDT in liquid extracts and tinctures reached 0.45 ng/g, while DDE amounted to 3.07 ng/g [84].

Systematic exposure to pollutants can pose a threat to public health even at levels that do not exceed the hygienic standards. Accordingly, assessment of the risk to human health should include the total estimate of the integrated impact of inhaled polluted air and also the entry of substances with drinking water and food [85, 86].

Due to the lack of opportunities to conduct research on such a wide range of objects, recommendations have been made for the analysis of POPs in air, water, breast milk, and human blood [87]. In epidemiological practice, the 4.4-DDE/4.4-DDT ratio in blood and breast milk is usually used to estimate how long ago DDT entered the body: the higher the value of this ratio, the lower the concentration of the initial 4.4-DDT was and the longer the exposure lasted [88].

The ratio of concentrations of α- and γ-HCH is used to estimate the time period since the entry of pesticides into the ecosystem. A ratio value of more than unity indicates a long-term presence of OCPs in the environment; a value below unity, i.e. the dominance of the γ-isomer, is characteristic of *fresh* entry [88]. Relatively large amounts of the β-isomer compared to other HCH forms are evidence of the long use and substantial microbiological biotransformation of the substance [89].

Long-term observations have also shown that serum levels of certain organochlorine pollutants, in particular β-HCH, are closely related to biomarkers of carbohydrate and lipid metabolism, e.g., the concentration of the hormone leptin and the degree of insulin resistance in adolescents living in Chapaevsk, a well-known focus of chemical hazard [90].

Table 15.2 Concentrations of organochlorine compounds in biological fluids and materials from residents of various regions of Russia and former Soviet republics

Region	HCH	DDT	References
Breast milk, ng/g lipid weight			
Armenia	0.0106[15.1]	0.0054[15.1]	[43]
Arkhangelsk Oblast	0.2–3.2	1037–1098	[60]
Baikal Region (Usolye-Sibirskoye)	0.5–16	566	[49]
Republic of Belarus	2.22[2]	7.66[2]	[12]
Vladimir Oblast	–[8]	–	[50]
Far East	76	12.9	[50, 61]
Irkutsk Oblast	0.7–4.2	521	[13]
Murmansk	2–3	900	[54]
Republic of Buryatia	0.45–18	660	[62]
Samara Oblast	115–196	207–244	[58]
Tyumen Oblast	–	–	[49]
Chelyabinsk Oblast	–	–	[63]
Chechen Republic	–	–	[63]
Chukotka Autonomous Okrug	0.1–5.6	204–418	[60]
Blood			
Far East (Primorsky Krai)	90–950[3]	6.37[2]	[8, 50]
Republic of Crimea	–	–	[64]
Murmansk Oblast	1.05[2]	6.37[2]	[53]
Russian Arctic	–	1.5–4.7[2]	[14]
Samara Oblast (Chapaevsk)	–	–	[56]
Urine			
Far East (Primorsky Krai)	110–160[6]	70–490[6]	[8, 50]
Sperm			
Kyrgyzstan	0.0001–0.014[15.1]	0.0004–0.053[15.1]	[45]
Placenta			
Kyrgyzstan	–	–	[16]
Abortion tissues			
Kyrgyzstan	–	–	[18]
Subcutaneous adipose tissue			
Ukraine	–	3100[5]	[65, 66]
Liver			
Ukraine	–	820[5]	[67]
Brain			
Uzbekistan	–	–	[68]
Saliva			

(continued)

Table 15.2 (continued)

Region	HCH	DDT	References
Krasnodar	–	–	[69]
Hair			
Kazan	–	$2.8–28.0^7$	[70]

[15.1] mg/L; [2] μg/L; [3] ng/L; [4] mg/mL; [5] μg/kg wet weight; [6] pg/L; [7] ng/mg wet weight; [8] not detected

15.4 Russian Studies of POP Effects on the Human Body

Some researchers note the pronounced bioaccumulation potential of POPs [91]. Their combined effect on organs is a simple summation of the POP contents that would be observed with their separate entries.

There are many signs and symptoms of chronic intoxication that depend on route of toxicants' entry into the body [92]:

• skin manifestations such as chloracne, hyperpigmentation, and hyperkeratosis;
• hepatic syndrome such as liver fibrosis, damage to pancreas, increased blood levels of transaminase and triglycerides, elevated cholesterol level, and digestive disorders (vomiting, nausea, defecation disorders, intolerance to alcohol and fatty foods);
• cardiovascular syndrome such as shortness of breath, palpitations, myocardiodystrophy, and arterial hypotension;
• respiratory syndrome such as lesions of the upper respiratory tract, chronic toxic bronchitis, which is characterized by diffuse atrophy of the mucous membrane, low ventilation rates, and increased viscous resistance.

15.4.1 Skin Cover (Derma)

Chloracne is a well-known specific skin lesion in the form of increased pigmentation, pathological pore dilation, and hypersensitivity in persons who have been exposed to organochlorine compounds [93].

15.4.2 Cardiovascular System

A chronic exposure of the human body to POPs increases the prevalence and aggravates diseases of the cardiovascular system, e.g., coronary insufficiency, hypertension, and vascular atherosclerosis [94]. In 1986, a study of the atherogenic properties of POPs showed characteristic patterns of disturbance of lipid metabolism that may accelerate the progression of and aggravate atherosclerosis [95]. Subsequently, in 1990, this hypothesis was confirmed in the studies by Gadalina: pollutants caused biochemical changes characteristic of early manifestations of atherosclerosis, with more pronounced atherogenic effect caused by smaller doses of xenobiotic [96].

POPs have a direct positive inotropic effect on the myocardium [97, 98]. According to biochemical studies, an increase in lactate dehydrogenase (LDH) activity in the myocardium and a decrease in LDH levels in blood serum, which indicates certain disorders and rearrangement of bioenergetic processes in cardiomyocytes, occur on day 90 of POP intoxication. Damage to mitochondrial membranes (both external and internal) entails disturbance of oxidative phosphorylation, ATP utilization by myofibrils, etc. These changes, in turn, lead to a reduction in the contraction energy of the myofibrillar apparatus and a disturbance of excitation associated with myocardium contraction and relaxation. The energy deficiency is compensated by the development of giant mitochondria forms and hypertrophy of myofibrils [95].

15.4.3 Nervous System

Disorders of the nervous system are characteristic of effects of all POPs. In case of chronic poisoning, various functional and dynamic disorders of the nervous system are observed: diffuse lesions (encephalo-polyneuropathy) with scattered small-focal organic symptoms; a complete set of symptoms of vegetative asthenic syndrome in the form of decreased performance, insomnia, increased irritability, lability of blood pressure, hyperhidrosis, hyperreflexia, etc. The most severe pathological changes in case of organochlorine poisoning occur in the central nervous system structures, which are largely similar to the clinical manifestations of encephalitis with a predominant damage to the subcortical region [98]. In severe forms of intoxication, hypothalamic syndrome may develop (hyperglycemia, arterial hypertension, and obesity) [98].

It should be taken into account that some POPs have a long-term neurotoxic effect. Surveys of the cognitive functions in the populations of three ecologically unfavorable villages of Talgar District showed that almost 70% of local residents had a low level of resilience and decreased focus of attention on information received, a significant and substantial decrease in performance due to the dominance of the inhibition process in the central nervous system, which is characteristic of overtiredness or asthenization [39].

15.4.4 Immune System

Persistent suppression of the immune system activity is characteristic of most xenobiotics. This effect is observed even under exposure to small, low-toxic doses [99]. The negative impact of POPs on the human immune system, leading to atypical forms of various infectious and non-communicable diseases accompanied by an increased risk of autoimmune pathologies, has also been confirmed [99, 100].

Intoxication with pollutants has been reported to induce lymphopenia that affects all types of lymphocytes (T-, B-, NK-cells). After a short-term exposure to xenobiotics, changes in immunity persist for four months or more. With a long-term exposure to pesticides, disorders in the immune system may persist for 2–10 years [99].

A study by Israilova [101] showed that people living in areas of intensive application of pesticides manifest persistent and profound disorders of the immune system, leading to protracted pneumonia. Accordingly, pathologies of the respiratory system are indicators of an ecologically unfavorable air environment [102].

Long-term international studies in the Arctic region have convincingly proven the immunodeficiency states induced by POP poisoning of adult population whose diet includes fish and meat of marine animals. Infants that receive toxicants with their mother's milk also suffer [103].

15.4.5   Excretory System

In the regions where POPs are actively used, impairment of kidney function is more frequently recorded in the form of a decrease in glomerular filtration and inhibition of nitrogen excretion function. Dysmetabolic nephropathies increase multifold, while anomalies of the urinary system organs and pathological changes in the biochemical parameters of urine become increasingly common [104]. The incidence of infectious and inflammatory kidney diseases in areas of intensive use of pesticides or direct exposure to them is 12.5%; in the zone of medium intensity of use, 7.9%; in conditionally *clean* zones, 1.6% [104, 105]. A clinical and epidemiological survey of humans residing in the regions of pesticide use has revealed signs of acute infectious and inflammatory kidney diseases with severe outcomes in 6.5% of cases [106].

The histological and cytoenzymatic changes in the nephron structures suggest the entry of pesticides into the endothelium of glomerular capillaries during blood filtration and into the epithelium of tubules during reabsorption of ultrafiltrate. Thus, HCH has a greater nephrotoxic effect associated with a pronounced inhibition of enzymes involved in infiltration/reabsorption/secretion processes in nephrons [107].

15.4.6   Male Reproductive Health

The interest in finding the relationship between POPs and male infertility has arisen due to the fact that almost all anthropogenic pollutants exert a gonado- and embryotoxic effect associated with their hormone-like properties, due to which these substances are referred to as hormone-like xenobiotics, or *environmental hormones* [108, 109]. Many researchers associate male infertility with the action of pesticides, in particular DDT [110]. In POP-contaminated areas, the incidence of male infertility is 2–2.5-fold higher (primary male infertility constitutes 8–10%) than in ecologically clean areas. OCPs were recorded from semen of patients with male infertility in 75.4% of cases versus 4.3% in the control. The infertility incidence correlates with the POP content of semen, which suggests POPs to be one of the probable causes of male infertility [111].

Low rates of POP removal contribute to a buildup of concentrations of these substances with age. However, the pattern of xenobiotic removal varies between men and women [57]. In the male body, more POPs are accumulated during life. Dudarev notes a higher POP content of blood from men compared to that from women of the same age groups [112]. This may be explained by a higher proportion of meat and fish in the diet of men, as well as by the fact that the female body can partially remove OCCs with breast milk and partially transfer the stored substances to offspring [7, 8].

When the male body is exposed to POPs, the spermatogenesis processes become disturbed [45, 113]. With further increase in contamination, the percentage of pathological semen increases, and the number of patients with a complete lack of spermatozoa in the ejaculate (azoospermia) grows almost five-fold.

15.4.7 Gastrointestinal Tract

It has been noted that the hematosalivar barrier is not a barrier for many toxic agents. A direct relationship has been found for α- and γ-HCH levels between blood serum and oral fluid. Further comparative studies have shown that the saliva from the parotid gland contains these isomers at higher concentrations than the oral fluid and correlates with the concentration of the pesticide in the serum at a higher correlation coefficient. DDT and its metabolites were also found in mixed saliva [69]. Residents of rural areas exposed to pesticides to a significant degree were diagnosed with caries in 78.3%, periodontal disease in 82.2%, stomatitis in 62.7%, and pathological tooth wear in 22.2%; in a control area, these diseases were detected in 48.3, 15.4, 6.4, and 0%, respectively [114].

The effect of POPs on the liver can be manifested in the form of toxic hepatitis that causes a pathological bile composition and the cholestasis syndrome. Toxic pancreatitis is also possible. Damage to the pancreas is associated with increased pressure in the pancreatic duct system due to the dyskinesia of the gastrointestinal tract (GIT) and dystonia of the sphincter of Oddi [102].

Omarova has suggested certain mechanisms of POP exposure. A direct damaging effect on the gastrointestinal mucosa leads to inflammatory, erosive, and ulcerative changes. Damage to the mucous membrane causes impaired motility with dyskinesia of the stomach, intestines, and biliary tract. The sphincter of Oddi dystonia leads to a disturbance of the outflow of secretion from the pancreas. Another mechanism of involvement of GIT in the pathological process is the effect of POPs directly on the nervous system with the development of autonomic dysfunctions and disorders of gastrointestinal motility [102].

15.4.8   Cancer Diseases

The period of culmination in the use of POPs (1980–1995) showed that the prevalence rates of cancer diseases increase as the environment becomes saturated with pesticides [115].

Studies conducted in Kursk Oblast [116] showed a relationship between the increase in the incidence of stomach and breast cancer and the amount of pesticides applied to the soil. In another region, in rural areas of southern Kyrgyzstan, OCPs are one of the major causes of breast cancer in multiparous women [117]. Based on the facts collected, it was proposed to consider the breast cancer incidence as a sensitive criterion for environmental pollution by pesticides [1]. An analysis of the prevalence of acute leukemia in various regions of Tajikistan showed the large-scale use of chemical fertilizers to have been a key factor in the development of this pathology [118]. The employment of parents in agriculture and the associated exposure to pesticides during production activity can also increase the risk of cancer in their children [119].

15.4.9   Female Reproductive Health and Pregnancy

Variations in values of the reproductive health parameters are indicative of the state of the environment and also characterize mutagenicity and embryotoxicity of the factors and their potential to suppress the adaptive mechanisms of the body [120]. Significant changes in the health of pregnant women exposed to a polluted environment are reported for Chelyabinsk Oblast (Chelyabinsk, Magnitogorsk, and Karabash) and Sverdlovsk Oblast (Kirovgrad and Nizhny Tagil). The health of women living near chemical plants in Tambov, Irkutsk, and other regions also suffers [14, 16, 121].

A study of combined effects of pesticides on the reproductive function of women employed in workshops with seed protectants showed a 3.5-fold higher incidence of pregnancy complications than in the control group. A histological examination revealed destructive changes in the placenta (infarctions, vascular obliteration, etc.) that reduce the placental functional efficiency, which becomes one of the causes of perinatal pathologies [122].

Pregnant women's exposure to pesticides increases the incidence of pregnancy and childbirth complications, induces stillbirth, abnormalities in newborns, and leads to deterioration of children's health [93, 123]. In 100% of the examined pregnant women in Ukraine, DDT was detected in peripheral blood at concentrations from 0.28 (in Poltava) to 6 µg/L (in Kiev) [44]. The most severe complications during pregnancy, childbirth, and in the postpartum period were recorded simultaneously with the detection of aldrin and DDT [93].

OCPs are capable of penetrating the placental barrier [82, 124]. An analysis of stillbirths showed a medium DDT level in the fetus' subcutaneous adipose tissue (3100 µg/kg), which differed little from the level in the subcutaneous adipose tissue in operated adults (4330 µg/kg) [65]. The DDT concentration in the liver of stillborn infants averaged at 820 µg/kg [69]. HCH also overcomes the placental barrier [6]. Pesticides increase the release of catecholamines from synaptic nerve endings and penetrate the placental barrier, which causes structural and metabolic disorders in placental tissues, thus, having a toxic effect on the fetus and provoking miscarriages [125].

In the Kyrgyz Republic, OCPs were recorded from the placenta (where their concentration reached 2.27 mg/kg) in 39.2% of pregnant women. The higher the OCP content of placenta was, the more frequently pathologies were observed in pregnant women. A correlation was found between the OCP detection in placenta and the gynecological complications in the postpartum period [16].

In 30.5% of cases, xenobiotics were detected in abortion tissues collected after termination of pregnancy (within the first two months before the placenta formation); in 5.33%, a high OCP concentration was recorded; in 2.66%, there were embryo pathologies (congenital anomalies, CA) (Table 15.3) [18].

Evident differences were also observed after childbirth. If HCH was detected in women's blood, the percentage of asphyxia in their newborns doubled (12.0%). Primiparous women (with HCH in the blood) delivered a significantly larger proportion of infants with developmental anomalies (2.56% vs. 0.15% in the control) [126].

High levels of pollutants cause negative changes in the demographic trends. Thus, with a higher pesticide pressure, the mortality rate increases ($r = 0.82$, $p < 0.05$), while the birth rate drops ($r = -0.67$, $p = 0.046$) [91].

The parameters of newborns' health such as, in particular, prevalence of congenital anomalies are considered the most sensitive indicators for assessing the health status of a population and the effect of environmental factors on it. It has been found that congenital pathologies are associated with the effect of OCPs on pregnant women and women of reproductive age [18]. The primary incidence of congenital anomalies and diseases of the digestive and endocrine systems in infants correlates with the area of arable lands treated with pesticides and mineral fertilizers in the region [125]. To reduce the intake of pesticides, Khamidov [126] recommended increasing the time interval between pregnancies to three years, with "the conception desirable in the winter and spring seasons of the year."

Table 15.3 Level of pesticides in abortion tissues analyzed [18]

OCP concentration in abortion tissues	Number of samples analyzed	Of these, OCP were detected in	
		Number	%
Total analyzed	75	23	30.6
Of these, OCPs were up to 0.1 mg/kg	75	4	5.33
Of these, CAs were found	75	2	2.66
Of these, OCPs were over 0.1 mg/kg	75	17	22.6

15.4.10 Breast Milk

A specific problem that pediatrics currently faces is the risk of residual pesticide level in foods, especially in baby food. The younger the child, the greater the relative dose of pesticide that he/she receives in conditions equivalent to those for adults [102].

The number of children with diseases of the ENT organs (chronic pharyngitis, trophic rhinitis, sinusitis, laryngitis, acute and chronic tonsillitis, and otitis media) increases in areas with elevated levels of POPs. An increase in the prevalence of inflammatory diseases of the middle ear is observed in Eskimo infants fed breast milk and children eating marine fish and meat of marine mammals containing elevated concentrations of pesticides [103].

Breast milk has a number of advantages as an observation object, since breast milk sampling is a non-invasive technique, free of usual technical and instrumental problems. Levels of POPs in breast milk are considered an indicator of their pressure on the maternal body [127]. Mothers with DDT detected in their breast milk were more likely to deliver low-weight and premature born infants ($26.5 \pm 2.7\%$) than mothers whose milk was free of this xenobiotic ($13.1 \pm 3.7\%$) [68]. This has been confirmed by a study of Zastenskaya with co-author [128] who show a pronounced negative relationship between the degree of DDT contamination of breast milk and the infant's birth weight.

A high DDT content of breast milk was reported for women from Kazakhstan, which is generally characteristic of the former USSR republics [129, 130]. According to the results of an OCP monitoring in breast milk samples of parturient women from the Republic of Armenia, the rate of detection of the major pollutants (γ-HCH and DDE) reached 100%; the rate of DDT detection increased compared to 2009 and amounted to 71.4%. However, detectable residual OCP amounts turned out to be by an order of magnitude higher than in 2009 (in mg/L): γ-HCH, 0.0147; DDE, 0.0169; and DDT, 0.0039 [131].

In breast milk-fed newborns, the daily dose of PCBs can be 10–100-fold higher than in their mothers. For newborns, whose metabolic potencies are not yet fully developed, the major mechanism for removing PCBs from the body is regular steatorrhea (fatty stool) [132].

Mamontova with co-authors [13] note that mother's diet has an effect on the PCB level in breast milk. Traditionally, residents of the settlements on the Lake Baikal coast use fat of Baikal seals and Baikal oilfish in their diet. The PCB concentrations in this fat are much higher than in commercial fish species. In the breast milk from female residents of the Lake Baikal coast, the level of PCBs is comparable only to that in the breast milk from women of the Faroe Islands, whose diet includes mainly fat and meat of marine fish, mammals, and birds, and also to that in the breast milk from female residents of the city of Serpukhov who worked at the local transformer plant where a technical mixture of PCBs (locally called *sovol*) was used in production.

15.5 Conclusion

Human is at the top of the food chain and, therefore, faces the danger posed by POPs to the greatest extent. In this regard, assessment of the impact of xenobiotics on the health of human population is still relevant.

Pesticides entering the human body via migration and translocation chains can have a mutagenic effect, increase the number of point mutations and chromosomal aberrations in somatic and germ cells, induce neoplasms, spontaneous miscarriages and perinatal fetal death, congenital anomalies, and infertility. According to the widely accepted practice, assessment of pesticide mutagenicity is based on identification of the mutagenic properties of certain active components that are part of formulations. The issue of necessity to further study formulations with several active substances seems extremely relevant.

The culmination of the research on acute effects that pesticides exert on the human health took place in the 1960s and 1970s. In the 1980s–1990s, studies were published on pesticides' chronic effects. In recent years, the number of such studies has decreased sharply. Today, biomonitoring is either not carried out in Russian cities or is very irregular, which does not provide a full view of the current status of POPs all over the country. New publications concerning the impact of POPs on public health studied by modern methods are almost absent.

Thus, there is an urgent need to develop at the legislative level and implement a system of continuous state-governed biomonitoring studies including analysis of the environment, living organisms, and human as a bioindicator of accumulation, biotransformation, and long-term effects of POPs.

References

1. Gichev YP (2003) Environment pollution and ecology-related human pathology. State Public Scientific and Technical Library of the Siberian Branch of the Russian Academy of Sciences, Novosibirsk
2. Dudarev AA, Odland JO (2017) Human health in connection with arctic pollution—results and perspectives of international studies under the Aegis of AMAP. Ekologiya cheloveka (Human Ecology) 24:3–14. https://doi.org/10.33396/1728-0869-2017-9-3-14
3. Inelova ZA, Nurzhanova AA, Zhamabalinova RD et al (2010) Phytocenosis bioindication of soils contaminated with pesticides (Talgar district, Almaty region). Eurasian J Ecol 29
4. Melnikov NN (1974) Chemistry and technology of pesticides. Khimiya, Moscow
5. Rakitsky VN, Nikolaeva NI (2001) Morphofunctional criteria of environmental factors effect on an organism. Meditsina, Moscow, Russia
6. UNEP (2020) Stockholm Convention on Persistent Organic Pollutants
7. Tsygankov VY, Boyarova MD, Kiku PF, Yarygina MV (2015) Hexachlorocyclohexane (HCH) in human blood in the south of the Russian Far East. Environ Sci Pollut Res 22:14379–14382. https://doi.org/10.1007/s11356-015-4951-3
8. Tsygankov VY, Khristoforova NK, Lukyanova ON et al (2017) Selected Organochlorines in human blood and urine in the South of the Russian far east. Bull Environ Contam Toxicol 99:460–464. https://doi.org/10.1007/s00128-017-2152-0

9. Tsygankov VY, Yarygina MV, Lukyanova ON et al (2019) Trace concentrations of organochlo-
    rine compounds in biological liquids of the Russian far east residents. Ekologiya Cheloveka
    (Human Ecology) 2019:15–19
10. Tsygankov VY (2019) Organochlorine pesticides in marine ecosystems of the Far Eastern
    Seas of Russia (2000–2017). Water Res 161:43–53. https://doi.org/10.1016/j.watres.2019.
    05.103
11. Revich BA, Shelepchikov AA (2008) The prevalence of male infertility in residents living
    in conditions of the pollution of the environment by organochlorine pesticides. Gigiena I
    Sanitariia 26–32
12. Zastenskaya IA (2009) Organochlorine pesticides: biological monitoring and environmental
    monitoring in evaluation of newborn health impact. Health Environ 540–547
13. Mamontova EA, Tarasova EN, Kuz'min MI et al (2010) The levels of stable organic pollutants
    in the breast milk of women living in the Irkutsk region [Article in Russian]. Gig Sanit 35–38
14. Dudarev AA, Chupakhin VS (2014) Influence of exposure to persistent toxic substances (PTS)
    on pregnancy outcomes, gender ratio and menstrual status in indigenous females of Chukotka.
    Gigiena I Sanitariia 36–40
15. Zastenskaya IA, Piven NP, Kochubinsky VV, Kochubinski AV (2014) Influence of polychlo-
    rinated biphenyls and heavy metals on immune system indexes in experimental study. Toxicol
    Rev 28–31
16. Toichuev RM (2015) The effect of organochlorine pesticide content in placenta on the course
    of pregnancy and childbirth. Gigiena I Sanitariia 106–108
17. Amirova ZK, Speranskaya OA (2016) New persistent organic supertoxicants and their impact
    on human health. Eco-soglasie, Moscow, Russia
18. Darbishev EP, Yeshiev AM (2017) Impact of organochlorous pesticides on the birth of children
    with congenital pathology in Osh region. In: Collection of scientific papers based on the results
    of the international scientific and practical conference. Innovative center for the development
    of education and science, Samara, Russia, pp 96–99
19. Dudarev AA (2018) Public health practice report: water supply and sanitation in Chukotka
    and Yakutia, Russian Arctic. Int J Circumpolar Health 77:1423826. https://doi.org/10.1080/
    22423982.2018.1423826
20. NPO "Tayfun" (2017) The state of pollution by pesticides of environmental objects of the
    Russian Federation in 2016, Yearbook. FGBU «NPO «Tayfun», Obninsk
21. Tsygankov VY, Boyarova MD, Lukyanova ON (2015) Chemical and ecological aspects
    of persistent organic pollutants, textbook. Admiral Nevelskoy Maritime State University,
    Vladivostok, Russia
22. Semerenko SA (2015) Ecology and plant protection. Oil Crops 103–137
23. Chernykh AM (2003) Threats to human health when using pesticides (Review). Gigiena I
    Sanitariia 25–28
24. Larionov KV (2008) Distribution of pesticides in the ecosystem of the Krasnodar Territory
    and minimization of their impact on the environment. Abstract of the dissertation for the
    degree of candidate of chemical sciences, Kuban State University
25. Chernogaeva GM (2019) Review of the state and environmental pollution in the Russian
    Federation for 2018. Federal Service for Hydrometeorology and Environmental Monitoring
    (Roshydromet), Moscow
26. Danilov-Danilyan VI (1999) Order dated April 13, 1999 No. 165 "About Recommendations
    for the purposes of inventory in the territory of the Russian Federation of production facilities,
    equipment, materials using or containing PCBs, as well as PCB-containing wastes"
27. Danilov-Danilyan VI (1999) Order dated February 23, 1999 No 76 "On conducting an
    inventory of production facilities, equipment, materials using or containing polychlorinated
    biphenyls (PCBs), as well as PCB-containing wastes on the territory of the Russian Federation"
28. Speranskaya O, Tsittser O (2004) Persistent organic pollutants: a review of the situation in
    Russia. IPEN (International Pollutants Elimination Network)
29. Korpakova IG, Korotkova LI, Larin AA et al (2015) Availability of persistent organochlo-
    rine pesticides and polychlorbiphenyls in the water area of the licensed Site of Llc "NC
    "Priazovneft" in the Azov Sea. Environmental protection in the oil and gas industry

30. Government of the Russian Federation (2014) Resolution dated July 30, 2014 No. 720 "About measures to ensure the fulfillment by the Russian Federation of the obligations stipulated by the Stockholm Convention on Persistent Organic Pollutants of May 22, 2001"
31. Sobol M (2004) Time to act. IPEN (International Pollutants Elimination Network)
32. Lukyanova ON, Tsygankov VY, Boyarova MD, Khristoforova NK (2014) Pesticide biotransport by Pacific salmon in the northwestern Pacific Ocean. Doklady Biol Sci 456:188–190. https://doi.org/10.1134/S0012496614030089
33. Lukyanova ON, Tsygankov VY, Boyarova MD, Khristoforova NK (2015) Pacific salmon as a vector in the trasnsfer of persistent organic pollutants in the Ocean. J Ichthyol 55:425–429. https://doi.org/10.1134/S0032945215030078
34. Pavlov AV (1986) Handbook of pesticides: hygiene of application and toxicology. Urojay, Kiev, Ukraine
35. Saleh MA, Afify AMR, Ragab A et al (1996) Breast milk as biomarker for monitoring human exposure to environmental pollutants. In Biomarkers for agrochemicals and toxic substances. In: Biomarkers for agrochemicals and toxic substances. ACS Symposium series, USA
36. Krauthacker B, Reiner E, Votava-Raić A et al (1998) Organochlorine pesticides and PCBs in human milk collected from mothers nursing hospitalized children. Chemosphere 37:27–32. https://doi.org/10.1016/S0045-6535(98)00035-6
37. Liderman EM, Zabelin MV (2018) Analysis of ecological conditions of conditionality of the population health in the southern federal district in aspect of the organization of the regional system of medico-social rehabilitation. Healthc Educ Secur 7–20
38. Rakitsky VN (2015) Prognostic risk of pesticides' toxic effects in workers. Russ J Occup Health Ind Ecol 5–7
39. Kapysheva UN, Bakhtiyarova SK, Zhaksymov BI (2019) Effects of long-term pesticidal pollution environmental human health. Int J Appl Basic Res
40. Ruder AM, Waters MA, Butler MA et al (2004) Gliomas and farm pesticide exposure in men: the upper midwest health study. Arch Environ Health Int J 59:650–657. https://doi.org/10.1080/00039890409602949
41. Ivanov AV (2005) Human health status in the areas with intensive use of pesticides. Gig Sanit 24–27
42. Barkatina EN, Pertsovsky AL, Murokh VI (2002) The content of residual quantities of organochlorine pesticides in staple foods, breast milk and adipose tissue of residents of Belarus. Storage and processing of farm products 29–32
43. Tadevosyan NS, Muradyan SA, Tadevosyan AE et al (2012) Monitoring of environmental pollution in Armenia and certain issues on reproductive health and cytogenetic status of organism. Gigiena I Sanitariia 48–51
44. Demchenko FV (1989) Hygienic aspects of biomonitoring of organochlorine pesticides. Dissertation for the degree of candidate of biological sciences
45. Mirzakulov DS, Eshbaev AA, Mirzokulov SS, Kalmatov RK (2016) Features of influence of organochlorous compounds and their metabolites on the state of fertility of men living in the osh region of Kyrgyzstan. Mod Prob Sci Educ
46. Ushakov AA, Turbinskiy VV, Paschenko IG, Katunina AS (2015) Hygienic assessment of habitat adverse social and sanitary factors in the Altai Krai. Health Risk Anal 50–61. https://doi.org/10.21668/health.risk/2015.4.07
47. Polder A, Gabrielsen GW, Odland JØ et al (2008) Spatial and temporal changes of chlorinated pesticides, PCBs, dioxins (PCDDs/PCDFs) and brominated flame retardants in human breast milk from Northern Russia. Sci Total Environ 391:41–54. https://doi.org/10.1016/j.scitotenv.2007.10.045
48. Lyzhina AV, Buzinov RV, Unguryanu TN, Gudkov AB (2012) Chemical contamination of food and its impact on population health in Arkhangelsk region. Ekologiya cheloveka (Human Ecology) 19:3–9. https://doi.org/10.17816/humeco17398
49. Mamontova EA, Tarasova EN, Mamontov AA (2017) Ecological and hygienic assessment of the consequences of persistent organic compounds pollution of the industrial town (by the Example of Usol'e-Sibirskoe): II. food, human tissues, health risk assessmen. Environ Chem 26:41–52

50. Yufit S, van Leeuwen R, Malisch R, Samsonov DP (2002) Contamination of human milk with PCDDs, PCDFs and PCBs in two Russian cities. Organohalogen Compd 56
51. Tsygankov VY, Boyarova MD, Lukyanova ON, Khristoforova NK (2017) Bioindicators of organochlorine pesticides in the Sea of Okhotsk and the Western Bering Sea. Arch Environ Contam Toxicol 73:176–184. https://doi.org/10.1007/s00244-017-0380-2
52. Dudarev AA, Chupakhin VS, Ivanova ZS, Lebedev GB (2012) Specificity of exposure of the indigenous dwellers of coastal and Inland Chukotka to Dichlorodiphenyltrichloroethane. Gigiena I Sanitariia 15–20
53. Dudarev AA, Chupakhin VS, Ivanova ZS, Lebedev GB (2012) Peculiarities of exposure to polychlorinated biphenyls (PCBs) in the indigenous population of the coastal and Mainland Chukotka. Gigiena I Sanitariia 22–28
54. Dudarev AA, Dushkina EV, Sladkova YN et al (2016) Exposure levels of persistent organic pollutants (POPs) among population of the Pechenga District in the Murmansk region. Toxicol Rev 2–9. https://doi.org/10.36946/0869-7922-2016-3-2-9
55. Shushpanov AG (2011) Influence of the pesticides forbidden to application on health of man. Technol Merchandizing Innov Foodstuff 73–77
56. Revich BA, Sergeev OV, Hanser R (2006) Dioxins, furans and PCB in blood of tecnagers in the town of Chapayevsk first outcome of a perspective epidemiological study. Toxicol Rev 2–8
57. Revich BA, Sergeyev OV, Shelepchikov AA (2012) Innovative environmental and epidemiologic technologies of assessment of dioxins impacts on childrens health. Ekologiya cheloveka (Human Ecology) 19:42–49. https://doi.org/10.17816/humeco17455
58. Sergeyev O, Shelepchikov A, Denisova T et al (2008) Contamination of human milk with PCDDs, PCDFs and PCBs in two Russian cities. Organohalogen Compd 70
59. Lazareva NV, Lineva OI (2017) Interdependent pathogenetic risks of the influence of ecotechnological factors on human somatic and reproductive health. Med Almanac
60. Arctic Monitoring and Assessment Programme (2004) AMAP assessment 2002: persistent organic pollutants in the Arctic. AMAP Arctic Monitoring and Assessment Programme, Oslo
61. Tsygankov VY, Gumovskaya YP, Gumovskiy AN et al (2020) Bioaccumulation of POPs in human breast milk from south of the Russian Far East and exposure risk to breastfed infants. Environ Sci Pollut Res 27:5951–5957. https://doi.org/10.1007/s11356-019-07394-y
62. Tsydenova OV, Sudaryanto A, Kajiwara N et al (2007) Organohalogen compounds in human breast milk from Republic of Buryatia, Russia. Environ Pollut 146:225–232. https://doi.org/10.1016/j.envpol.2006.04.036
63. Amirova ZK, Shahtamirov IY (2012) PCDD/Fs and PCBs-WHO in plasma and breast milk of residents of the Chechen Republic. South Russia Ecol Dev 7:125–129
64. Moskovchuk OB, Moskovchuk KM, Demchenko VF, Evstafyeva HV (2012) Correlations between the content of pesticides in venous blood and immune state of puerperants. Tavricheskiy Mediko-Biologicheskiy Vestnik 15:176–178
65. Komarova LI (1969) Carriage of DDT and some aspects of its influence on the body. Abstract of the dissertation for the degree of candidate of medical sciences
66. Vaskovskaya LF (1971) Chemical and biological characteristics of the accumulation and distribution of DDT in the human body. Dissertation for the degree of candidate of biological sciences
67. Kuzminskaya UA (1975) Biochemical characteristics of liver subcellular cultures under the influence of pesticides: on the mechanism of action of organochlorine and carbamate pesticides. Abstract of the dissertation for the degree of doctor of medical sciences
68. Azizova OM (1981) Morphological changes in the brain in chronic pesticide intoxication. Abstract of the dissertation for the degree of doctor of medical sciences
69. Korot'ko GF (2006) The salivadiagnostics—renaissance of non-invasive technologies. Kuban Sci Med Bull
70. Cuong LP, Evgenev MI, Gumerov FM et al (2011) Environmental monitoring of pesticides in the hair of Vietnamese living in Da Nang (Vietnam) and Kazan (Russia). Bull Kazan Technol Univ 31–37

71. Paramonchik VM (1968) The functional state of the liver in persons exposed to certain organochlorine compounds in the conditions of their production. Abstract of the dissertation for the degree of candidate of medical sciences

72. Nuritdinova F (1974) The state of the organ of vision in persons with pesticide intoxication working in agriculture in Uzbekistan: clinical and experimental study. Abstract of the dissertation for the degree of doctor of medical sciences

73. Yusupova FD (1988) Issues of occupational health and early clinical, functional and biochemical changes in the skin under the influence of certain pesticides. Abstract of the dissertation for the degree of candidate of medical sciences

74. Perkhurova VP (1974) Research on the toxicology of karbofos and the combined action of intermediate products of its production. Abstract of the dissertation for the degree of candidate of medical sciences

75. Shafeev MS (1978) Influence of some pesticides and their combinations on the indicators of immunity and nonspecific reactivity of the organism. Abstract of the dissertation for the degree of candidate of medical sciences

76. Ruzybakeev RM (1987) Immunodeficiency states in chronic pesticide intoxication and the problem of their correction: clinical and experimental study. Abstract of the dissertation for the degree of doctor of medical sciences

77. Saraimanova ZS (1971) Influence of some pesticides on the genital area (clinical and experimental study). Abstract of the dissertation for the degree of candidate of medical sciences

78. Sattarova SS (1981) The influence of some pesticides on the reproductive system, reproductive function, intrauterine development, the development of the fetus and offspring. Abstract of the dissertation for the degree of candidate of medical sciences

79. Krasnyuk EP (1961) Clinical picture and treatment of chronic intoxication in DDT production workers. Abstract of the dissertation for the degree of candidate of medical sciences

80. Atabayev ST (1975) Hygienic study of environmental objects when using pesticides in a hot climate and health-improving and preventive measures. Abstract of the dissertation for the degree of doctor of medical sciences

81. Khakimov AM (1975) The state of the vestibular analyzer in workers with organochlorine and organophosphorus pesticides (clinical and experimental study). Abstract of the dissertation for the degree of candidate of medical sciences

82. Rosivaya L, Sokolai A (1983) Transplacental transfer of pesticides into the human embryo. Czech Med 1–7

83. Onikienko FA (1966) The state of some aspects of carbohydrate metabolism and oxidative processes when exposed to the body of certain organochlorine insecticides. Abstract of the dissertation for the degree of candidate of biological sciences

84. Gravel IV (2005) Evaluation of the transition of Ddt and its metabolites into liquid extracts and tinctures from medicinal plant raw materials. Tradit Med 28–31

85. Zhurba OM, Taranenko NA (2007) Social-hygienic aspects of determining the residual amounts of chloroganic pesticides in food products. Bulletin of the East Siberian Scientific Center of the Siberian Branch of the Russian Academy of Medical Sciences 56–58

86. Zakharenkov VV, Kislitsyna VV (2014) Prioritization of environmental measures based on risk assessment for the health of the population of an industrial city. Adv Curr Nat Sci 12–15

87. Barr DB, Barr JR, Driskell WJ et al (1999) Strategies for biological monitoring of exposure for contemporary-use pesticides. Toxicol Ind Health 15:168–179. https://doi.org/10.1191/074823399678846556

88. Rovinsky FY, Voronova LD, Afanasev MI et al (1990) Background monitoring of ground ecosystems contamination by organochlorine compounds. Gidrometeoizdat, Leningrad

89. Galiulin RV, Galiulina RA (2008) Ecological and geochemical assessment of "Fingers" of persistent organochlorogenic pesticides in the soil–surface water system. Agrochemistry 52–56

90. Burns JS, Williams PL, Korrick SA et al (2014) Association between chlorinated pesticides in the serum of prepubertal Russian boys and longitudinal biomarkers of metabolic function. Am J Epidemiol 180:909–919. https://doi.org/10.1093/aje/kwu212

91. Shumeiko AY (2004) Ecological assessment of the interaction of pesticides and radiation in the agroecosystems of the Bryansk region. Abstract of the dissertation for the degree of candidate of agricultural sciences, Bryansk State University
92. Lotkov VS (2000) Clinical and pathogenetic features of the chronic effects of chlorinated hydrocarbons on the respiratory organs and other body systems (experimental clinical study). Abstract of the dissertation for the degree of doctor of medical sciences, Samara State Medical University
93. Fedorov LA, Yablokov AV (1999) Pesticides—a toxic blow to the biosphere and human. Nauka, Moscow, Russia
94. Boykulov MC (2004) Comparative characteristics of the aorta of rats in the norm and under the influence of pesticides. Morphology 126:22
95. Shitskova AP, Nikolaeva NI, Gadalina ID (1986) Hygienic assessment of the cardiotoxic effect of some pesticides. Gigiena I Sanitariia 4–7
96. Gadalina ID (1990) Some methodological approaches to assessing the cardiotoxic effect of pesticides, taking into account the age factor. Gigiena I Sanitariia 77–78
97. Akhmedov BK, Saliev KK (2000) Peripheral blood parameters in rats with chronic pesticide intoxication. Med J Uzbekistan 89–90
98. Azovskova TA, Vakurova NV, Lavrentiev NE (2014) Occupational intoxication with pesticides. SamLuxPrint LLC, Samara, Russia
99. Khaitov RM, Pinegin BV, Istamov KhI (1995) Ecological immunology. Russian Federal Research Institute of Fisheries and Oceanography (VNIRO), Moscow, Russia
100. Ovchinnikova EL, Rezanova NV, Brusentsova AV (2003) Monitoring the quality and safety of foodstuffs as a component of social and hygienic monitoring. Sibir'–Vostok 31–36
101. Israilova M (1992) Clinical and immunological features of the course and treatment of protracted pneumonia in patients living in the zone of pesticide use. Abstract of the dissertation for the degree of candidate of medical sciences
102. Omarova ZM (2010) Effect of pesticides on children's health. Russ Bull Perinatol Pediatr 55:59–64
103. Donaldson S, Øyvind Odland J, Allard B (2016) AMAP assessment 2015: human health in the Arctic. Arctic monitoring and assessment programme, Oslo, Norway
104. Latypova RI (1971) Functional state of the kidneys in persons working with a complex of chlorine and organophosphorus pesticides. Med J Uzbekistan 19–23
105. Allazov SA (1992) Acute infectious and inflammatory diseases of the kidney under the influence of pesticides (clinical and experimental study). Abstract of the dissertation for the degree of doctor of medical sciences
106. Sitdikova ME, Allazov SA, Sayapova DR (2010) The effect of organochlorine compounds on some urological diseases. Kazan Med J 91:372–374
107. Rasulov MT, Shakhnazarov MA, Shakhnazarov AM, Magomedgadzhiev BG (2017) Morphological and histoenzymatic characteristics of the kidneys under chronic exposure to pesticides hexachlorocyclohexane, chlorophos and copper sulfate. In: VII scientific and practical conference in memory of Professor S.A. Abusueva. Dagestan State Medical University, Makhachkala, pp 242–245
108. Nikitin AI (2008) Environmental factors and the human reproductive system (responsibility to future generations), second edition. ELBI-SPb, St. Petersburg, Russia
109. Nikitin AI (2009) Hormone-like pollutants of the biosphere and their impact on human reproductive function. Biosfera 1:218–249
110. Moshansky VF, Kagan SA, Tektinsky OL (1987) Differential diagnosis of two forms of necrospermia. Urol Nephrol 52:57–59
111. Toichuev RM, Mirzakulov DS, Payzildaev TR (2015) The prevalence of male infertility in residents living in conditions of the pollution of the environment by organochlorine pesticides. Gigiena I Sanitariia 99–101
112. Dudarev AA (2009) Persistent polychlorinated hydrocarbons and heavy metals in Arctic biosphere: the main regularities of exposure and reproductive health of indigenous people. Biosfera 1:186–202

113. Ter-Avanesov GV (2004) Problems of male reproductive health, practical guidance. Telsy, Moscow, Russia

114. Zhumatov U (1982) The state of the oral cavity organs under the influence of organochlorine and phosphorus pesticides on the body. Abstract of the dissertation for the degree of candidate of medical sciences

115. Abdullaev RB, Matkarimova DS, Duchanov SB (2012) Aral crisis: problems of ecological culture and health. Urgench, Uzbekistan

116. Adamovich VL (1986) Ecological tactics of using pesticides in agriculture and ways to prevent harmful consequences. Russia, Bryansk

117. Paizova ZM, Toychuev RM (2012) The influence of environmental pollution with organochlorine pesticides on the development of breast cancer in women, depending on the number of births, in the conditions of the South of Kyrgyzstan. J Sci Art Health Educ XXI Century 14:160

118. Mustafakulova NI, Melikova TI, Mustafakulova NS (2015) Risk factors and clinical features of leukemiasin the Republic of Tajikistan. Avicenna Bull 17:67–70. https://doi.org/10.25005/2074-0581-2015-17-1-67-71

119. Solenova LG (2011) Cancer risk factors in children and approaches to the prevention of their impact. Pediatrics 90:120–126

120. Kuz'min DV (2007) Comparative analysis of reproductive health indices in women living in proximity to an aluminum work area. Gig Sanit 13–15

121. Ailamazyan EK, Belyaeva TV, Vinogradova EG (1996) The influence of the environmental situation on the reproductive health of women. A new look at the problem. Bull Russ Assoc Obstet Gynecol 13–16

122. Verzhanskiy PS (1979) Influence of combined pesticides (fentiuram, homecin, etc.) on reproductive function and prevention of complications in women working at a chemical enterprise. Abstract of the dissertation for the degree of candidate of medical sciences

123. Yudaev AI (2010) Polychlorinated biphenyls and dioxins—superecotoxicants of the XXI century. Energiia: ekonomika, tekhnika, ekologiia 60–65

124. Khudoley VV, Mizgirev IV (1996) Environmentally hazardous factors. Publishing House, St. Petersburg, Russia

125. Khamitova RY, Mirsaitova GT (2014) The population morbidity in conditions of lasting moderate application of pesticides. Health Care Russ Fed 58:38–42

126. Khamidov MK (1986) Childbearing function, the outcome of pregnancy and childbirth for the fetus and newborn in women engaged in the cultivation and cultivation of cotton. Abstract of the dissertation for the degree of doctor of medical sciences

127. Puerto C, Fernandez AMD, Jimenez MA et al (1990) Levels of DDT and its metabolites in the human biological environment. Hyg Sanitation 73–75

128. Zastenskaya IA, Kochubinsky VV (2010) Potential role of epidemiological researches and biological monitoring in hygienic rate setting. Actual Probl Transp Med

129. Goncharuk EI, Sidorenko GI, Golubchikov MV, Prokopovich AS (1990) Using the mother-fetus-newborn system to study the combined action of pesticides and other chemicals. Gigiena I Sanitariia 4–7

130. GN 1.2.3111-13 (2014) GN 1.2.3111-13. Hygienic standards for the content of pesticides in environmental objects (list)

131. Tadevosyan NS, Tadevosyan AE, Dzhandzhapanyan AN et al (2012) Issues of accumulation and detection of some persistent organic pollutants in rural residents of Armenia. Vestnik of Kazakh National Medical University 212

132. Ginsberg G, Hattis D, Sonawane B (2004) Incorporating pharmacokinetic differences between children and adults in assessing children's risks to environmental toxicants. Toxicol Appl Pharmacol 198:164–183. https://doi.org/10.1016/j.taap.2003.10.010

Chapter 16
Comparison of POP Accumulation Between Human Organs, Tissues, and Organ Systems with Focus on Sex-Related Differences: First Investigations in Russia

Abstract The present study aimed to consider the accumulation of persistent organic pollutants (POPs) in various human organs and tissues with focus on sex-related differences. POPs were determined in samples of organs and tissue systems from a man (63 years of age) and a woman (57 years of age). The concentration ranges of OCPs ($\sum$HCH + $\sum$DDT) and PCBs were 14.6–1398.1 (with an average of 254.3 ± 7.8) and 42.4–359.2 (198.0 ± 6.7) ng/g lipid weight, respectively. The $\sum$HCH and $\sum$DDT concentrations ranged within 1.1–1295.6 and 0–472.5 ng/g, with average values of 172.5 ± 49.9 and 81.7 ± 18.8 ng/g, respectively. $\sum$HCH in the man's organs amounted to 221.7 ± 66.1 ng/g; in the woman's organs, to 250.9 ± 49.9 ng/g. As our study showed, the main organs and tissues accumulating the major part of POPs are the blood, hair, spleen, liver, and brain.

Keywords HCHs · DDTs · PCBs · Human organs and tissues · Sex-related differences

16.1 Introduction

Chemical compounds affecting the human health and environment quality are becoming an increasingly global concern [1, 2]. In this respect, human biological monitoring can help evaluate the state of the environment on the basis of human body's response to environmental changes that is manifested in the form of various pathologies [3–5].

Persistent organic pollutants (POPs) including polychlorinated biphenyls (PCBs) and organochlorine pesticides (OCPs) are among the most hazardous chemical compounds. These pollutants are highly resistant, subject to bioaccumulation up food chains, can be transported over long distances, and have a wide range of harmful effects [6–9].

Numerous studies that detected toxicants in human tissues and organs, considered the mechanisms of their entry and accumulation levels, and assessed their effects on the human body still cannot answer the question as to what actually happens in organs and tissues exposed to POPs [10–13].

Biological fluids such as breast milk, urine, and blood, which provide the home-ostatic functions in the body, are often used as markers to identify effects of environmental factors on humans [14]. The exposure biomarkers are also considered as indicators of presence of an exogenous substance in the body (Biomarkers in risk assessment: Validity and validation. IPCS. Environmental health criteria.). Blood, while being a reliable biological material for determining long-term exposure to various toxic compounds, nevertheless, does not reflect their short-term effects [15]. The biomarkers of POP exposure most commonly mentioned in the literature are HCH, DDT, and PCB [16].

In view of the above facts, the present study aimed to investigate the accumulation of POPs in various human organs and tissues with focus on sex-related differences.

16.2 Comparison of POP Accumulation Between Human Organs

POPs were determined in various organs and tissue systems of a man (63 years of age) and a woman (57 years of age) (Table 16.1). The biological samples were collected at several health-care facilities. Informed written consent was obtained from both participants of the study. The personal information is protected by the law "On the bases of health care of citizens of the Russian Federation" and by medical confidentiality. The frozen samples (–20 °C) from them were delivered to the laboratory of the Far Eastern Federal University.

Organochlorine compounds were found in all the tissue and organ samples analyzed. The level of POPs ($\sum$HCH + $\sum$DDT + $\sum$PCB) in the samples ranged from 103.6 to 1695.3 ng/g lipid weight (l.w.). The concentration ranges of OCPs ($\sum$HCH + $\sum$DDT) and PCBs were 14.6–1398.1 (with an average of 254.3 ± 7.8) and 42.4–359.2 (198.0 ± 6.7) ng/g, respectively. The $\sum$HCH and $\sum$DDT concentrations ranged within 1.1–1295.6 and 0–472.5 ng/g, with average values of 172.5 ± 49.9 and 81.7 ± 18.8 ng/g, respectively. $\sum$HCH in the man's organs amounted to 221.7 ± 66.1 ng/g; in the woman's organs, to 250.9 ± 49.9 ng/g. As our study showed, the main organs and tissues accumulating the major part of POPs are the blood, hair, spleen, liver, and brain.

HCH was detected in 100% of the samples. The level of α-HCH ranged from 2.9 to 1106.6 (with an average of 167.6 ± 56.08 ng/g l.w. ng/g). α-HCH was below the detection limits in the hair, kidneys, and subcutaneous tissue of both sexes, in the woman's striated musculature and gallbladder, and in the man's testes. β-HCH was not found only in the man's hair and testes. In the rest of the samples, the levels of the pollutant varied from 1.1 to 191.7 ng/g. γ-HCH was detected at concentrations from 1.1 to 312.9 ng/g. The toxicant was below the detection limits in the man's colon, subcutaneous tissue, and kidneys, and in the woman's striated musculature,

Table 16.1 Lipid content (%) of the tissues and organs analyzed

Organ/tissue analyzed		% lipid weight
Hair	♂	1.20
	♀	1.55
Brain	♂	2.66
	♀	2.59
Subcutaneous tissue	♂	63.53
	♀	68.28
Muscles	♂	1.19
	♀	1.88
Lungs	♂	1.29
	♀	1.26
Liver	♂	5.80
	♀	2.57
Gallbladder	♂	14.76
	♀	6.03
Colon	♂	29.06
	♀	44.31
Kidneys	♂	4.50
	♀	2.23
Spleen	♂	1.04
	♀	0.78
Blood	♂	0.42
	♀	0.56
Mammary glands	♀	78.82
Prostate	♂	2.62
Testes	♂	7.01

mammary gland, and gallbladder. δ-HCH was below the detection limits in the lungs, gallbladder, colon, mammary gland, prostate, and testes. Its level in the other organs varied from 0.8 to 80.0 ng/g.

DDT metabolites were detected in 92% of the samples. DDTs below the detection limits were found in the woman's hair and liver. Mainly o,p'- and p,p'-DDE were identified. The o,p'-DDE concentration ranged from 0.2 to 107.3 ng/g l.w.; the p,p'-DDE concentration, from 12.0 to 472.5 ng/g. No significant differences were observed between the man's and woman's samples in the DDE content: 80.5 versus 72.3 ng/g, respectively. o,p'-DDT was recorded only from the man's colon and omentum at concentrations from 3.2 to 5.2 ng/g; p,p'-DDT, from the woman's omentum at a concentration of 4.0 ng/g; p,p'-DDE, from the woman's omentum and subcutaneous tissue and the man's colon at concentrations from 2.8 to 19.1 ng/g.

Pathoanatomical studies conducted in the 1970s showed that the OCP content in organs with pathologies was 2–threefold higher than normal [17]. Establishing whether a person has been exposed to pesticides is a challenge. Pesticides, which regularly enter the human body at minimum doses with water, food, and air, gradually build up in all organs and organ systems, thus, causing chronic poisoning. In the body, up to 97% of toxicants are accumulated in the bone marrow, lymph nodes, spleen, and adipose tissue [18].

The predominance of α- and β-HCH isomers in human organs indicates a long period since the contamination of the environment. Emissions of α-HCH may occur from hazardous waste burial sites, and also from stockpiles and residues of lindane production [6].

POPs enter the human body mainly with food [19, 20]. Data collected through monitoring of various representatives of biota, including humans, show significant sorption of HCH from the environment, which indicates its bioavailability. The biological half-life of this group of pollutants is 0.7–154 days. The probability of its involvement in food chains depends on the level of lipophilicity of HCH isomers ($-\log KO/W = 7.61$–8.88). Its permissible daily intake is estimated only at 0.001 mg/kg of human body weight [21].

DDT can be spread both by air (from areas that were actively treated in the past and from those where DDT is currently applied) and by water (with anti-fouling paints and dicofol), and also due to its ability to biomagnify [22]. According to our previous studies [6, 7], DDT and its metabolites are much less commonly distributed in the Russian Far East than HCH, probably, because the latter found significantly wider application in the agriculture of the region [23]. This fact may explain the lower concentrations of DDT in human organs and tissues compared to those of HCH.

The evident predominance of HCH content in the organs and tissues of the woman was expected to occur also in the blood. Nevertheless, we observed much higher levels of this pollutant in the man's compared to woman's blood: 1296 versus 303 ng/g l.w., respectively. This may be due to the cyclical changes in the woman's body, and also the ability of POPs to be transmitted through the placenta and excreted with breast milk during lactation [24]. Furthermore, the monthly menstrual blood loss does not allow xenobiotics to build up in this rapidly regenerating tissue.

Polychlorinated biphenyls were found in 100% of samples. The levels of the PCB congeners 143, 155, and 207 were below the detection limits. The lower-chlorinated PCB 28 and PCB 52 were detected in 57 and 86% of samples, respectively. The higher-chlorinated PCB 101 was detected in a range of 0.7–2.8 ng/g l.w. only in the woman's organs (mammary gland, colon, omentum, and subcutaneous tissue). The concentration ranges of PCBs 118, 153, 138, and 180 were 1.9–113.2, 23.7–165.3, 15.6–120.5, and 25.6–64 ng/g, respectively.

PCB 118 may indirectly indicate where this xenobiotic has come from. Its major source is the soil. The practice of burning dry vegetation to free up an arable area is often used in agriculture. In this case, excessive accumulations of chlorine may cause certain PCB congeners to form, and the accumulated pollutants to be transported with biomass-burning aerosol particles [25, 26]. PCB 118 was found in the organs of both the man and the woman. Accordingly, it can be assumed that foods made from plants

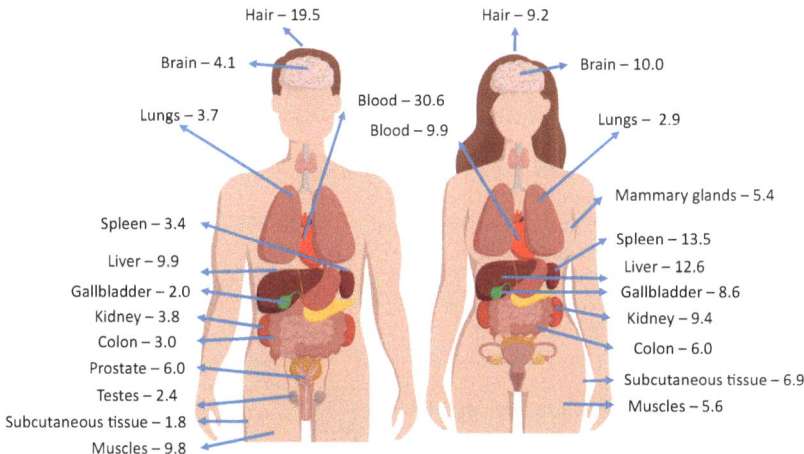

Fig. 16.1 POPs content (%) of the organs in the male and female body

growing in such areas or meat from animal fed these plants can be the responsible for the appearance of this toxicant in the human body (RF Government Order of 2015 no. 1316-r). Furthermore, lubricants, inks, dyes, cement additives, flame retardants, pesticides, glue, sealing fluids, etc. can also be sources of PCBs [27].

According to our preliminary studies (Fig. 16.1), the POP content of lung tissues was comparable between the woman and the man (190.5 vs. 206.9 ng/g l.w., respectively). The man's muscles contained greater amounts of this toxicant (545.2 ng/g) than the woman's ones (375.2 ng/g). In the rest of the analyzed organs, the POP concentration in the man was significantly lower than in the woman. Thus, the concentration in the woman's liver was 1.5-fold, in the colon twofold, in the brain and kidneys threefold, in the spleen and subcutaneous tissue fourfold, and in the gallbladder almost fivefold higher. However, the POP concentration in the man's blood was 2.5-fold higher than that in the woman's blood (1695.3 vs. 662.9 ng/g, respectively), while in the hair it was 1.5-fold higher (1078.4 vs. 604.5 ng/g, respectively).

16.3 Conclusion

Thus, our study has shown that POPs are more actively accumulated in women's organs. Their concentration in blood may not correlate with the actual accumulation of pollutants in certain human organs and tissues. A biological material commonly used for human biological monitoring is hair [28]. In our study, we have convincingly demonstrated that the POP content of hair correlates well with the level of these pollutants in blood. Hair analysis is a non-invasive approach to assessing the accumulation of POPs in the human body that can simplify biomonitoring.

References

1. Khamidullina KK (2015) Preventive toxicology tasks in provision of safe chemical regulations. Occup Med Hum Ecol 280–286
2. Shilov VV, Markova OL, Kuznetsov AV (2019) Biomonitoring of influence of harmful chemicals on the basis of the modern biomarkers. Literature review. Gigiena i sanitariia 98:591–596. https://doi.org/10.47470/0016-9900-2019-98-6-591-596
3. Bobun II (2011) Regional standardization of water chemical substances in case of the Arkhangelsk region. Gigiena i sanitariia 91–95
4. Buzinov RV, Kiku PF, Unguryanu TN et al (2016) From Pomorye to Primorye: socio-hygienic and environmental problems of public health, monograph. Publishing House of the Northern State Medical University, Arkhangelsk
5. Chashchin MV, Chashchin VP, Chashchin VN (2012) The main trends in the concentration of persistent toxic substances in the blood of the indigenous population of the Arctic. Hum Ecol 19:3–7. https://doi.org/10.17816/humeco17459
6. Donets MM, Tsygankov VY, Kulshova VI et al (2020) Food safety of bivalves from the south Vietnam: organochlorine compounds and heavy metals as risk factors for human health. Med Acad J 20:45–48. https://doi.org/10.17816/MAJ34285
7. Donets MM, Tsygankov VY, Boyarova MD et al (2020) Organochlorine compounds in flounders of genus Hippoglossoides Gottsche, 1835 from the Far Eastern seas of Russia. mbj 5:29–42. https://doi.org/10.21072/mbj.2020.05.1.04
8. Tsygankov VY, Gumovskaya YP, Gumovskiy AN et al (2020) Organic chlorine compounds in breast milk of women in the south of the Russian Far East. Ekologiya Cheloveka (Human Ecology) 12–18. https://doi.org/10.33396/1728-0869-2020-4-12-18
9. Tsygankov VY, Gumovskaya YP, Gumovskiy AN et al (2020) Bioaccumulation of POPs in human breast milk from south of the Russian Far East and exposure risk to breastfed infants. Environ Sci Pollut Res 27:5951–5957. https://doi.org/10.1007/s11356-019-07394-y
10. Bedi JS, Gill JPS, Aulakh RS et al (2013) Pesticide residues in human breast milk: Risk assessment for infants from Punjab, India. Sci Total Environ 463–464:720–726. https://doi.org/10.1016/j.scitotenv.2013.06.066
11. Černá M, Krsková A, Čejchanová M, Spěváčková V (2012) Human biomonitoring in the Czech Republic: an overview. Int J Hyg Environ Health 215:109–119. https://doi.org/10.1016/j.ijheh.2011.09.007
12. Chen M-W, Santos H, Que D et al (2018) Association between organochlorine pesticide levels in breast milk and their effects on female reproduction in a Taiwanese population. Int J Environ Res Public Health 15:931. https://doi.org/10.3390/ijerph15050931
13. Waliszewski SM, Caba M, Herrero-Mercado M et al (2012) Organochlorine pesticide residue levels in blood serum of inhabitants from Veracruz, Mexico. Environ Monit Assess 184:5613–5621. https://doi.org/10.1007/s10661-011-2366-2
14. Yusa V, Millet M, Coscolla C, Roca M (2015) Analytical methods for human biomonitoring of pesticides. A review. Anal Chim Acta 891:15–31. https://doi.org/10.1016/j.aca.2015.05.032
15. Jakubowski M (2011) Chapter 3E. Lead. In: Knudsen L, Merlo DF (eds) Issues in toxicology. Royal Society of Chemistry, Cambridge, pp 322–337
16. Lu D, Jin Y, Feng C et al (2017) Multi-analyte method development for analysis of brominated flame retardants (BFRs) and PBDE metabolites in human serum. Anal Bioanal Chem 409:5307–5317. https://doi.org/10.1007/s00216-017-0476-6
17. Gaponyuk EI (1977) Residual content of pesticides in environmental objects and their biological significance. In: Proceedings of IEM. Gidrometeoizdat, Moscow, pp 65–68
18. Chupak VV (2013) Pesticides problems in the ecology and toxicology. Scientific Notes of Orel State University 338–344
19. Malykh KA, Poroshin KV (2018) Determination of pesticide residues in chicken eggs. Bull Mod Res 290–291

20. Rakitskii VN, Doan NH, Fedorova NE et al (2020) Safety of imported agricultural products: pesticide residues. Health Care Russ Fed 64:150–157. https://doi.org/10.46563/0044-197X-2020-64-3-150-157

21. Amirova ZK, Speranskaya OA (2016) New persistent organic supertoxicants and their impact on human health. Eco-soglasie, Moscow, Russia

22. Liu R, Tan R, Li B et al (2015) Overview of POPs and heavy metals in Liao River Basin. Environ Earth Sci 73:5007–5017. https://doi.org/10.1007/s12665-015-4317-7

23. Lin T, Hu Z, Zhang G et al (2009) Levels and mass burden of DDTs in sediments from fishing harbors: the importance of DDT-containing antifouling paint to the coastal environment of China. Environ Sci Technol 43:8033–8038. https://doi.org/10.1021/es901827b

24. Arctic Monitoring and Assessment Programme (2016) AMAP assessment 2015. Arctic Monitoring and Assessment Programme, Oslo, Norway

25. Urbaniak M (2007) Polychlorinated biphenyls: sources, distribution and transformation in the environment—a literature review. Acta Toxicologica 15:83–93

26. Bao L-J, Maruya KA, Snyder SA, Zeng EY (2012) China's water pollution by persistent organic pollutants. Environ Pollut 163:100–108. https://doi.org/10.1016/j.envpol.2011.12.022

27. Cui S, Fu Q, Li Y-F et al (2016) Levels, congener profile and inventory of polychlorinated biphenyls in sediment from the Songhua River in the vicinity of cement plant, China: a case study. Environ Sci Pollut Res 23:15952–15962. https://doi.org/10.1007/s11356-016-6761-7

28. Human biomonitoring: facts and figures. WHO Regional Office for Europe, Copenhagen (2015)

Chapter 17
Persistent Organic Pollutants in Residents of Coastal Areas in the Russian Far East

Abstract A number of countries implement national programs for monitoring persistent organic pollutants (POPs) in the human body. In Russia, monitoring is currently conducted in several regions. As regards the Russian Far East, only preliminary data on the accumulation of POPs in the human body have been published to date. In this chapter, we provide the results of a survey of POP levels in the blood and breast milk of residents of the Russian Far East.

Keywords POPs · Biomonitoring · Blood · Breast milk · Far East

17.1 Introduction

The problem of POP accumulation in the human body is extremely urgent, since these substances have potential teratogenic, carcinogenic, hormonal, neurological, and immunological effects [1–3]. Presumably, about 90% of all pollutants enter the body with food and only 10% with the air breathed and via skin contact [4].

Despite numerous studies on detection of toxicants in human tissues and organs, the mechanisms of entry and accumulation of pollutants, as well as their effects, remain insufficiently understood or largely as theories [5–9]. Although bans and restrictions on the use have been imposed in most countries, the negative effects of these compounds on organisms continue due to their persistence and retention in the biosphere.

The south of the Russian Far East is a zone of developed agriculture where organochlorine pesticides (OCP) were applied to farmlands until the ban. Furthermore, the region is adjacent to China where these substances are still used to control crop pests and vectors of infectious diseases.

National programs for monitoring organochlorine compounds (OCC) in the human body are implemented in various countries, e.g., the Czech Republic [6, 7] and the Republic of Korea [9]. Similar studies are also conducted in Russia [10–14].

© The Author(s), under exclusive license to Springer Nature Switzerland AG 2023
V. Tsygankov, *Persistent Organic Pollutants in the Ecosystems of the North Pacific*,
Earth and Environmental Sciences Library,
https://doi.org/10.1007/978-3-031-44896-6_17

17.2   POPs in Blood of Far East Residents

Biological fluids that provide homeostatic functions of the body such as, primarily, urine and blood are often analyzed to identify effects of environmental factors on humans [15].

In 2019, we collected samples from residents aged 18–30 yr. The numbers of men and women were 35 and 41, respectively. The lipid content of blood ranged from 0.03 to 0.33% (with a mean value being $0.17 \pm 0.06\%$).

Organochlorine compounds were found in 63 people (about 83% of the sample). α-, β-, γ-, δ-HCH, DDE, and PCB 52 were detected in 41, 92, 60, 52, 27, and 40% of samples, respectively.

The total POP concentrations ranged from 56 to 2475.8 (mean 886.4 ± 572) ng/g lipid weight (l.w.).

The total HCH concentrations were within a range of 16.9–1130.2 (mean 414.8 $\pm$ 282) ng/g l.w. The levels of α-, β-, γ-, δ-HCH ranged from 9.9 to 1015.7 (mean 375 $\pm$ 234), from 7 to 494.8 (121.8 $\pm$ 109), from 13.8 to 292.4 (99.1 $\pm$ 72), and from 19.4 to 468.2 (143.1 $\pm$ 98), respectively (Fig. 17.1).

The levels and distribution of certain HCH isomers indicate a long time since contamination (long-term exposure) ($\alpha + \beta$), and also the active degradation of lindane ($\gamma < \delta$) (Fig. 17.2).

As Fig. 17.2 shows, the sum of α- and β-isomers reaches almost 70%, i.e. the ecosystem has been exposed to this pesticide for a long time, and it has already degraded by 2/3 into metabolites.

Of the DDT metabolites, we detected only DDE in the blood within a concentration range of 1.9–213.1 (mean 92 $\pm$ 68) ng/g l.w. This fact indicates the lack of fresh contamination and the degradation of the initial DDT. Of PCB congeners, we recorded only PCB 52 whose concentrations ranged from 28.9 to 483.4 (110.9 $\pm$ 122) ng/g l.w. This congener is water-soluble (therefore, its presence in the blood is

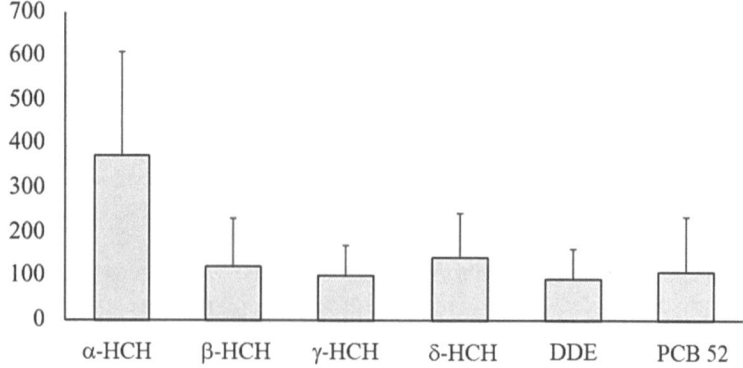

Fig. 17.1 Average concentrations of certain POPs in blood samples from residents of Primorsky Krai, ng/g l.w.

Fig. 17.2 Proportions of
HCH isomers in the blood
samples from residents of
Primorsky Krai

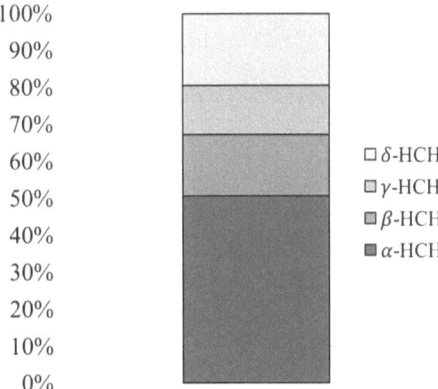

expected) and is subsequently excreted from the body with urine. However, as our initial studies showed, only picograms of toxicants are removed with urine, while the remaining portion is *deposited* in adipose tissues [16, 17].

Concentrations of organochlorine compounds, which are lipophilic, show a relationship with the lipid content (Fig. 17.3). Although this relationship is not statistically confirmed, POP concentrations, nevertheless, show a general tendency to increase with higher lipid content.

We did not find any statistically significant age- and sex-related differences in POP accumulation, probably, because the 10-year age span is not sufficient to assess time trends of POP accumulation (Fig. 17.4).

Hexachlorocyclohexane was used in the mid-twentieth century as a technical mixture with the following proportions of isomers: α-HCH, 55–70%; β-HCH, 5–14%; and γ-HCH, 9–13% [18]. It is still applied in China, India, and other countries due to the lack of cost-effective and environmentally friendly alternative methods to control insect vectors of infections [9]. Since 1990, the production of pesticides in China has increased and now amounts to 2.2×10^6 t/yr [19]. The ban on the use of DDT in the Russian agriculture was imposed in 1971, and the ban on HCH in 1990 [11, 14].

The data obtained confirmed the lack of environmental risk to human health in the region. The conclusion was drawn on the basis of a comparison with the international permissible threshold values of OCCs in blood: the plasma DDT concentration of 200 μg/L [1]; in the whole blood, the HCH and DDT concentrations within ranges from 0.3 to 0.9 μg/L and from 1.5 to 31 μg/L, respectively [20].

A comparison of the data that we obtained (Table 17.1), converted to ng/g l.w. (concentration range, from 56 to 2476; average OCC concentration, 886; and median, 780 ng/g l.w.), with the results of other authors showed that the lower OCC content in the blood samples from residents of the south of the Russian Far East is, nevertheless, higher in median than that in the blood from residents of the UK and Japan.

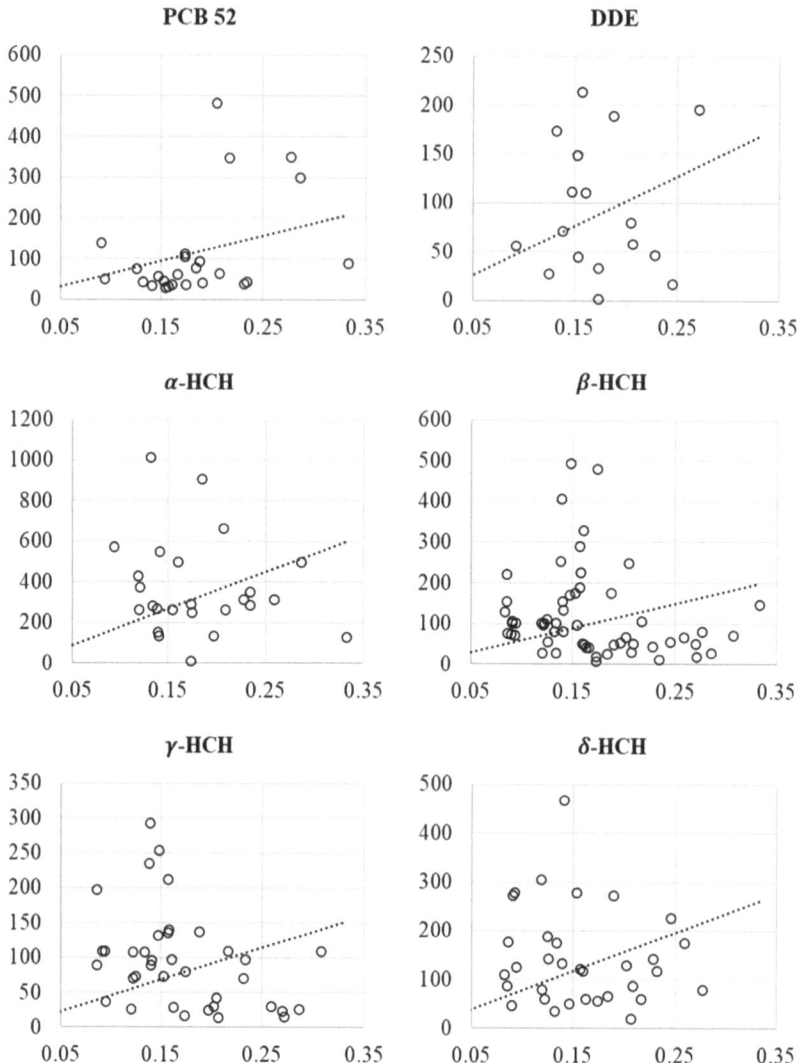

Fig. 17.3 Relationship between the concentrations of certain POPs (X-axis, ng/g l.w.) and the lipid content (Y-axis, %) for the blood samples from residents of Primorsky Krai

Thus, POPs are present in the blood of residents at noticeable concentrations in up to 83% of the statistical sample. This is probably explained by the fact that the developing countries continue to use POPs because of the lack of safer alternatives.

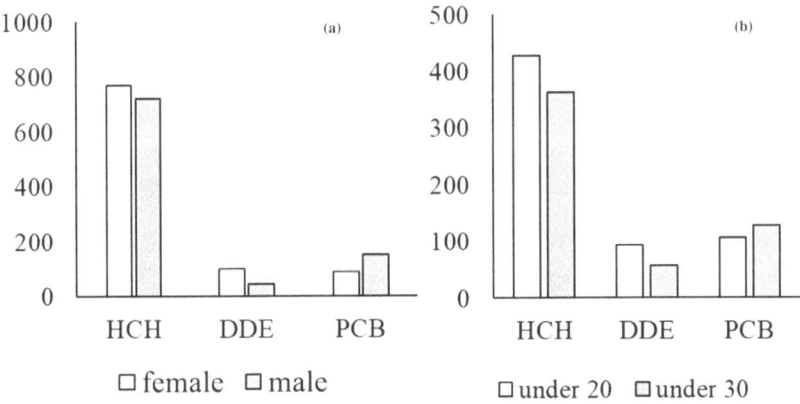

Fig. 17.4 Average POP concentrations in the blood samples from Primorsky Krai residents of different sexes (**a**) and ages (**b**), ng/g l.w.

Table 17.1 Concentrations of organochlorine compounds in blood from residents of different world's regions, ng/g l.w.

Region	Range	Mean value	Median	References
Primorsky Krai, Russia	56–2476	886	780	Present study
Mexico	1600–182,600	23,100	13,800	[8]
Romania	446–47,120	–	–	[21]
Spain	–	6187	4123	[22]
Great Britain	10–2720	–	115	[23]
Switzerland	–	887	–	[5]
Japan	–	151	210	[24]

17.3 POPs in Breast Milk of Far East Residents

According to the WHO recommendations, one of the most reliable indicators of the impact of POPs on human health is their content in the blood and women's milk [25].

17.3.1 Primorsky Krai

An ecological and analytical study of breast milk samples from female residents of the south of the Russian Far East (Primorsky Krai), which involved 29 women in 2017 and 37 in 2018, revealed POPs in 100% of samples. This survey was a stage of the regular POP monitoring within the framework of the Plan for the Fulfillment of Russian Federation's Obligations under the Stockholm Convention on Persistent

Organic Pollutants (Order No. 529 of October 3, 2017 "On Approval of the Plan for the Fulfillment of the Russian Federation's Obligations under the Stockholm Convention on Persistent Organic Pollutants").

The women were from 20 to 49 yr of age. The age groups of 20–29, 30–39, and 40–49 yr in 2017 numbered 10, 11, and 5, respectively; for three samples, the age of mothers was unknown. In 2018, the sizes of the age groups were 10, 25, and 2 persons, respectively.

The OCC level (ΣHCH + ΣDDT + ΣPCB) in the samples ranged from 23 to 878.3 (with a mean value of 151.4) ng/g l.w. (Table 17.2). The concentration ranges of OCPs (ΣHCH + ΣDDT) and PCBs were 2.8–291 and 3.2–720 ng/g l.w., respectively. The average values of OCP and PCB concentrations were 80.1 and 74.9 ng/g l.w., respectively.

The total content of organochlorine pesticides is formed by the sum of HCH isomers and DDT and its metabolites. The concentrations of ΣHCH and ΣDDT were within ranges of 2.84–291 and 1.1–83 ng/g l.w., respectively. The average values

Table 17.2 Levels of organochlorine compounds (OCC) in the breast milk samples from female residents of Primorsky Krai, ng/g l.w.

Toxicant	2017			2018		
	Age groups			Age groups		
	20–29	30–39	40–49	20–29	30–39	40–49
α-HCH	–	–	–	0.6 ± 0.1	1.2 ± 0.2	0.7 ± 0.03
β-HCH	36.3 ± 15.4	66.6 ± 14.2	47.4 ± 6.6	62.2 ± 13.5	91.5 ± 11.7	114.1 ± 72.2
γ-HCH	–	–	–	7.9 ± 2.2	7.6 ± 2.2	–
o,p'-DDT	–	2.3 ± 0.2	–	–	–	–
p,p'-DDT	10.2 ± 2.2	2.2 ± 0.01	–	–	–	–
o,p'-DDD	–	–	–	–	–	–
p,p'-DDD	–	–	–	–	–	–
o,p'-DDE	–	–	–	9.3 ± 0.5	8.9 ± 1.5	–
p,p'DDE	17.7 ± 2.3	3.5 ± 0.2	–	–	9.1 ± 2.1	–
PCB 28	–	2 ± 0.2	–	14.5 ± 6.0	4.8 ± 1.1	–
PCB 52	47.2 ± 6.6	16.8 ± 5.4	11.9 ± 4.1	26.4 ± 11.4	12.3 ± 3.2	12.6 ± 3.3
PCB 101	12.8 ± 3.3	9.7 ± 3.9	–	3.05 ± 0.8	11.9 ± 2.2	–
PCB 118	49.9 ± 24.4	23.6 ± 5.1	20.8 ± 8.1	6.7 ± 0.8	9.1 ± 1.2	11.2 ± 2.1
PCB 138	48.3 ± 22.5	38.1 ± 15.2	15.2 ± 2.8	7.9 ± 1.4	15.5 ± 2.5	26.5 ± 14.2
PCB 143	–	–	–	–	–	–
PCB 153	49.2 ± 21.5	28.8 ± 6.4	32.7 ± 11.2	7.6 ± 1.1	16.7 ± 2.4	23.3 ± 10.4
PCB 155	25.0 ± 9.8	2.3 ± 0.03	–	16.8 ± 4.8	25.6 ± 12.5	–
PCB 180	–	12.8 ± 3.9	–	–	12.9 ± 2.5	–
PCB 207	–	–	–	–	–	–

Note –, below the detection limits

of ΣHCH and ΣDDT concentrations were 76 and 12.9 ng/g l.w., respectively. All HCH isomers were found in the breast milk. The most frequently detected form was β-HCH. The concentrations of α-, β-, and γ-HCH were within ranges of 0.3–6.6, 2.8–290, and 0.8–26 ng/g l.w., respectively. DDT and its metabolites were below the detection limits in 50% of samples. Mainly o,p' and p,p'-DDE were identified (in 29 and 23% of samples, respectively). Their concentrations ranged from 0.7 to 22 and from 0.5 to 29 ng/g l.w., respectively. o,p'-DDT, o,p'-DDD, and p,p'-DDD were detected in two samples where their concentrations amounted to 3.62 and 1.1 ng/g, 2.2 and 1.4 ng/g, 34.4 and 25.7 ng/g l.w., respectively. p,p'-DDT was found in 11% of samples (within 1.8–27.6 ng/g).

Polychlorinated biphenyls (PCB) were found in almost all samples. Concentrations of PCBs 143 and 207 were below the detection limit. The level of lower chlorinated PCBs 28 and 52 (whose molecules contain up to four chlorine atoms) were within ranges of 1–35 and 1.8–130 ng/g l.w., respectively. The concentration ranges of higher chlorinated PCBs 101, 118, 138, 153, 155, and 180 were 1.8–95, 2.5–253, 2.9–169, 3.4–163, 2.3–49, and 7–19 ng/g l.w., respectively.

In 2017, we analyzed a total of 29 breast milk samples. The OCC level in them ranged from 22.6 to 878.3 (with an average of 144.1) ng/g l.w. The concentration ranges of OCPs and PCBs were 2.8–158 and 16.3–720.3 ng/g l.w., respectively; the average OCP and PCB concentrations were 53.8 and 97.4 ng/g l.w., respectively. The ΣHCH and ΣDDT concentrations ranged within 2.8–158 and 1.4–83.4 ng/g; the average values of ΣHCH and ΣDDT concentrations were 51 and 17.9 ng/g l.w.

In 2018, we analyzed a total of 37 breast milk samples. The OCC level in the samples ranged from 23.7 to 412.5 (with an average of 157.11) ng/g l.w. The ranges of OCP and PCB concentrations were 10.8–291.1 and 3.2–177.5 ng/g l.w., respectively; the average OCP and PCB concentrations were 99.2 and 57.89 ng/g l.w., respectively. The ΣHCH and ΣDDT concentrations ranged within 10.8–291.1 and 1.1–22.1 ng/g l.w.; the average values were 92.9 and 10.6 ng/g l.w., respectively.

An assessment of correlation between the concentrations of OCCs and lipids showed no relationship. The total OCP level in 2018 was higher than in 2017 ($p = 0.035$). HCH, as the most detectable component among the compounds under consideration, was found in all samples in 2018 and in 84% of samples in 2017. The HCH and DDT concentrations in 2018 were significantly higher than in 2017 ($p = 0.016$ and $p = 0.008$) (Figs. 17.5 and 17.6). This fact indicates a possible use of OCCs in the Far East, as well as in southern China and India.

According to the Ministry of Health of the Russian Federation, substantial amounts of DDT are currently stockpiled across the territories of Primorsky Krai, Khabarovsk Krai, and other regions of the Asian part of Russia. The results of our research clearly show that the total DDT concentration is decreasing. The decrease in the level of DDT and its metabolites indicates a reduction or lack of use of pollutants of this group in agriculture.

The total PCB concentration in 2018 was lower than that in 2017 (Fig. 17.7). This is probably related to the decommissioning of PCB-containing equipment and a reduction in the use of such disposal technologies as incineration of household and industrial waste. Also, the results show that the lower chlorinated congeners account

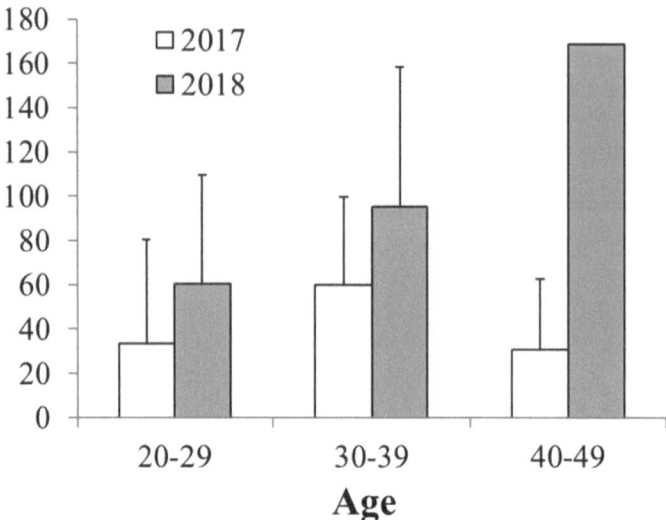

Fig. 17.5 Average levels of hexachlorocyclohexane (HCH) isomers in breast milk samples, ng/g l.w.

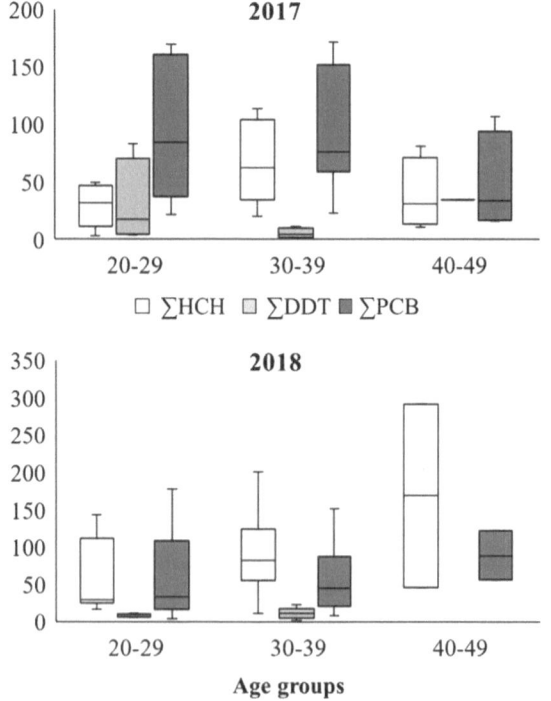

Fig. 17.6 Total levels of HCH, DDT, and PCBs (median) in the breast milk samples in 2017 and 2018, ng/g l.w.

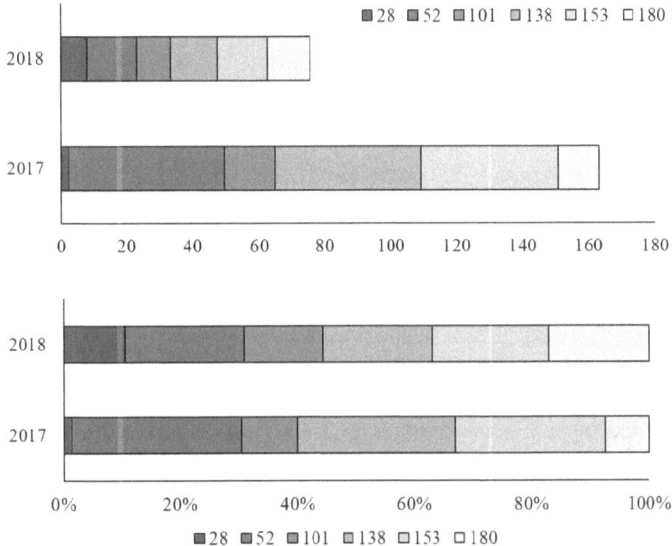

Fig. 17.7 Concentrations (ng/g l.w.) and proportions (%) of lower and higher chlorinated indicator PCBs

for 30% of the total PCB concentration (Fig. 17.7). As is known, the solubility in water decreases with increasing chlorine content [26]. Accordingly, lower chlorinated PCBs are excreted faster by the urinary system. Higher chlorinated (heavy) PCBs are accumulated by body's lipids and are almost not removed with the urine. This feature indicates the transfer of these compounds with mother's milk to infant, which increases possible health risks.

17.3.2 Chukotka Autonomous Okrug

In 2019, we carried out an ecological and analytical survey of breast milk samples from a total of 26 women residing in the Chukotka Autonomous Okrug (CAO). The women were from 15 to 44 yr of age. The age groups *under 30 years* and *over 30 years* numbered 15 and 10, respectively. For one of the samples, the age of the woman was unknown.

Organochlorine compounds (OCC) were found in all samples. The total OCC level ($\sum$HCH + $\sum$DDT + $\sum$PCB) in the samples ranged from 12.9 to 620.6 (with a mean value of 110.2 $\pm$ 142.7) ng/g l.w. The concentration ranges of OCPs ($\sum$HCH + $\sum$DDT) and $\sum$PCB were 7.1–275.4 and 1.1–430.8 ng/g l.w.; the average concentrations were 41.3 $\pm$ 51.9 and 74.1 $\pm$ 112.4 ng/g l.w., respectively.

The levels of $\sum$HCH and $\sum$DDT ranged from 4.8 to 162.3 (with a mean value of 31.1 $\pm$ 30.5) and from 1.5 to 113.1 (13.7 $\pm$ 25.8) ng/g l.w., respectively. α-, β-, γ-, and δ-HCH were detected in the breast milk samples. Among them, β-HCH was most frequently found in all the samples analyzed, and α-HCH was the least frequently occurring form. The concentrations of α-, β-, γ-, and δ-HCH ranged within 9.2–32.2, 1.3–162.3, 1.4–11.8, and 0.8–6.8 ng/g l.w., respectively. The mean levels were as follows: α-HCH, 20.5 ± 9.6; β-HCH, 21.5 ± 30.6; γ-HCH, 4.2 ± 2.8; and δ-HCH, 2.4 ± 1.7 ng/g l.w. Among the DDT metabolites, p,p'-DDT, o,p'-DDD, and p,p'-DDD were not detected. o,p'-DDT and o,p'-DDE were found in two samples at the following concentrations: 33.8 and 56.0 ng/g l.w. in one and 3.0 and 9.4 ng/g l.w. in the other. The most frequently occurring metabolite was p,p'-DDE, whose levels ranged from 1.5 to 57.1 (mean 9.8 ± 12.9) ng/g l.w.

PCBs were detected in almost all samples. However, the congeners 155, 101, and 207 were below the detection limit in all the samples analyzed. Two samples contained PCBs 28 and 143 at the following concentrations: 3.5 and 30.8 in one and 3.6 and 45.0 ng/g l.w. in the other. PCB 180 was found in three samples: 26.8, 60.2, and 75 ng/g l.w. The concentration ranges of PCBs 52, 118, 153, and 138 were 1.1–9.8 (with a mean value of 5.2 ± 2.8), 3.3–74.4 (18.8 ± 23.6), 4.3–203.5 (43.6 ± 58.4), and 4.4–84.1 (24.1 ± 26.1) ng/g l.w., respectively.

The $\sum$OCC concentrations in the breast milk of women under 30 yr of age were within a range of 12.9–459.2 (with a mean value of 99.8 ± 124.9) ng/g l.w. The levels of $\sum$OCP and $\sum$PCB ranged from 9.9 to 51.4 (28.8 ± 14.8) and from 1.1 to 430.8 (81.1 ± 128.9) ng/g l.w., respectively. The levels of $\sum$HCH and $\sum$DDT were within ranges of 5.2–52.1 (24.8 ± 14.6) and 1.5–11.6 (5.4 ± 3.1) ng/g l.w., respectively.

The $\sum$OCC level in the breast milk of women over 30 yr of age ranged from 221 to 620.6 (123.8 ± 168.4) ng/g l.w. The concentrations of $\sum$OCP and $\sum$PCB ranged from 7.1 to 275.4 (58.0 ± 76.0) and from 9.4 to 345.2 (65.8 ± 94.5) ng/g l.w., respectively. The $\sum$HCH level was within a range of 4.8–162.3 (39.4 ± 43.2) ng/g l.w.; the $\sum$DDT level, within 2.3–113.1 (24.8 ± 37.6) ng/g l.w.

A statistical analysis did not show correlations between the concentrations of lipids and OCCs in breast milk. A comparison of the levels of pollutants in samples from different age groups revealed significant differences only in β-HCH level (Fig. 17.8), with its level being significantly higher in women over 30 yr ($p = 0.046$).

The higher β-HCH levels in women over 30 compared to younger ones (Figs. 17.8 and 17.9) may be associated with the degradation of lindane in the human body with age. Among the DDT metabolites, p,p'-DDE was most frequently detected, which indicates a long period since the contamination of the environment by the initial pollutant and its degradation. The higher levels of p,p'-DDE in the breast milk of women over 30 yr (Fig. 17.8) (statistically insignificant) most likely indicate biomagnification and degradation of DDT.

The PCB congeners were dominated by higher chlorinated PCBs 118, 153, and 138 (Figs. 17.10 and 17.11). The levels of PCB 52 in breast milk were comparable in women from both age groups. PCB 143 was found only in the breast milk samples from younger women. The average levels of PCBs 153 and 138 tended to increase in women under 30 yr (statistically non-significant).

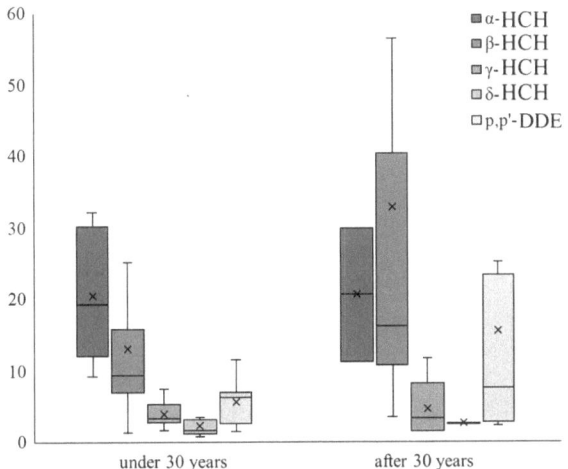

Fig. 17.8 Average levels (median) of HCH isomers and DDE in the breast milk samples from CAO women, ng/g l.w.

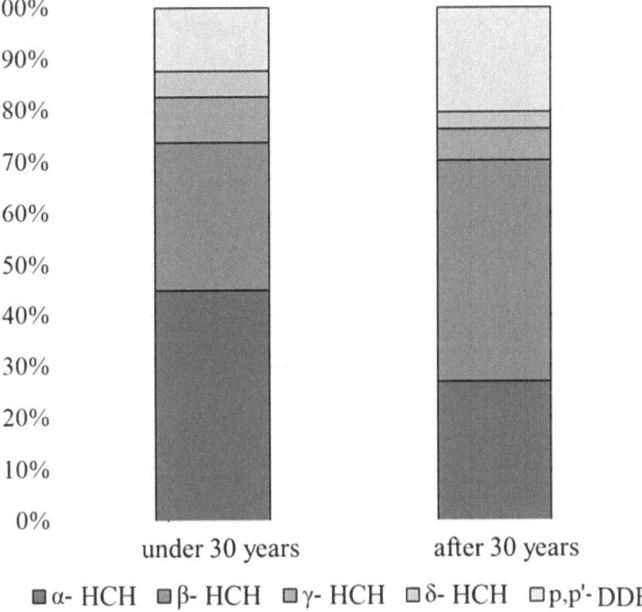

Fig. 17.9 Age–frequency distribution of OCPs in the breast milk samples from women of two age groups, %

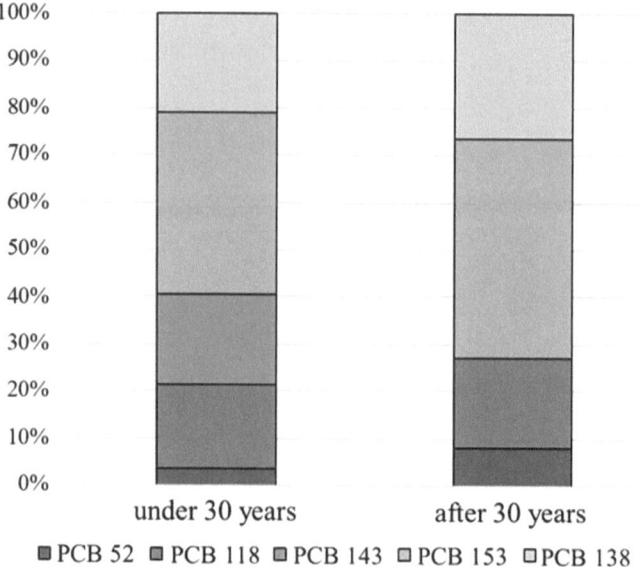

Fig. 17.10 Age–frequency distribution of detected PCB congeners in the breast milk samples from women of two age groups, %

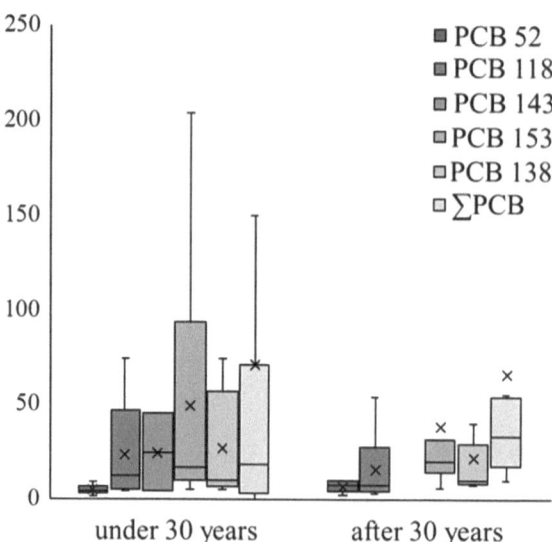

Fig. 17.11 Average concentrations (medians) of PCB congeners in the breast milk samples from CAO women, ng/g l.w.

The higher PCB concentrations in the breast milk from younger women may be explained by the greater number of births in the older age group before the study. As known, the major routes of POP removal from the woman's body are transfer through the placenta (mother-to-fetus transfer of toxicants) and excretion with breast milk during lactation [27, 28]. Most likely, the older and more mature women had given more than one birth and, therefore, the concentration of PCBs in their breast milk was lower.

A comparison of our data with the results of POP surveys by researchers from Russia and other countries has shown (Table 17.3) that the total of $\sum$HCH in breast milk from residents of the Chukotka Autonomous Okrug (CAO) is lower than in samples from Primorsky Krai, the Republic of Buryatia (Russia), Australia, Bangladesh, Iran, Libya, Tunisia, Turkey, India, Vietnam, Japan, and China and higher than in samples from Irkutsk Oblast, Transbaikalia (Russia), Mexico, Croatia, Tanzania, Norway, and Taiwan. The HCH concentrations in the samples from CAO are similar to those from Poland, South Korea, and the USA. The $\sum$DDT levels are comparable in CAO, Primorsky Krai (Russia), and Croatia, where levels of pollutants are not higher than 20 ng/g l.w.

In the case of Primorsky Krai, which is a developed agricultural territory, the detectable levels of pesticides may be associated with the extensive application of HCHs in farmlands and the accumulation of pollutants in plant- and animal-derived foods through the general biomagnification of xenobiotics [29–31].

The Chukotka Autonomous Okrug, due to its harsh natural conditions, is a poorly developed region in terms of agriculture. In this regard, the use of plant protection chemicals was not widespread in CAO. However, a study by Dudarev [50] provides data from regional veterinary services which indicates the intensive use of OCP-containing (in particular DDT and HCH) chemicals in reindeer ranching in the 1960s and 1970s. To protect deer from botfly, animals were sprayed with solutions of these substances. The POP content in them was not reliably documented, but the available information suggests concomitant soil pollution, which still affects local residents of the region. Moreover, low temperatures significantly reduce the evaporation rate and act as a "cold trap" for POPs, which leads to the accumulation of volatile and semi-volatile pollutants in northern latitudes [50]. This obviously explains why OCPs are still found in representatives of the local population.

The total PCB concentrations in breast milk from women of Primorsky Krai and Chukotka Autonomous Okrug (CAO) are almost on the same level with those reported for some countries of the Asia–Pacific region, while in such countries as Norway and Russia (in particular the Republic of Buryatia and Irkutsk Oblast), the levels of these pollutants are higher. The indicator PCBs (28, 52, 101, 138, 153, and 180) in CAO account for more than 40% of total PCBs. In Primorsky Krai, such pollutants account for about 26%. Most likely, this proportion indicates the impact of high shipping activities and the local effect of transformers, capacitors, hydraulic systems, and other PCB-containing equipment operated in the region. Another serious source of polychlorinated biphenyls in the northern part of the Far East could be the so-called *northern delivery* of fuels, lubricants, and other technical liquids. It turned areas around many villages into landfills of abandoned barrels still containing residues of

Table 17.3 Levels of POPs in breast milk samples from various regions of Russia and other countries (ng/g l.w.)

Region	Year	ΣHCH	ΣDDT	ΣPCB ind	ΣPCB	References
Chukotka[**]	2019	20	6.3	24	58	Present study
Primorsky Krai[**]	2017–2018	76	13	20	78	Present study
Irkutsk Oblast[*]	1997–2009	4.3	534	155	267	[12]
Transbaikalia[*]	1997–2009	2.5	1122	106	2125	[12]
Buryatia[*]	2003–2004	810	660	–	240	[32]
Australia[**]	2002–2003	33.36	359	–	–	[33]
Bangladesh[**]	2010	58[2]	2033	–	–	[34]
Iran[*]	2007	1660	1930	690	–	[35]
China[**]	2006–2010	173[2]	655[3]	–	–	[36]
Colombia[*]	2012	–[***]	203[3]	–	–	[37]
South Korea	2011	24 ± 16	114 ± 67	–	14.2 ± 11.8	[38]
Libya[*]	2007	70	220[1]	–	–	[39]
Mexico	2004–2014	11 ± 19[2]	972 ± 828	–	–	[40]
Poland	2003	20 ± 6	1195 ± 475	–	115 ± 111.3	[41]
United States	2004	18.9 ± 19	65 ± 75	–	–	[42]
Northern Tanzania[**]	2012	1.11	205	4.19	–	[43]
Tunisia	2002–2003	67 ± 2090	3863 ± 120	–	–	[44]
Turkey	2007–2008	160 ± 409	1407 ± 23	27 ± 12	–	[45]
Japan[*]	2008–2009	63	-	–	112	[36]
Vietnam[*]	2007–2008	140[2]	1200	–	84	[46]
India	2015–2016	47 ± 107	519 ± 1017	–	33 ± 68	[47]
Norway[*]	2002–2009	12.3[2]	167[1]	–	541.6	[48]
Croatia*	2011–2014	3.4	16.8	25	66	[49]

Note [1] p,p'-DDE and p,p'-DDT; [2] only β-HCH; [3] only p,p'-DDE; [*] mean value; [**] median; [***] no data

chemicals such as PCBs that gradually leak and enter soils, groundwater, rivers and seas [50]. Due to different toxicities of certain PCB congeners, the risk to human health from exposure to them may vary depending on the concentration and characteristics of environmental pollution. Infants, whose diet is normally based on breast milk, are particularly sensitive to the effects of POPs. This makes the continuation of environmental biomonitoring of pollutants even more important.

References

1. UNEP (1989) DDT and its derivatives: environmental aspects. World Health Organization, Geneva
2. Nicholson WJ, Landrigan PJ (1994) Human health effects of polychlorinated biphenyls. In: Schecter A (ed) Dioxins and health. Springer, US, Boston, MA, pp 487–524
3. Chashchin VP, Kovshov AA, Gudkov AB, Morgunov BA (2016) Socio-economic and behavioral risk factors for health disorders among the indigenous population of the Far North. Hum Ecol 23:3–8. https://doi.org/10.33396/1728-0869-2016-6-3-8
4. Tanabe S (2007) Contamination by persistent toxic substances in the Asia-Pacific region. Persistent Org Pollut Asia Sour Distrib Transp Fate 7:773–817. https://doi.org/10.1016/S1474-8177(07)07018-0
5. Wicklund Glynn A, Wolk A, Aune M et al (2000) Serum concentrations of organochlorines in men: a search for markers of exposure. Sci Total Environ 263:197–208. https://doi.org/10.1016/S0048-9697(00)00703-8
6. Černá M, Spěváčková V, Batáriová A et al (2007) Human biomonitoring system in the Czech Republic. Int J Hyg Environ Health 210:495–499. https://doi.org/10.1016/j.ijheh.2007.01.005
7. Černá M, Krsková A, Čejchanová M, Spěváčková V (2012) Human biomonitoring in the Czech Republic: an overview. Int J Hyg Environ Health 215:109–119. https://doi.org/10.1016/j.ijheh.2011.09.007
8. Waliszewski SM, Caba M, Herrero-Mercado M et al (2012) Organochlorine pesticide residue levels in blood serum of inhabitants from Veracruz, Mexico. Environ Monit Assess 184:5613–5621. https://doi.org/10.1007/s10661-011-2366-2
9. Choi W, Kim S, Baek Y-W et al (2017) Exposure to environmental chemicals among Korean adults-updates from the second Korean National Environmental Health Survey (2012–2014). Int J Hyg Environ Health 220:29–35. https://doi.org/10.1016/j.ijheh.2016.10.002
10. Mamontova EA, Tarasova EN, Mamontov AA et al (2007) The influence of soil contamination on the concentrations of PCBs in milk in Siberia. Chemosphere 67:S71–S78. https://doi.org/10.1016/j.chemosphere.2006.05.092
11. Mamontova EA, Tarasova EN, Kuzmin MI (2010) The content of persistent organic pollutants in the breast milk of women living in the Irkutsk region. Hyg Sanitation 1:35–38
12. Mamontova EA, Tarasova EN, Mamontov AA (2017) PCBs and OCPs in human milk in Eastern Siberia, Russia: levels, temporal trends and infant exposure assessment. Chemosphere 178:239–248. https://doi.org/10.1016/j.chemosphere.2017.03.058
13. Chashchin MV, Chashchin VP, Chashchin VN (2012) The main trends in the concentration of persistent toxic substances in the blood of the indigenous population of the Arctic. Hum Ecol 19:3–7. https://doi.org/10.17816/humeco17459
14. Lyzhina AV, Buzinov RB, Ungureanu TN, Gudkov AB (2012) Chemical contamination of food products and its impact on the health of the population of the Arkhangelsk region. Hum Ecol 12:3–9
15. Yasmeen H, Qadir A, Mumtaz M et al (2017) Risk profile and health vulnerability of female workers who pick cotton by organanochlorine pesticides from southern Punjab, Pakistan: health vulnerability of female workers who pick cotton. Environ Toxicol Chem 36:1193–1201. https://doi.org/10.1002/etc.3633
16. Tsygankov VY, Boyarova MD, Kiku PF, Yarygina MV (2015) Hexachlorocyclohexane (HCH) in human blood in the south of the Russian Far East. Environ Sci Pollut Res 22:14379–14382. https://doi.org/10.1007/s11356-015-4951-3
17. Tsygankov VY, Khristoforova NK, Lukyanova ON et al (2017) Selected organochlorines in human blood and urine in the South of the Russian far east. Bull Environ Contam Toxicol 99:460–464. https://doi.org/10.1007/s00128-017-2152-0
18. Braginsky LP, Komarovsky FY, Pischolka YK, Maslova OV (1980) Migration of persistent pesticides in freshwater ecosystems. In: Proceedings of the All-Union conference. Gidrometeoizdat, Leningrad, Russia, pp 226–231

19. Hu L, Zhang G, Zheng B et al (2009) Occurrence and distribution of organochlorine pesticides (OCPs) in surface sediments of the Bohai Sea, China. Chemosphere 77:663–672. https://doi.org/10.1016/j.chemosphere.2009.07.070

20. GHBC (2003) Aktualisierung der Referenzwerte Für PCB-138, -153, -180 im Vollblut sowie Referenzwerte für HCB, β-HCH und DDE im Vollblut. Bundesgesundheitsblatt—Gesundheitsforschung—Gesundheitsschutz 46:161–168. https://doi.org/10.1007/s00103-002-0545-6

21. Dirtu AC, Cernat R, Dragan D et al (2006) Organohalogenated pollutants in human serum from Iassy, Romania and their relation with age and gender. Environ Int 32:797–803. https://doi.org/10.1016/j.envint.2006.04.002

22. Porta M, de Basea MB, Benavides FG et al (2008) Differences in serum concentrations of organochlorine compounds by occupational social class in pancreatic cancer. Environ Res 108:370–379. https://doi.org/10.1016/j.envres.2008.06.010

23. Thomas GO, Wilkinson M, Hodson S, Jones KC (2006) Organohalogen chemicals in human blood from the United Kingdom. Environ Pollut 141:30–41. https://doi.org/10.1016/j.envpol.2005.08.027

24. Fukata H, Omori M, Osada H et al (2005) Necessity to measure PCBs and organochlorine pesticide concentrations in human umbilical cords for fetal exposure assessment. Environ Health Perspect 113:297–303. https://doi.org/10.1289/ehp.7330

25. Revich B, Sergeyev O, Shelepchikov A (2012) Innovative environmental and epidemiologic technologies of assessment of dioxins impacts on children's health. Hum Ecol 8:42–49. https://doi.org/10.17816/humeco17455

26. Lukyanova ON, Tsygankov VY, Boyarova MD (2018) Organochlorine pesticides and polychlorinated biphenyls in the Bering flounder (Hippoglossoides robustus) from the Sea of Okhotsk. Mar Pollut Bull 137:152–156. https://doi.org/10.1016/j.marpolbul.2018.10.017

27. IARC (2013) Polychlorinated biphenyls and polybrominated biphenyls. IARC Press, Lyon

28. Porpora M, Lucchini R, Abballe A et al (2013) Placental transfer of persistent organic pollutants: a preliminary study on mother-newborn pairs. IJERPH 10:699–711. https://doi.org/10.3390/ijerph10020699

29. Tsygankov VY, Boyarova MD, Lukyanova ON (2015) Bioaccumulation of persistent organochlorine pesticides (OCPs) by gray whale and Pacific walrus from the western part of the Bering Sea. Mar Pollut Bull 99:235–239. https://doi.org/10.1016/j.marpolbul.2015.07.020

30. VY, Tsygankov, Boyarova MD, Lukyanova ON, Khristoforova NK (2017) Bioindicators of organochlorine pesticides in the Sea of Okhotsk and the Western Bering Sea. Arch Environ Contam Toxicol 73:176–184. https://doi.org/10.1007/s00244-017-0380-2

31. Tsygankov VY, Lukyanova ON, Boyarova MD (2018) Organochlorine pesticide accumulation in seabirds and marine mammals from the Northwest Pacific. Mar Pollut Bull 128:208–213. https://doi.org/10.1016/j.marpolbul.2018.01.027

32. Tsydenova OV, Sudaryanto A, Kajiwara N et al (2007) Organohalogen compounds in human breast milk from Republic of Buryatia, Russia. Environ Pollut 146:225–232. https://doi.org/10.1016/j.envpol.2006.04.036

33. Mueller JF, Harden F, Toms L-M et al (2008) Persistent organochlorine pesticides in human milk samples from Australia. Chemosphere 70:712–720. https://doi.org/10.1016/j.chemosphere.2007.06.037

34. Bergkvist C, Aune M, Nilsson I et al (2012) Occurrence and levels of organochlorine compounds in human breast milk in Bangladesh. Chemosphere 88:784–790. https://doi.org/10.1016/j.chemosphere.2012.03.083

35. Dahmardeh Behrooz R, Esmaili Sari A, Bahramifar N et al (2009) Organochlorine pesticide and polychlorinated biphenyl residues in human milk from Tabriz, Iran. Toxicol Environ Chem 91:1455–1468. https://doi.org/10.1080/02772240902732472

36. Fujii Y, Ito Y, Harada KH et al (2012) Comparative survey of levels of chlorinated cyclodiene pesticides in breast milk from some cities of China, Korea and Japan. Chemosphere 89:452–457. https://doi.org/10.1016/j.chemosphere.2012.05.098

37. Rojas-Squella X, Santos L, Baumann W et al (2013) Presence of organochlorine pesticides in breast milk samples from Colombian women. Chemosphere 91:733–739. https://doi.org/10.1016/j.chemosphere.2013.02.026

38. Kim S, Eom S, Kim H-J et al (2018) Association between maternal exposure to major phthalates, heavy metals, and persistent organic pollutants, and the neurodevelopmental performances of their children at 1 to 2 years of age—CHECK cohort study. Sci Total Environ 624:377–384. https://doi.org/10.1016/j.scitotenv.2017.12.058

39. Zeinab HMA-T, Refaat GAEE, El-Dressi AY (2011) Organochlorine pesticide residues in human breast milk in El-Gabal Al-Akhdar, Libya. In: International proceedings of chemical, biological and environmental engineering, vol 3, pp 146–149

40. Chávez-Almazán LA, Diaz-Ortiz J, Alarcón-Romero M et al (2014) Organochlorine pesticide levels in breast milk in Guerrero, Mexico. Bull Environ Contam Toxicol 93:294–298. https://doi.org/10.1007/s00128-014-1308-4

41. Szyrwińska K, Lulek J (2007) Exposure to specific polychlorinated biphenyls and some chlorinated pesticides via breast milk in Poland. Chemosphere 66:1895–1903. https://doi.org/10.1016/j.chemosphere.2006.08.010

42. Johnson-Restrepo B, Addink R, Wong C et al (2007) Polybrominated diphenyl ethers and organochlorine pesticides in human breast milk from Massachusetts, USA. J Environ Monit 9:1205–1212. https://doi.org/10.1039/b711409p

43. Müller MHB, Polder A, Brynildsrud OB et al (2019) Prenatal exposure to persistent organic pollutants in Northern Tanzania and their distribution between breast milk, maternal blood, placenta and cord blood. Environ Res 170:433–442. https://doi.org/10.1016/j.envres.2018.12.026

44. Ennaceur S, Gandoura N, Driss MR (2007) Organochlorine pesticide residues in human milk of mothers living in northern Tunisia. Bull Environ Contam Toxicol 78:325–329. https://doi.org/10.1007/s00128-007-9185-8

45. Cok I, Yelken C, Durmaz E et al (2011) Polychlorinated biphenyl and organochlorine pesticide levels in human breast milk from the Mediterranean city Antalya, Turkey. Bull Environ Contam Toxicol 86:423–427. https://doi.org/10.1007/s00128-011-0221-3

46. Haraguchi K, Koizumi A, Inoue K et al (2009) Levels and regional trends of persistent organochlorines and polybrominated diphenyl ethers in Asian breast milk demonstrate POPs signatures unique to individual countries. Environ Int 35:1072–1079. https://doi.org/10.1016/j.envint.2009.06.003

47. Bawa P, Bedi JS, Gill JPS et al (2018) Persistent organic pollutants residues in human breast milk from Bathinda and Ludhiana districts of Punjab, India. Arch Environ Contam Toxicol 75:512–520. https://doi.org/10.1007/s00244-018-0512-3

48. Lenters V, Iszatt N, Forns J et al (2019) Early-life exposure to persistent organic pollutants (OCPs, PBDEs, PCBs, PFASs) and attention-deficit/hyperactivity disorder: a multi-pollutant analysis of a Norwegian birth cohort. Environ Int 125:33–42. https://doi.org/10.1016/j.envint.2019.01.020

49. Jovanović G, Romanić SH, Stojić A et al (2019) Introducing of modeling techniques in the research of POPs in breast milk—a pilot study. Ecotoxicol Environ Saf 172:341–347. https://doi.org/10.1016/j.ecoenv.2019.01.087

50. Dudarev AA (2009) Persistent polychlorinated hydrocarbons and heavy metals in the Arctic biosphere: the main patterns of exposure and reproductive health of indigenous people. Biosphere 1:186–202